LEE COLLEGE LIBRARY
BAYTOWN, TEXAS

DRAFTING SYMBOL SOURCEBOOK

DRAFTING SYMBOL SOURCEBOOK

DOUG WOLFF

McGraw-Hill
New York San Francisco Washington D.C. Auckland Bogotá
Caracas Lisbon London Madrid Mexico City Milan
Montreal New Dehli San Juan Singapore
Sydney Tokyo Toronto

Library of Congress Cataloging-in-Publication Data

Wolff, Doug.
Drafting symbol sourcebook / Doug Wolff.
p. cm.
Includes index.
ISBN 0-07-071332-4
1. Mechanical drawing—Notation. 2. Signs and symbols.
I. Title.
T353.W837 1999
604.2'42—dc21 98-31326
CIP

McGraw-Hill

A Division of The McGraw-Hill Companies

1 2 3 4 5 6 7 8 9 0 KGP/KGP 9 0 3 2 1 0 9 8

ISBN 0-07-071332-4

The sponsoring editor for this book was Zoe Foundotos, the editing supervisor was Paul R. Sobel, and the production supervisor was Modestine Cameron.

Printed and bound by Quebecor/Kingsport.

CONTENTS

INTRODUCTION

This book is intended as a tool in the interpretation and creation of engineering drawings. It arose from a need for a single reference to the more commonly used drafting symbols, that would relate to the often confusing reality of engineering documentation. It is not intended as a definitive or scholarly work.

Throughout the book some symbols are associated with letters in brackets, e.g. [A]. These letters refer to engineering organizations from whose standards those particular symbols are taken. These are the officially recognized symbols.

The letters and the organizations they refer to are:

[A] American National Standards Institute
[F] National Fire Protection Association
[G] U.S. Geological Survey
[I] Instrument Society of America
[IA] American Institute of Architects
[P] American Petroleum Institute

Symbols with the [P] designation were all taken from page 401 of the American Petroleum Institute Recommended Practice 35, "Oil Mapping Symbols", First Edition, 1957 and were reprinted courtesy of the American Petroleum Institute.

Symbols marked with (L) were created by Randy Lauresen, licensed surveyor, in San Mateo, California.

In the creation of engineering and architectural documents it is recommended that a key (or legend) be included, defining the particular symbols used in such documents.

Doug Wolff

ACKNOWLEDGMENTS

This book called upon the knowledge and resources of many people.

The following companies kindly supplied literature and information about their products:

A & K Railroad Material Inc., Salt Lake City, Utah
AAA Technology & Specialties Co. Inc., Houston, Texas
Advanced Valve Design, Whitehall, Pennsylvania
Akron Brass Inc., Wooster, Ohio
American Flow Control Corp., Birmingham, Alabama
Applied Electronics Company Inc., Boise, Idaho
Asahi/America Inc., Malden, Massachussetts
Becker Precision Equipment Inc., Elk Grove Village, Illinois
Ceel-co Corp., Lakewood, CO
Central Sprinkler Corp., Lansdale, PA
Charles Ross & Son Inc., Hauppauge, New York
Cherne Industries Inc., Minneapolis, Minnesota
Clow Valve Inc., Corona, California
Colton Industries, Inc., Buffalo, New York
Croll-Reynolds Inc., Westfield, New Jersey
Dixon Valve & Coupling Co. Inc., Chestertown, Maryland
Dorr-Oliver Corp., Milford, Connecticut
Draiswerke Inc., Mahwah, New Jersey
East Jordan Ironworks, East Jordan, Michigan
Edwards Signal Inc., Cheshire, Connecticut
Electro-Steam Corp., Alexandria, Virginia
Fluid Metering Inc., Syosset, New York
Fluitron Inc., Ivyland, Pennsylvania
Grainger Supply Inc., Chicago, Illinois
Habasit Belts & Conveyors, Inc., Atlanta, Georgia
Harrington Signal Corp., Moline, Illinois
Harrington Inc., Erie, Pennsylvania
Heyl & Patterson Inc., Pittsburgh, Pennsylvania
Hubbell Electric Products Inc., South Bend, Indiana
Hyde Products Inc., Cleveland, Ohio
Industrial Material Handling Corp., Albertville, Alabama
International Process Equipment Corp., Pennsauken, New Jersey
Jacobson Corp., Minneapolis, Minnesota
Jaeger Products Inc., Houston, Texas
Jaffrey (Global Industrial Technology), Woodruff, South Carolina

Jaygo Inc., Union, New Jersey
John C. Ernst Co. Inc., Dover, New Jersey
Joseph G. Pollard Co. Inc., New Hyde Park, New York
Komax Inc., Wilmington, California
Lithonia Lighting Corp., Conyers, Georgia
Marinetics Inc., Costa Mesa, California
Marion Mixers Inc., Marion, Iowa
MSC Industrial Supply Inc., Plainview, New York
National Time & Signal Corp., Oak Park, Michigan
NMP Corp., Tulsa, Oklahoma
Oceanic Electrical Manufacturing Co. Inc., Elizabeth, New Jersey
Ogantz Corp., Willow Grove, Pennsylvania
Omega Engineering Inc., Stamford, Connecticut
Omni Controls Technology Inc., Whitinsville, Massachussettes
Omni Fabricators Inc., Southampton, New Jersey
Paget Equipment Inc., Marshfield, Wisconsin
Patterson Industries Inc., Toronto, Ontario
Paul O. Abbe Inc., Little Falls, New Jersey
Potter-Roemer Inc., Atlanta, Georgia
Praher Corporation, Ontario, Canada
Prater Corporation, Cicero, Illinois
Red Head Brass Inc., Shreve, Ohio
Reheat Co. Inc., Danvers, Massachussetts
Schutte-Koerting Inc., Bensalem, Pennsylvania
Scott Equipment Inc., New Prague, Minnesota
Sharpsville Container Corp., Sharpsville, Pennsylvania
Silent Knight Corp., Maple Grove, Minnesota
Sloan Valve Co. Inc., Franklin Park, Illinois
Sturtevant Corp., Hanover, Massachussetts
Svedala Corp., Colorado Springs, Colorado
TAH Industries Inc., Robbinsville, New Jersey
Techna-flo Inc., Englewood, Colorado
Telelite Inc., Rochester, Minnesota
Terminal City Ironworks, Vancouver, British Columbia
The General Engineering Co. Inc., Frederick, Maryland
The Johnwood Co. Inc., Valley Froge, Pennsylvania
The Viking Corporation, Hasting, Michigan
Tyler Pipe Corp., Tyler, Texas
U.S. Filter Corp., Holland, Michigan
Universal Flow Monitors Inc., Hazel Park, Michigan
Waukesha Cherry-Burrell Inc., Delavan, Michigan

I am much indebted to the following people who very generously gave of their time and great professional knowledge:

Brent Hammond of Hammond Electrical Engineering (Electrical Components)
Bert Schaeffer of EUA Cogenex West (HVAC)
Carole Clynke, Boulder Medical Center (Nurse Call & Hospital Systems)
Dave Lowrey, Fire Control Engineer, City of Boulder (Fire Control)
Randy Laursen, Licensed Land Surveyor (Civil & Surveying)
Thomas R. Lofstrom, Electronic Cinematography Systems, Ltd. (Electronics)

A large thank you to Spirax Corporation of Allentown, Pennsylvania, for permission to use many of the line illustrations in their excellent Product Manual for steam systems.

Brian McNamara and James Kearns were valuable research assistants.

Last and not least, a thank you to Zoe Foundotos, my editor at McGraw-Hill, who shared my vision and enthusiasm for this project.

Doug Wolff
Broomfield, Colorado

ARCHITECTURAL SYMBOLS

ARCHITECTURAL REFERENCE SYMBOLS

NAME	SYMBOL	REMARKS
DRAWING TITLE	1/2"Φ TYP. (1) MAIN LEVEL FLOOR PLAN SCALE 1/4" = 1'-0"	PLACED CENTERED BENEATH DRAWING. IF DRAWING IS NOT SCALED, SCALE = NONE IF DRAWING IS AT A NON-STANDARD SCALE, NOTE AS "NOT TO SCALE."
ELEVATION REFERENCES	[1A] 1/2"Φ TYP.	TOP NUMBER IS DETAIL DRAWING NUMBER, BOTTOM NUMBER IS PAGE ON WHICH DETAIL DWG. APPEARS. -TYP.
	2 / A6	REFERENCING ONE ELEVATION VIEW. TRIANGLE POINTS TO ILLUSTRATED ELEVATION - TYP.
	2 / A2 A6	TOP NUMBER IS DETAIL DRAWING NUMBER, BOTTOM LEFT NUMBER IS PAGE WHERE DETAIL IS REFERENCED, LOWER RIGHT HAND NUMBER IS PAGE WHERE DETAIL APPEARS.
	3 / A7	REFERENCING TWO ELEVATIONS SHOWN IN ONE DRAWING.
	5 / A11	REFERENCING THREE ELEVATIONS SHOWN IN ONE DRAWING.
	3 / A7	REFERENCING FOUR ELEVATIONS SHOWN IN ONE DRAWING.
	2, A2, 3	REFERENCING TWO ELEVATION DRAWINGS ON ONE SHEET.
	5, A2, 6, 7	REFERENCING THREE ELEVATIONS ON ONE SHEET.
	1, 4, A2, 2, 3	REFERENCING FOUR ELEVATIONS ON ONE SHEET.

ARCHITECTURAL REFERENCE SYMBOLS

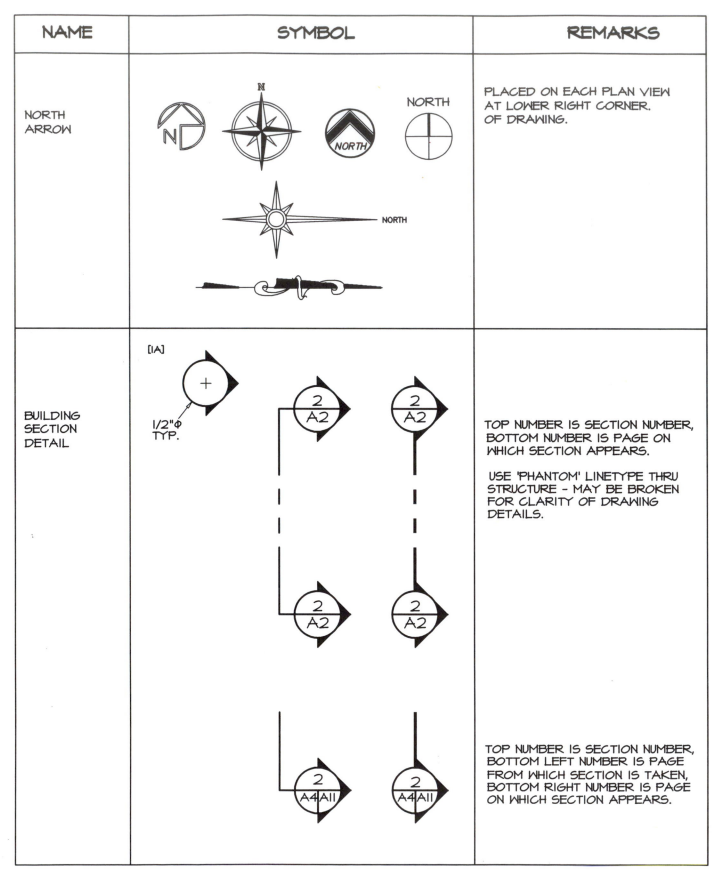

NAME	SYMBOL	REMARKS
NORTH ARROW	N NORTH NORTH NORTH	PLACED ON EACH PLAN VIEW AT LOWER RIGHT CORNER. OF DRAWING.
BUILDING SECTION DETAIL	[IA] 1/2"Φ TYP. 2 / A2	TOP NUMBER IS SECTION NUMBER, BOTTOM NUMBER IS PAGE ON WHICH SECTION APPEARS. USE 'PHANTOM' LINETYPE THRU STRUCTURE - MAY BE BROKEN FOR CLARITY OF DRAWING DETAILS.
	2 / A4 A11	TOP NUMBER IS SECTION NUMBER, BOTTOM LEFT NUMBER IS PAGE FROM WHICH SECTION IS TAKEN, BOTTOM RIGHT NUMBER IS PAGE ON WHICH SECTION APPEARS.

ARCHITECTURAL REFERENCE SYMBOLS

NAME	SYMBOL	REMARKS
WALL SECTION DETAIL	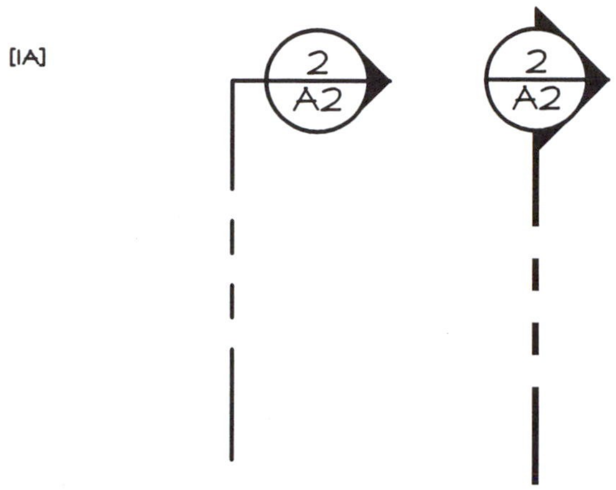 [IA] 2 A2 2 A2	TOP NUMBER IS DRAWING NUMBER, BOTTOM NUMBER IS PAGE ON WHICH SECTION APPEARS. USE 'PHANTOM' LINETYPE THRU STRUCTURE. SYMBOL SHOULD BE PLACED ON BOTH PLAN AND ELEVATION VIEWS.
DETAIL REFERENCE	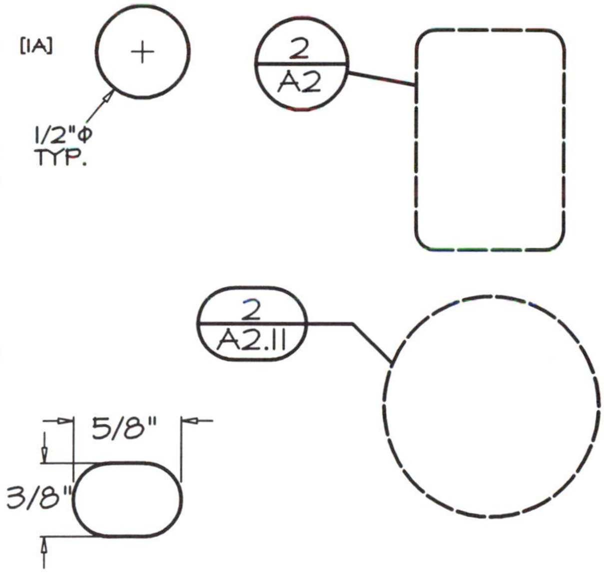 [IA] 1/2"ϕ TYP. 2 A2 2 A2.11 5/8" 3/8"	TOP NUMBER IS DETAIL NUMBER, BOTTOM NUMBER IS PAGE ON WHICH DETAIL APPEARS. DENOTE REFERENCED AREA WITH "DASHED" LINETYPE.
ROOM OR SPACE REFERENCE	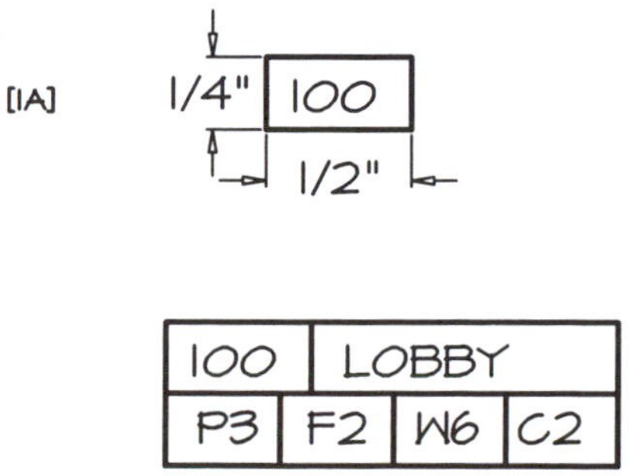 [IA] 1/4" 100 1/2" 100 LOBBY P3 F2 W6 C2	ROOM OR SPACE NUMBER. LOWER NUMBERS REFERENCE FINISH SCHEDULES, E.G. CARPET, WALL PAINT, ETC.

ARCHITECTURAL REFERENCE SYMBOLS

NAME	SYMBOL	REMARKS
EQUIPMENT NUMBER OR CEILING HEIGHT	[IA] 277 9'-0"	
DOOR KEY	[IA] 3/8" Φ TYP. 12	USED ON BOTH PLAN VIEW AND ELEVATIONS TO IDENTIFY DOOR SCHEDULE.
WINDOW KEY	[IA] 3/8" D	USED ON BOTH PLAN VIEWS AND ELEVATIONS TO IDENTIFY WINDOW ON SCHEDULE.
REVISION NUMBER	[IA] 3/8" 2	USE WITH 'CLOUD' TO OUTLINE RELEVANT AREA. ALSO USED TO KEY IN NOTES ON DRAWING.
MISCELLANEOUS REFERENCES	P-2 100 B6	USED ON BOTH PLAN VIEWS AND ELEVATIONS TO IDENTIFY FLOORING TYPES, PAINT TYPES, WALLPAPERS, ETC.

NAME	SYMBOL	REMARKS
COLUMN REFERENCE	[IA] 1/2"Φ TYP. A 1 GRID LINE	USE NUMBERS ON HORIZONTAL GRID LINES, LETTERS ON VERTICAL GRID LINES. USE 'CENTER' LINETYPE FOR GRID LINES THRU COLUMNS. RECTANGULAR AREAS DEFINED BY GRID LINES ARE REFERRED TO AS 'BAYS.'
ELEVATION REFERENCE	[IA] 1/4" Φ TYP.	USE ON PLAN VIEW TO LABEL ELEVATION, AND ON ELEVATION VIEW WITH CENTER LINE.
MATCH LINE	[IA]	SHADED HALF SHOWS PORTION DEPICTED. USE CENTER LINE TO DIVIDE AREAS. AREAS DEFINED ARE LABELED 'UNITS'.

DOORS

NAME	SYMBOL	ELEVATION	PERSPECTIVE	REMARKS
CASED OPENING		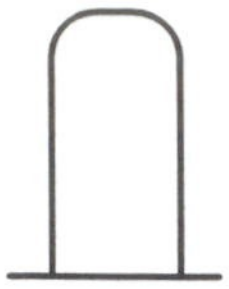	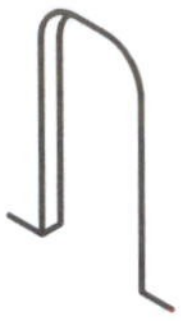	OPENING MAY BE SQUARED, ROUNDED, OR OTHWISE.
DOOR	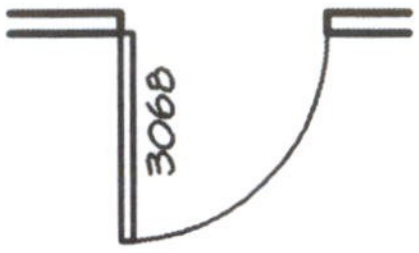 3068	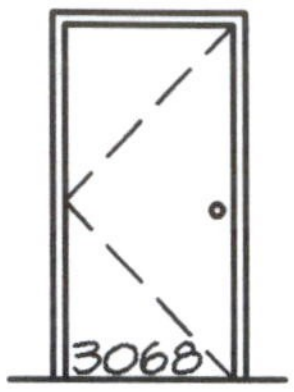 3068		DOOR SIZE MAY BE SHOWN DIRECTLY, AS SHOWN, BUT PREFERRED METHOD IS CIRCLE (SEE DOUBLE DOORS)
DOOR HANDING	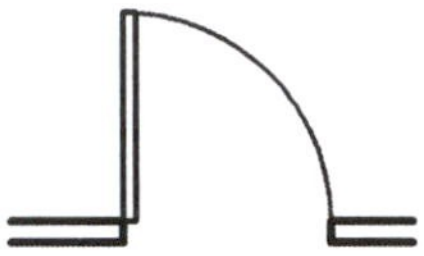 LEFT HANDED; 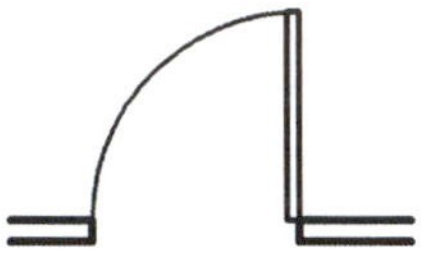 RIGHT HANDED; 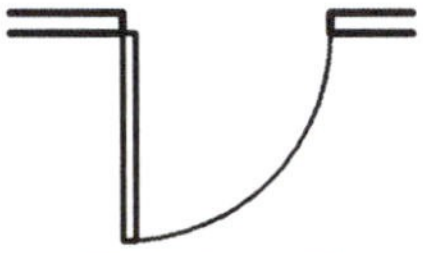 LEFT HANDED REVERSED; 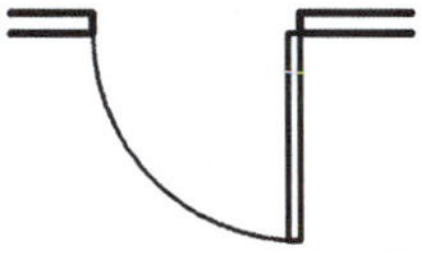 RIGHT HANDED REVERSED			
DOUBLE DOOR	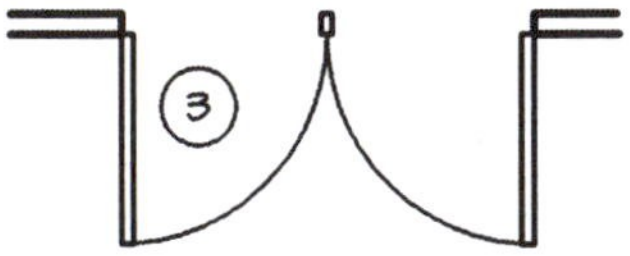 3	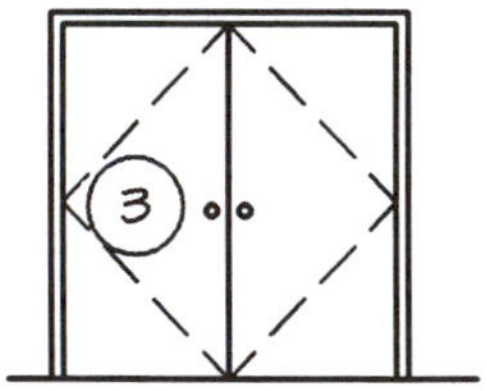 3	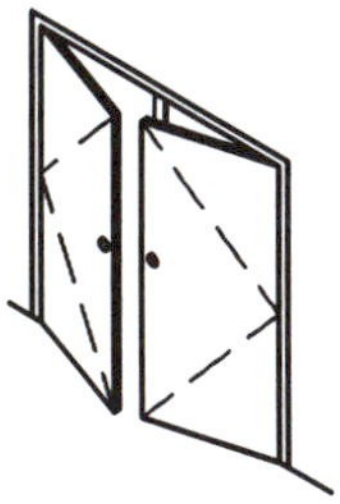	SHOWN WITH CENTER POST.
EXTERIOR DOOR	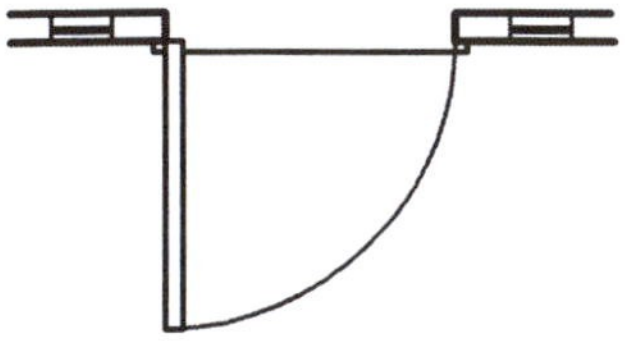	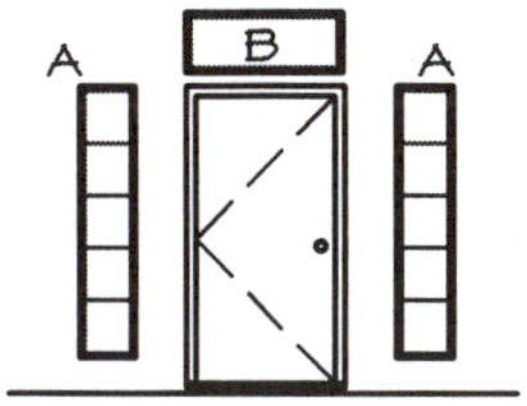 A B A		SHOWN WITH SIDE LITES (A) AND TRANSOM WINDOW (B).

NAME	SYMBOL	ELEVATION	PERSPECTIVE	REMARKS
EXTERIOR DOUBLE DOOR	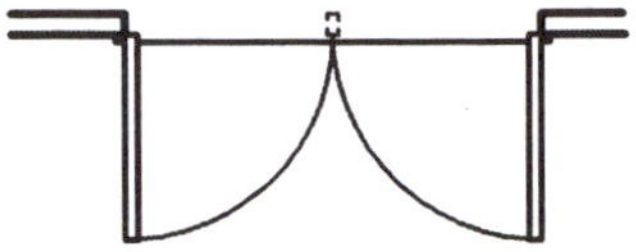	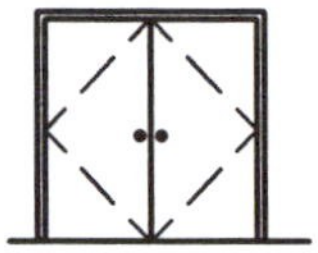	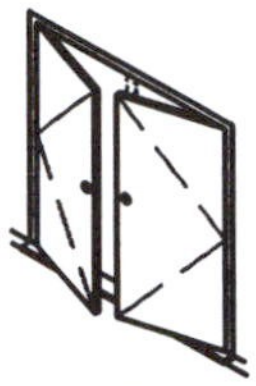	SHOWN WITH REMOVE-ABLE CENTER POST.
FIXED-LEAF DOUBLE DOOR		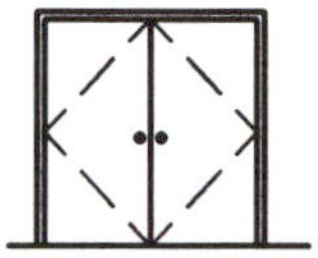	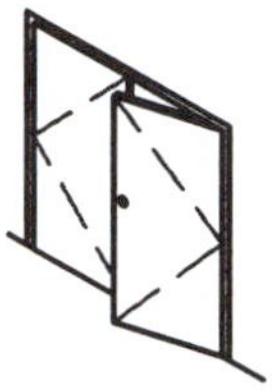	ONE DOOR NORMALLY IN CLOSED POSITION.
180 DEGREE DOOR	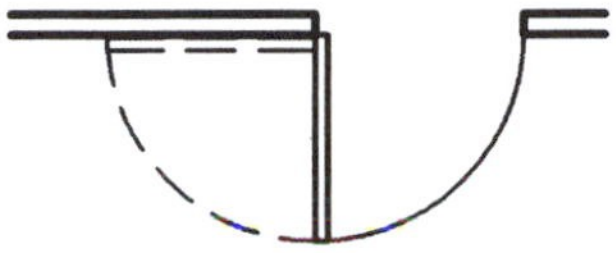	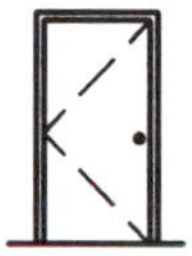		CLOSER HARDWARE ALLOWS FOR FULL OPEN.
180 SWINGING DOOR OR DOUBLE ACTING DOOR	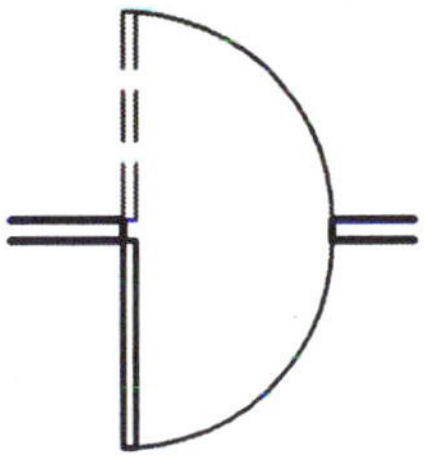	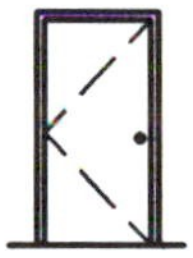		
DOUBLE LEAF DOOR		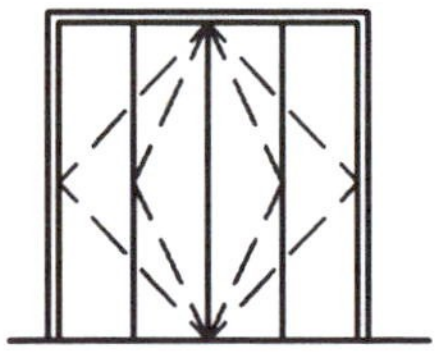	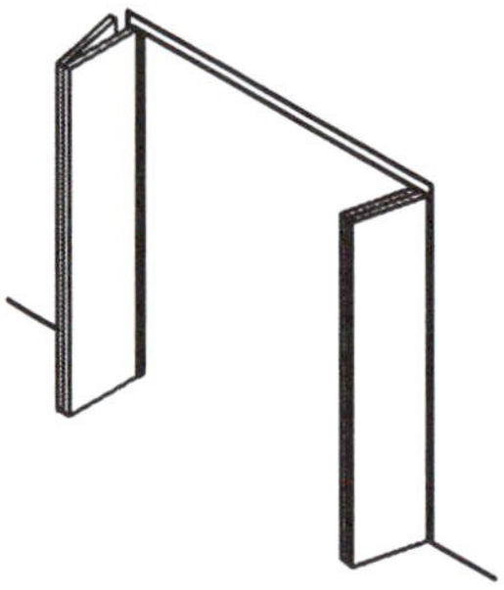	SOMETIMES LEAVES HOUSED IN ADJACENT WALL.
SLIDING DOORS		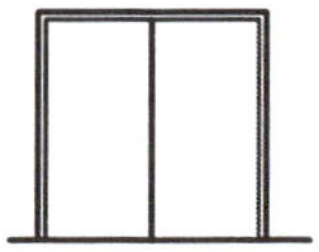		TYP. FOR CLOSETS

DOORS

NAME	SYMBOL	ELEVATION	PERSPECTIVE	REMARKS
DOUBLE ACTING DOUBLE DOORS	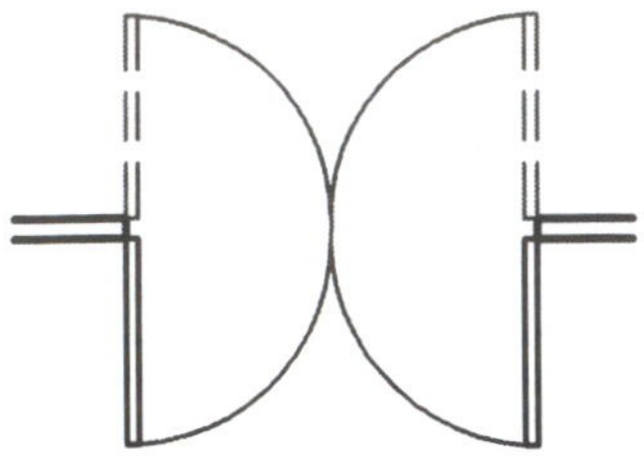	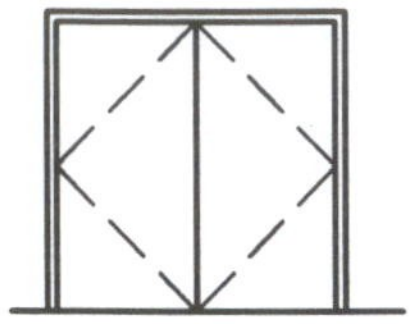	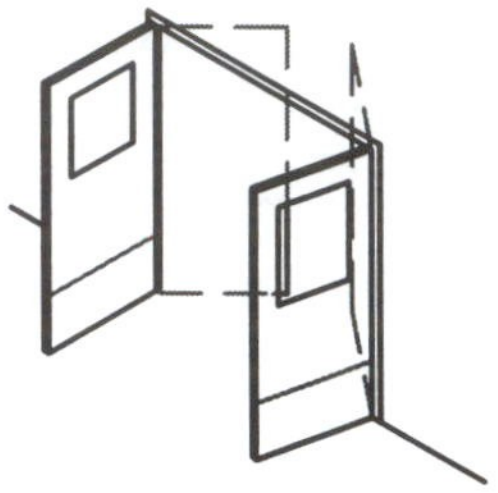	FOR EXAMPLE, ENTRY TO A KITCHEN AREA IN A RESTAURANT, OR STORAGE IN A SUPERMARKET.
OPPOSED DOUBLE DOORS OR IN AND OUT DOORS	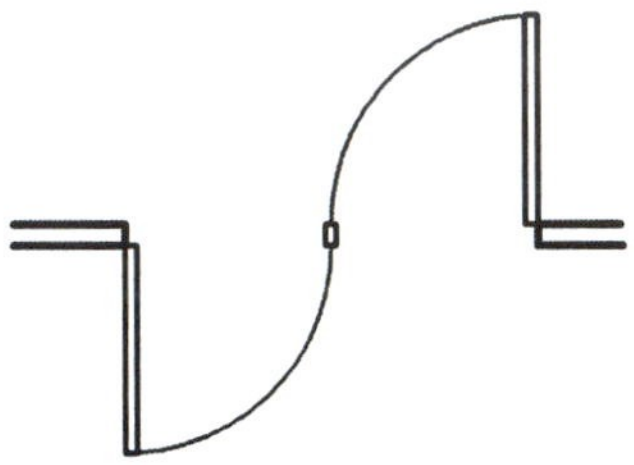	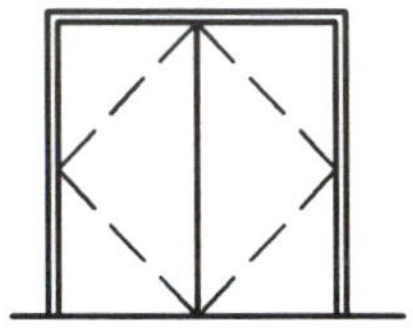	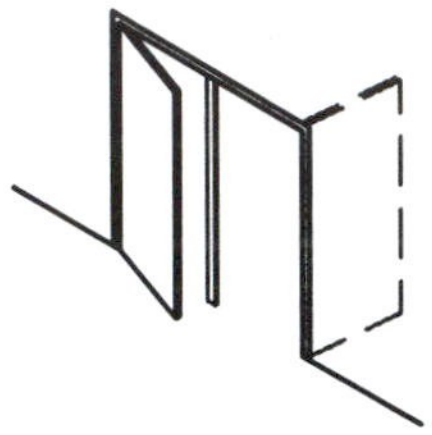	COMMONLY USED WITH SELF-CLOSERS IN IN HALLWAYS OF FACILITIES SUCH AS SCHOOLS & HOSPITALS WHERE THE HALLS ARE WIDE AND A FIRE BARRIER IS NEEDED.
POCKET DOOR		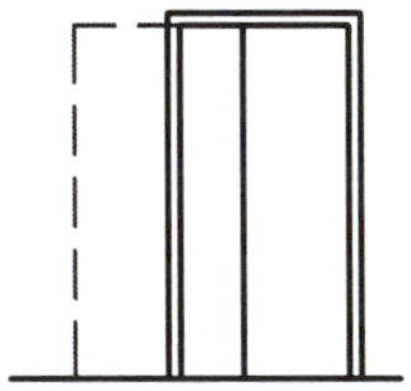	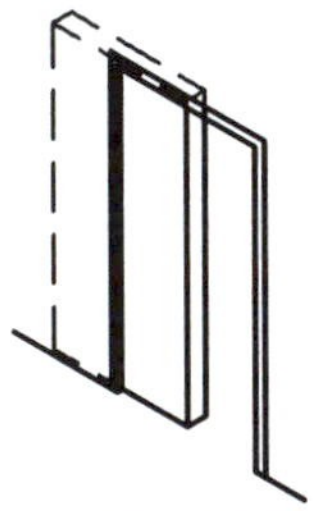	
DUTCH DOOR	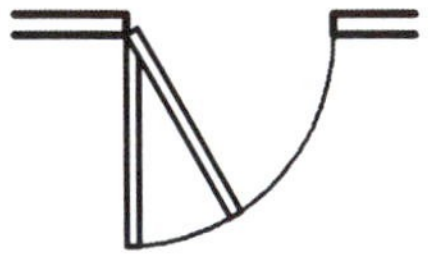	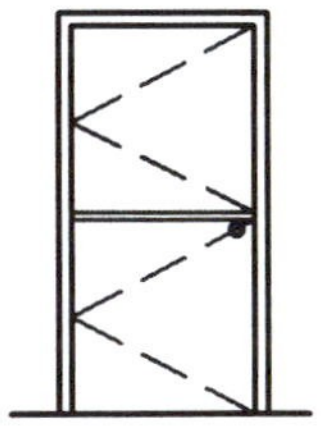		
SLIDING PATIO DOOR (GLASS)		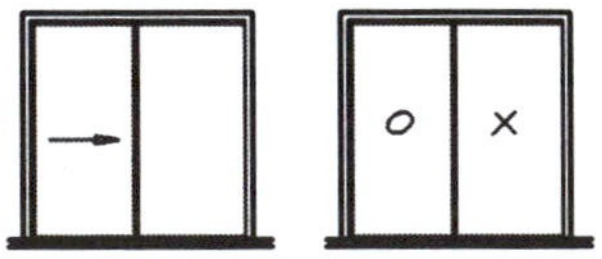		SHOWING TWO WAYS TO DENOTE SLIDING PANEL IN ELEVATION

DOOR

NAME	SYMBOL	ELEVATION	PERSPECTIVE	REMARKS
BI-FOLD DOORS		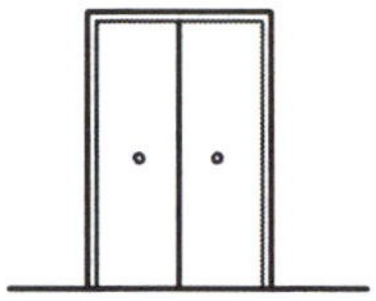	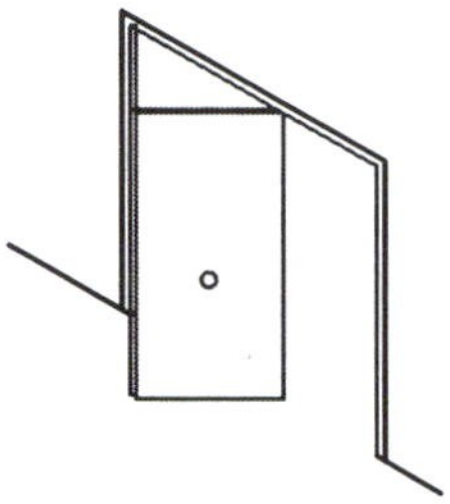	
DOUBLE BI-FOLD DOORS		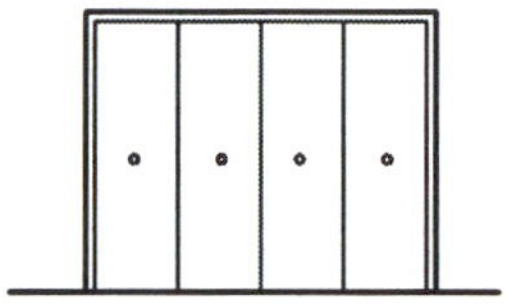	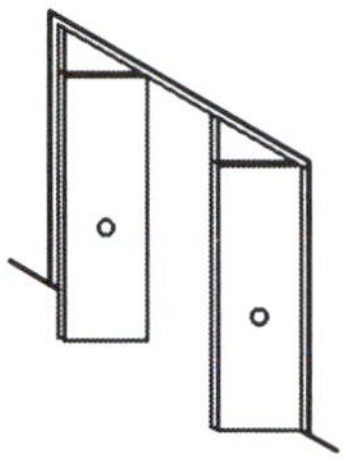	
ACCORD-ION DOOR			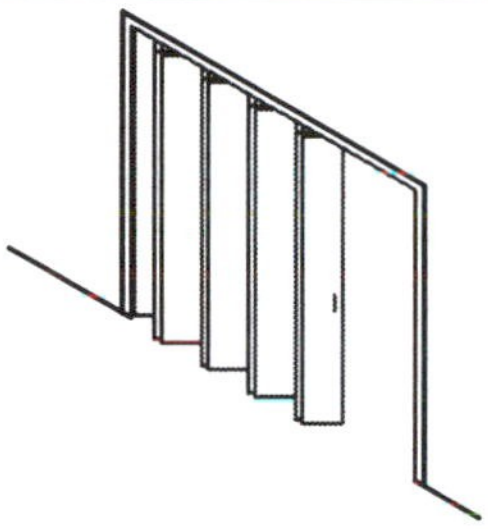	
REVOLV-ING DOOR	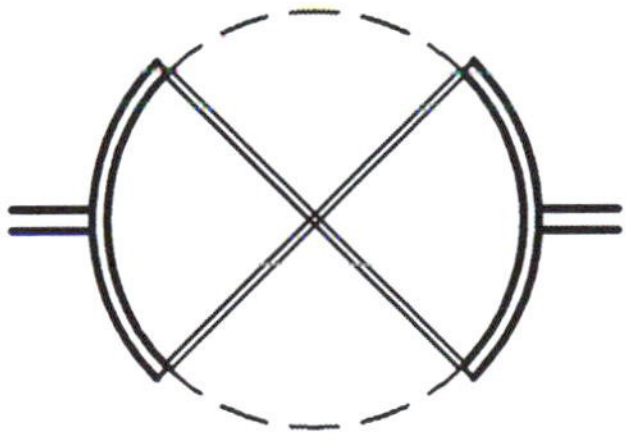	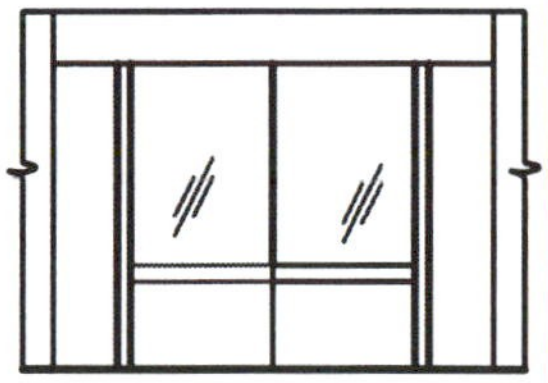	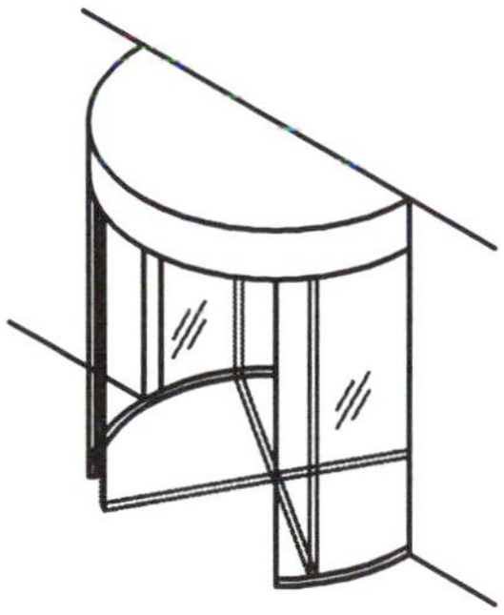	TYPICALLY DOORS AND WALLS OF GLASS. (PUSH BARS DELETED IN PERSPECTIVE ILLUSTRATION FOR CLARITY.)
GARAGE DOOR		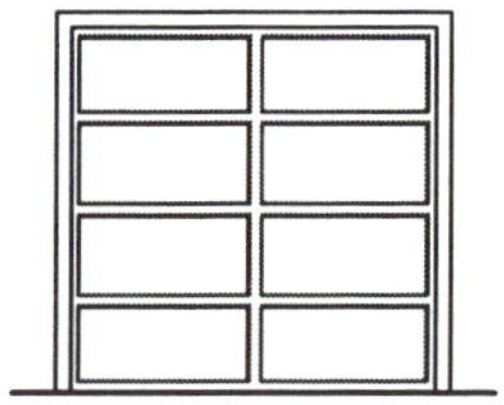	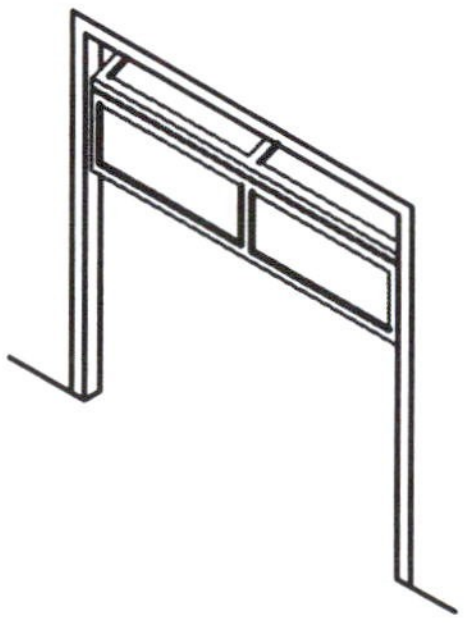	BOTH SINGLE UP-PIVOTING DOOR & HINGED SECTIONAL DOORS USE THE SAME SYMBOL.

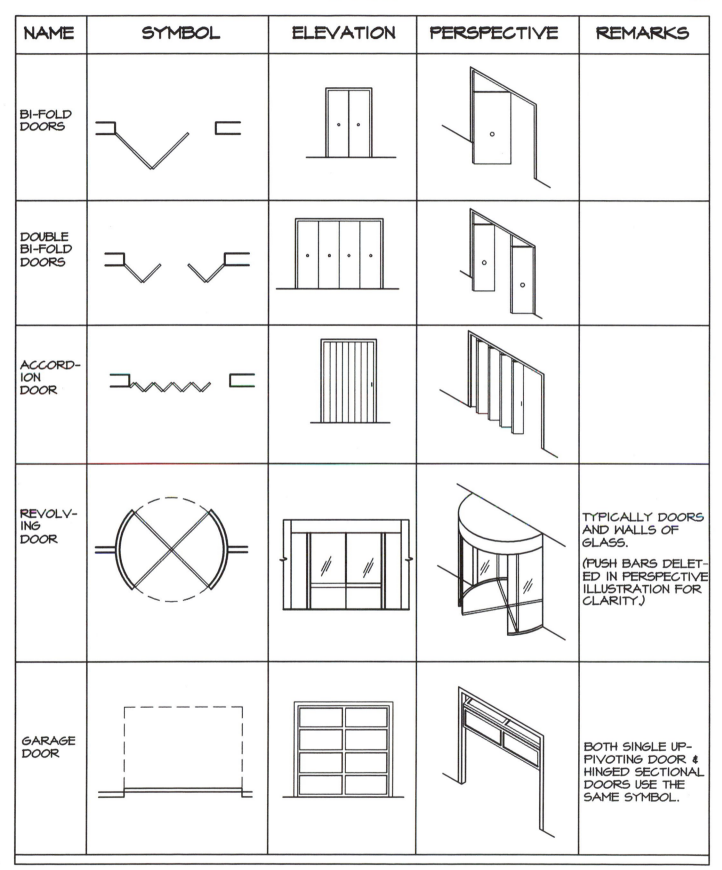

FARM BUILDINGS

NAME	SYMBOL	OBJECT	REMARKS
CORN CRIB	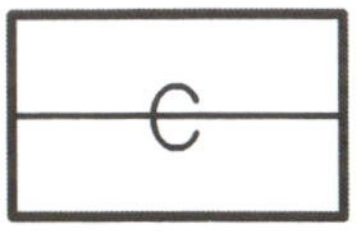 C	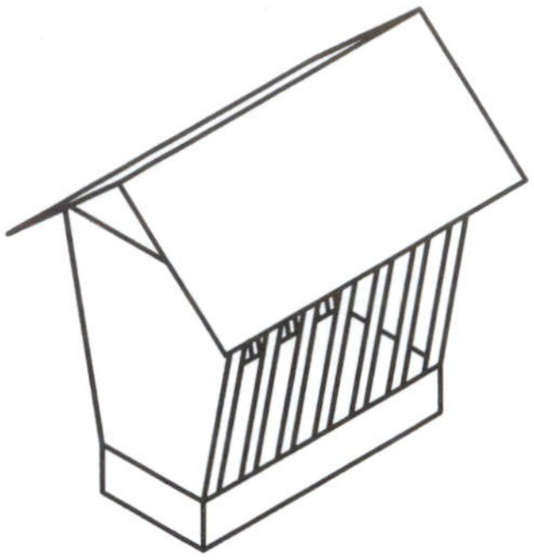	
HOUSE	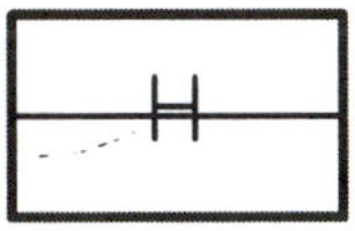 H		
BARN	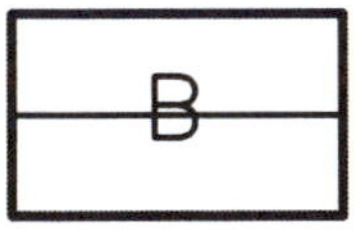 B		
SILO	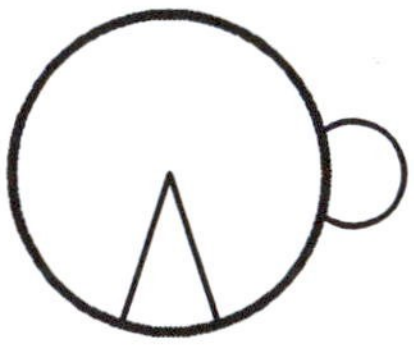 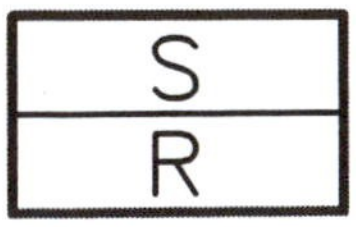	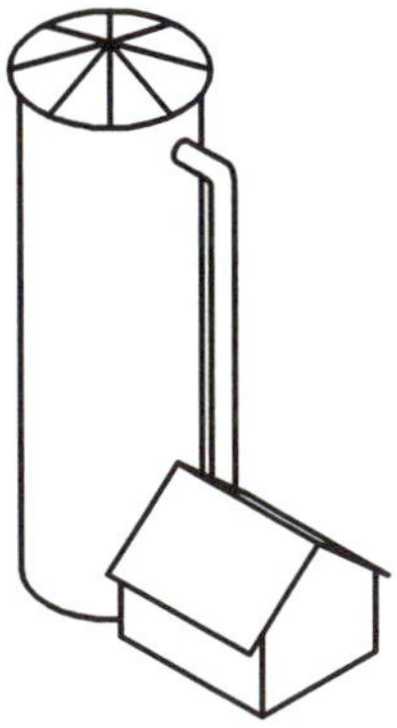	INDICATE FILLER DOOR & CHUTE LOCATION
SILO ROOM	S R		

NAME	SYMBOL	OBJECT	REMARKS
SHED OR SHELTER	S		
FEED BUNK	FB		
FEED STORAGE	F S		
FEED ROOM	F R		
GRANARY	G		
FEED GRINDER	F G		

NAME	SYMBOL	ELEVATION	REMARKS
MIXER	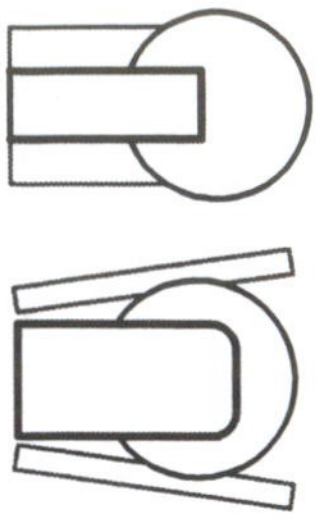	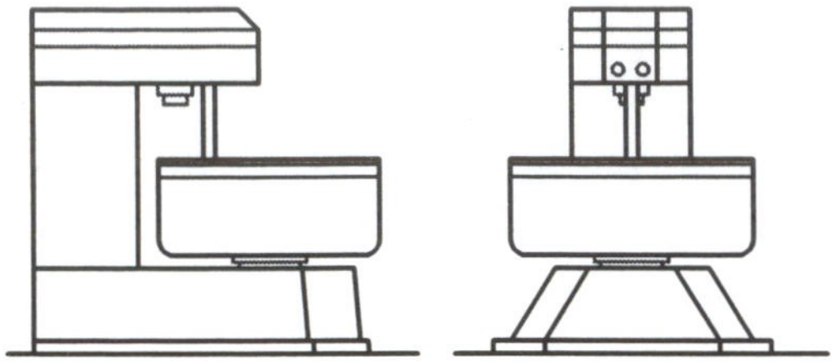	
SPRAY DOWN TABLE	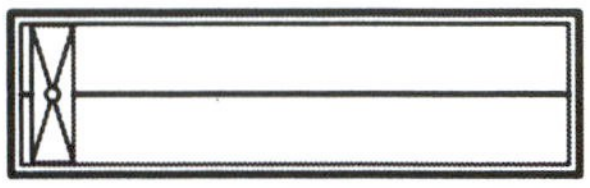		
BAGEL FORMER	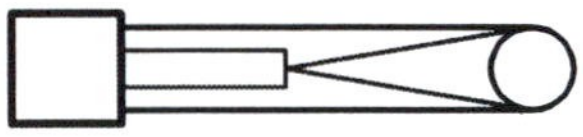	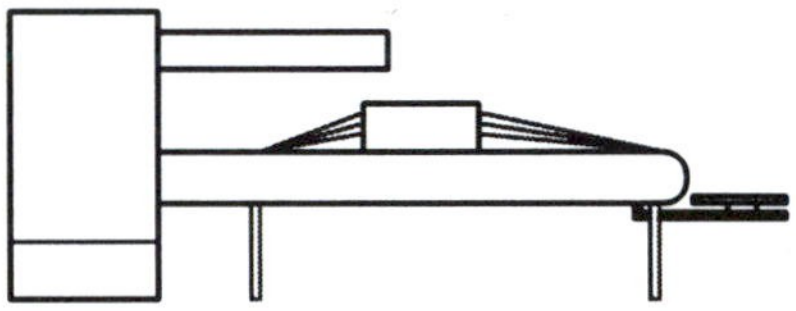	
ICE MAKER	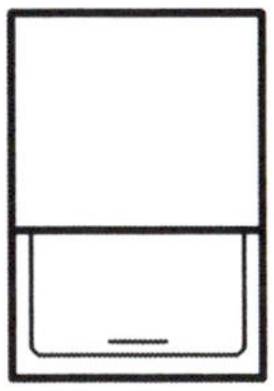	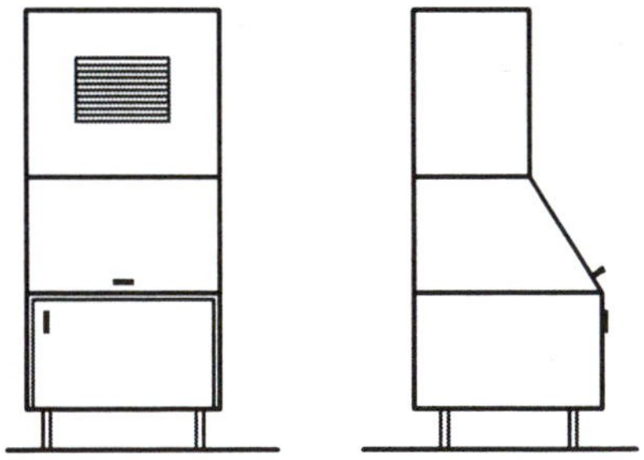	
SLICING MACHINE	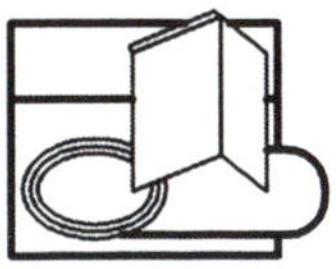		

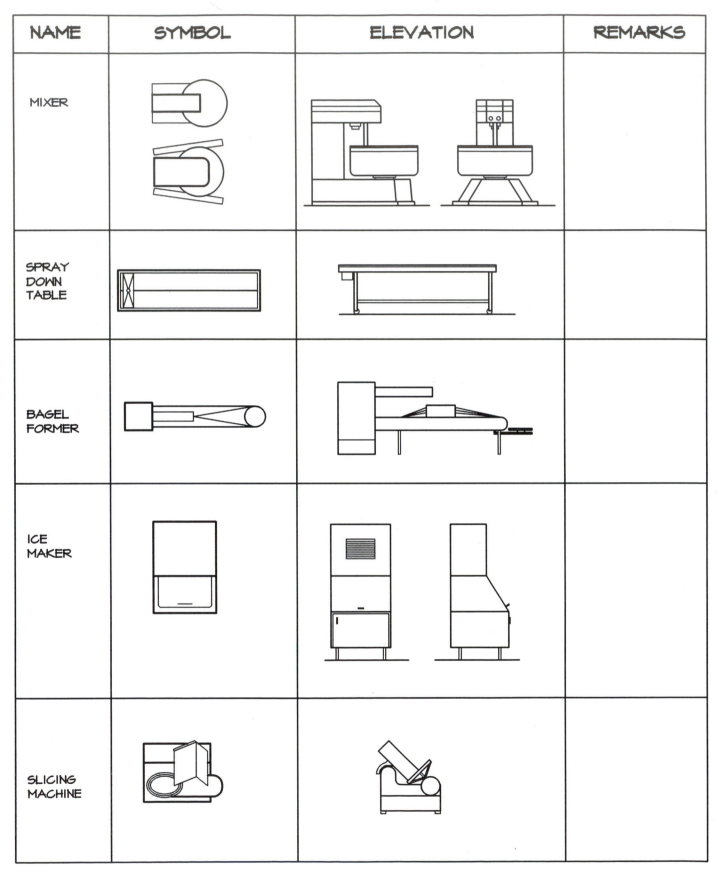

NAME	SYMBOL	ELEVATION	REMARKS
COFFEE WARMER (DOUBLE)	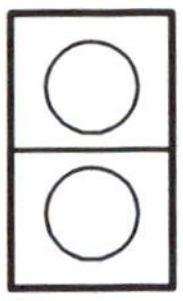	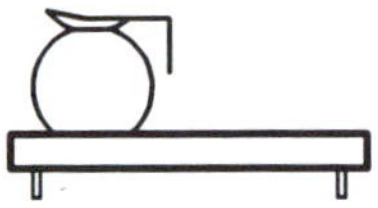	
TOASTER (CONVEYOR)	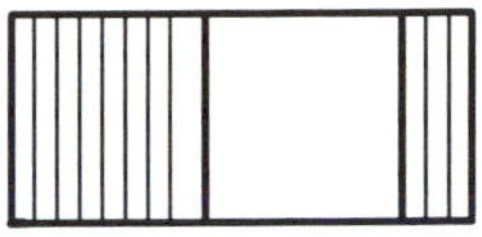	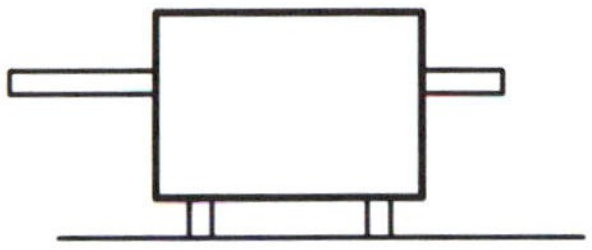	
SANDWICH UNIT (SMALL)	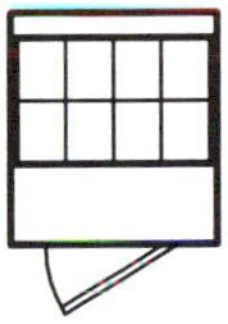	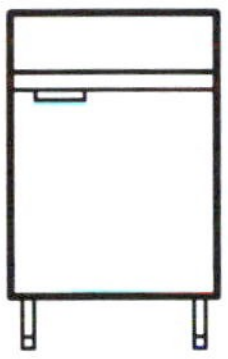	
SANDWICH UNIT (LARGE)	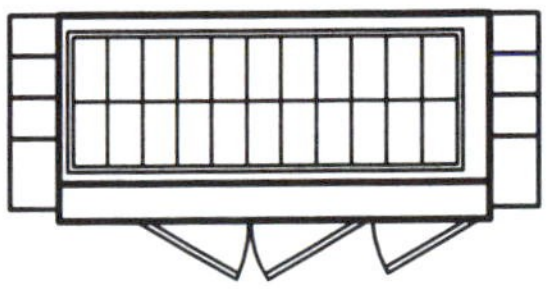	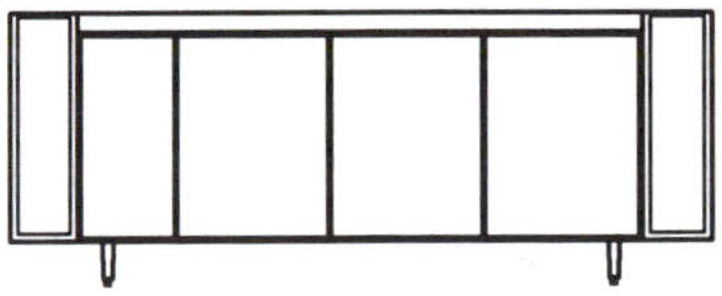	WITH ATTACHED CONDIMENT PANS
PIZZA OVEN	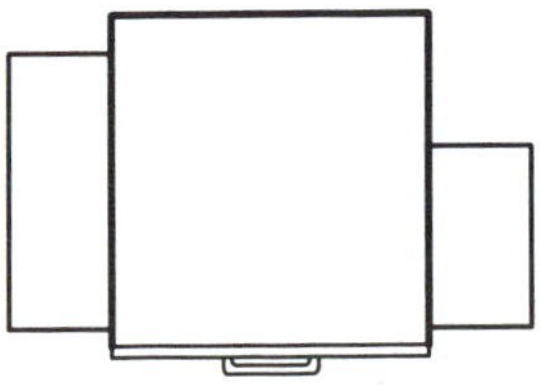		

NAME	SYMBOL	ELEVATION	REMARKS
REFRIG./ FREEZERS	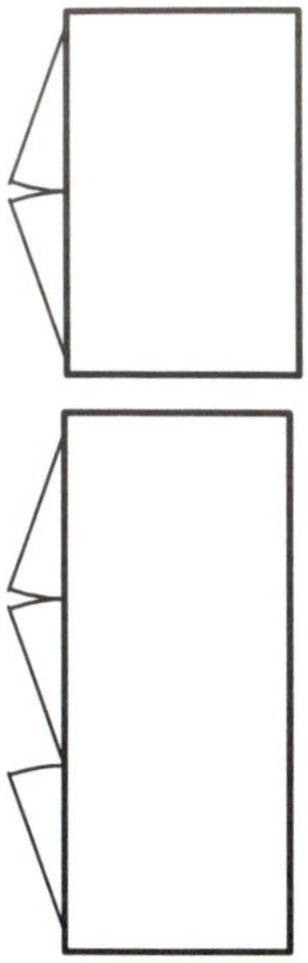		ELEVATIONS: UNDER COUNTER UNIT SHOWN ABOVE, FULL SIZE UNIT SHOWN BELOW.
KETTLE	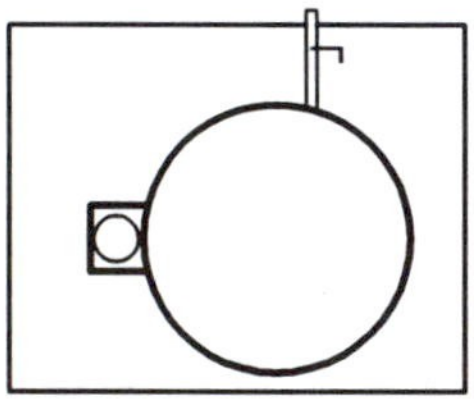		
TRASH	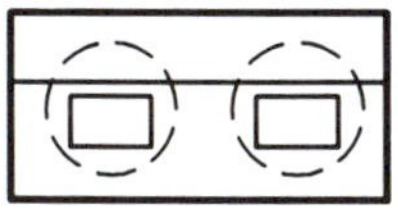	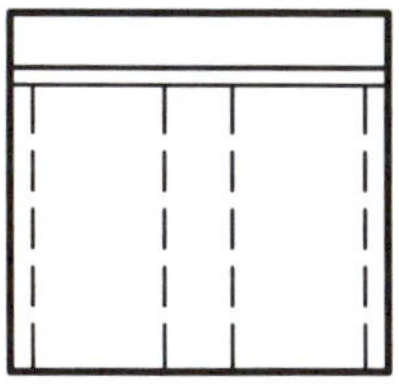	
AIR SCREEN REFRIG.	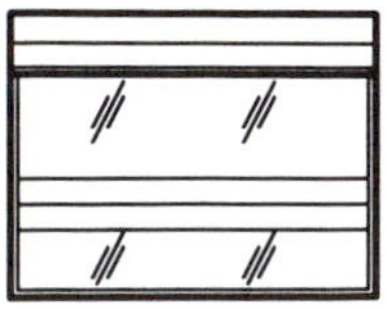	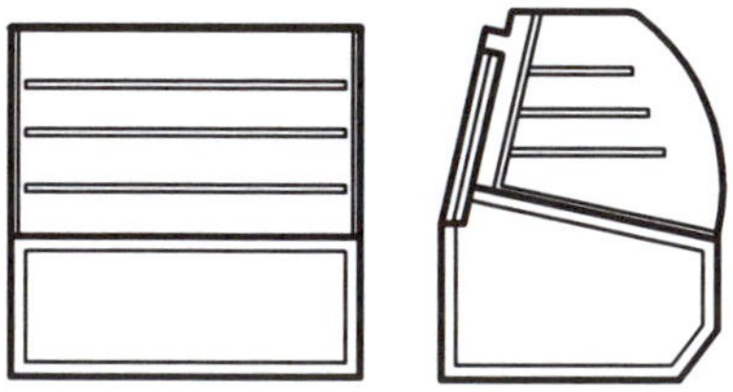	

NAME	SYMBOL	ELEVATION	REMARKS
DUAL COFFEE MACHINE			
COFFEE GRINDER		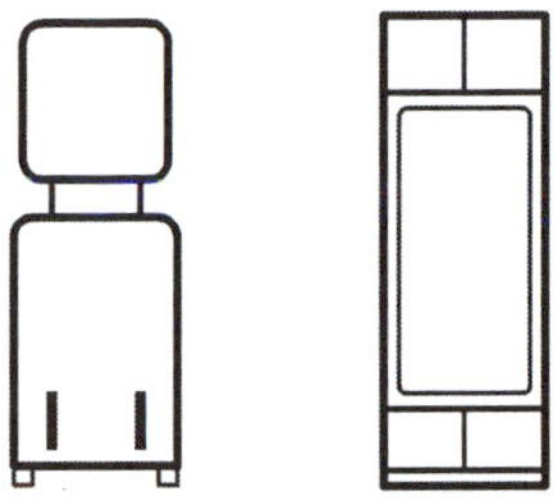	
TEA MACHINE	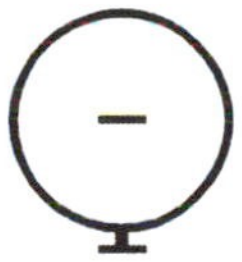	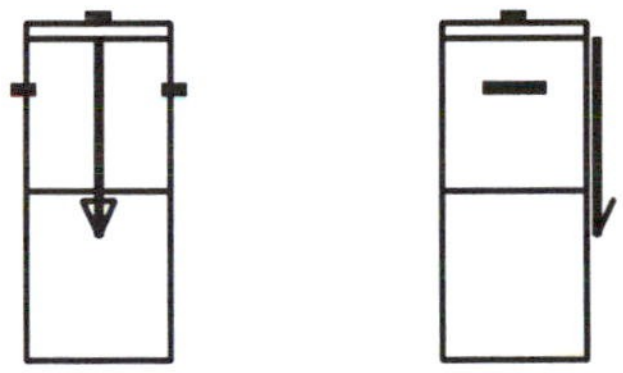	
SOFT DRINK DISPENSER		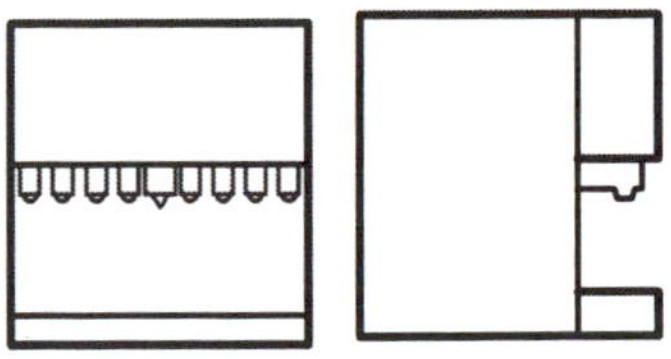	
CAPPUCINO MACHINE		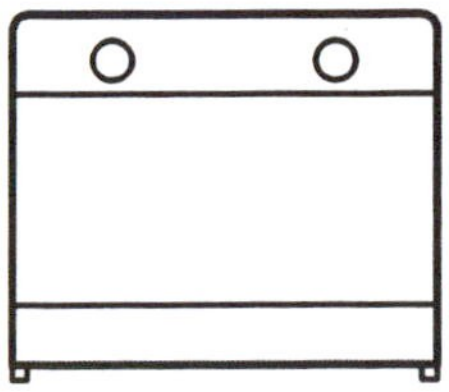	

NAME	SYMBOL	ELEVATION	REMARKS
ROLLING TABLE			
DISH WASHER			CORNER STYLE

WINDOWS

NAME	SYMBOL	ELEVATION	PERSPECTIVE	REMARKS
FIXED WINDOW	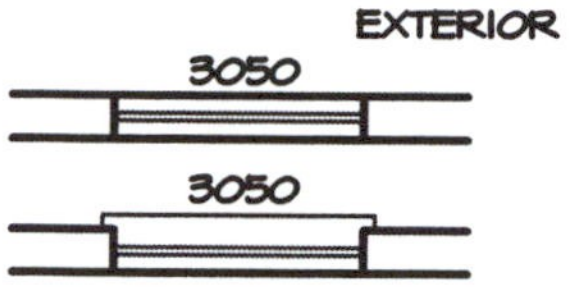 EXTERIOR 3050 3050	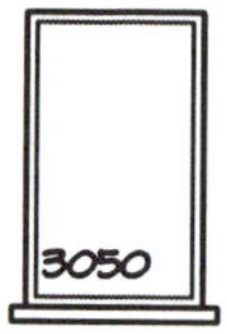 3050	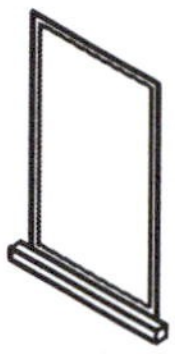	WINDOW SIZE MAY BE SHOWN MATHEMATICALLY BUT PREFERRED METHOD IS A LETTER IN A HEXAGRAM WHICH REFERENCES A WINDOW TABLE.
SINGLE HUNG WINDOW	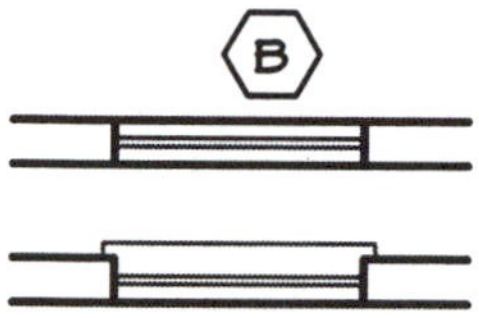 B	 B	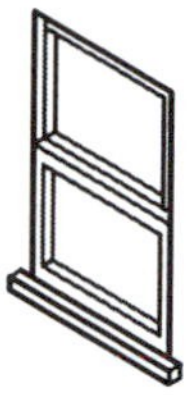	
DOUBLE HUNG WINDOW	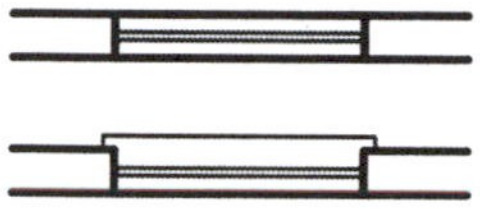	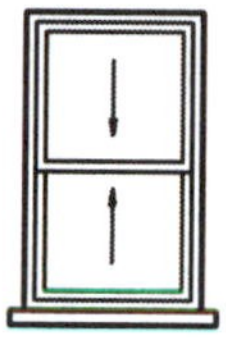	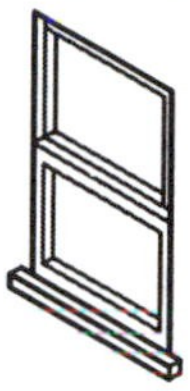	
SLIDING WINDOW	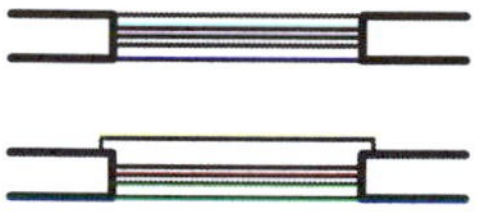			
CASEMENT WINDOW			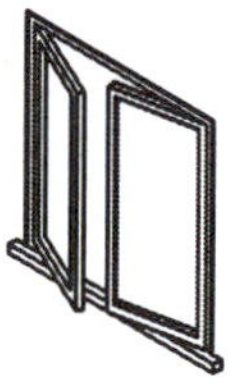	
AWNING WINDOW	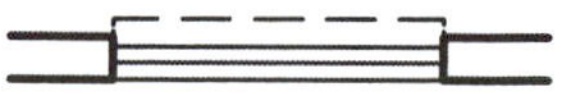 EXTERIOR	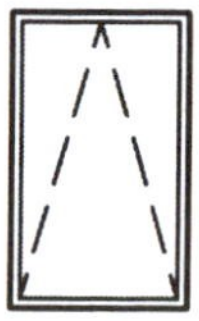	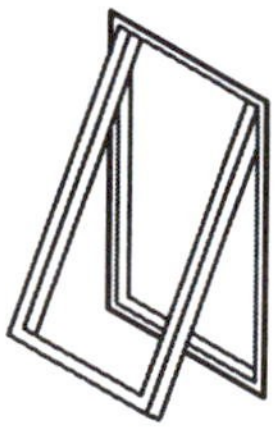	

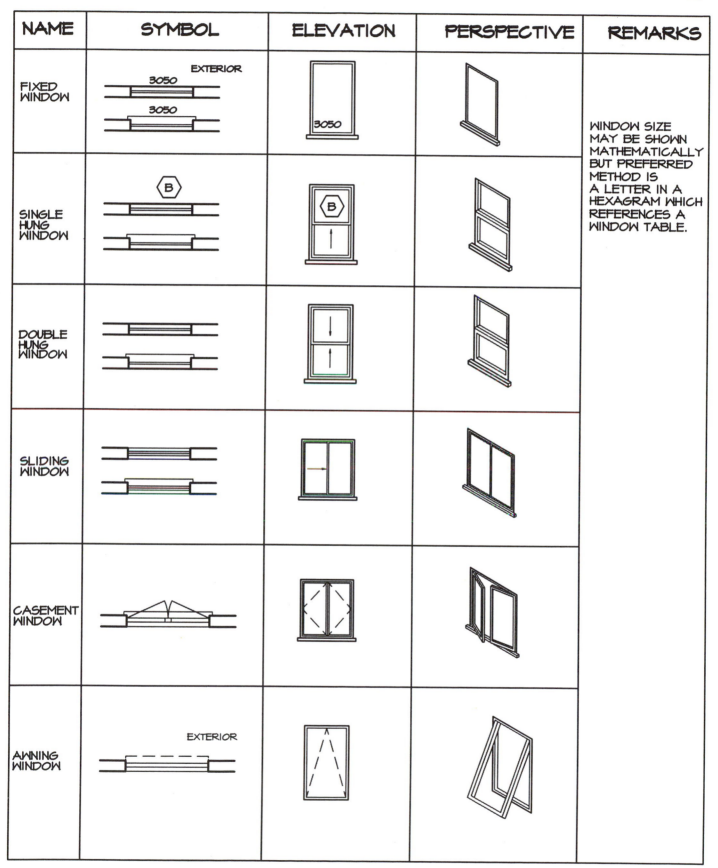

WINDOWS

NAME	SYMBOL	ELEVATION	PERSPECTIVE	REMARKS
HOPPER WINDOW	EXTERIOR 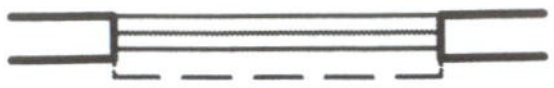		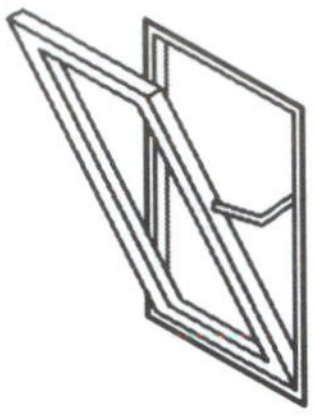	COMMONLY USED IN BATHROOMS.
HORIZON-TALLY PIVOTED WINDOW			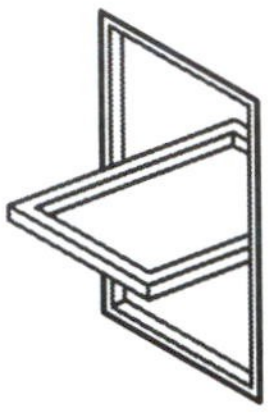	
COMMERC-IAL WINDOW	MULLIONS 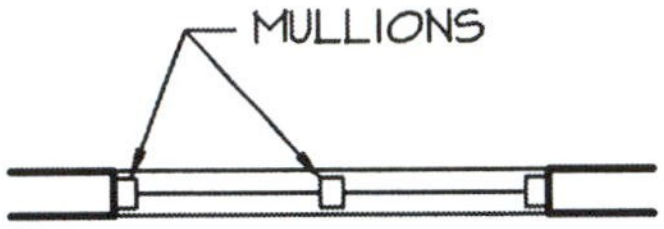			TYPICALLY BUILT ON SITE.
GLASS BLOCK WINDOW	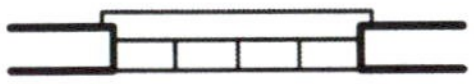	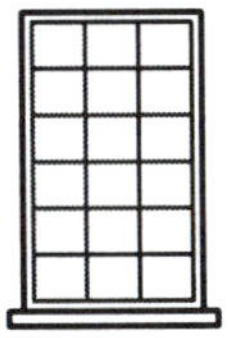		
JALOUSIE WINDOW	EXTERIOR 	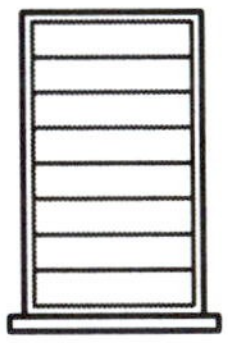	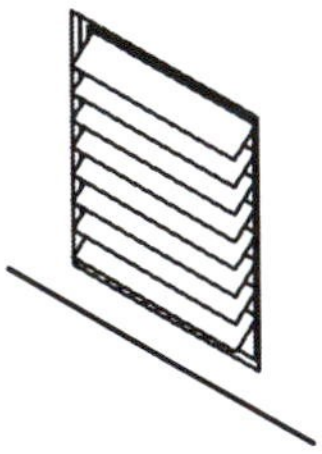	
STACKED WINDOWS	3030 FIXED ABOVE 3020 AWNING 			NOTE NOTATION.

GEOLOGICAL
&
TERRESTRIAL
SYMBOLS

NAME	SYMBOL	OBJECT	REMARKS
PLANT SYMBOLS			
			SAME SPECIES, DIFFERING SIZES
	a b		a = SPECIES b = SIZE

NAME	SYMBOL	OBJECT	REMARKS
VINE & SHRUB SYMBOLS		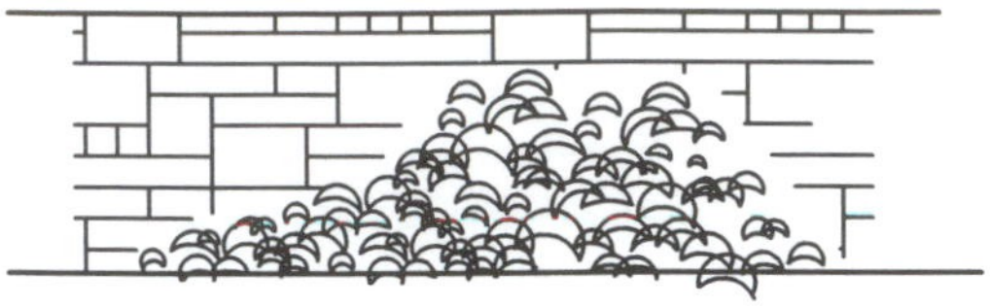	PARTHENOCISSUS QUINQUEFOLIA (VIRGINIA CREEPER)
	 W		WISTIRIA SINENSIS (JAPANESE WISTARIA)
TREE SYMBOLS	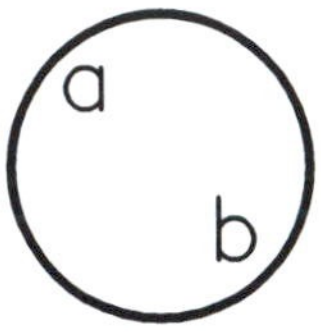 a b		a = SPECIES b = SIZE
	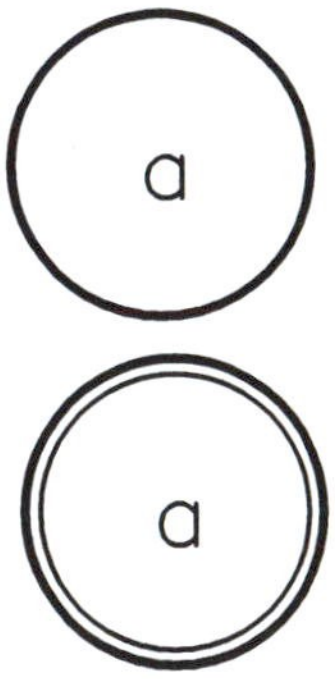 a a		SAME SPECIES, DIFFERENT HEIGHTS.

NAME	SYMBOL	OBJECT	REMARKS
CEMETERY	SMALL: [G] 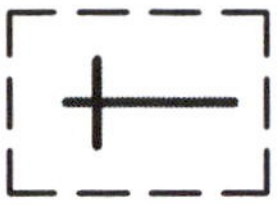 ; LARGE: [G]  Cem		HIDDEN LINETYPE
MAIL BOX	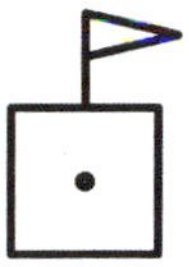		
STREET SIGN			
SCHOOL BUILDING			
CHURCH BUILDING			
GAGING STATION			

NAME	SYMBOL	ILLUSTRATION	REMARKS
CAMPGROUND	[G]		CAMPING MAY INCLUDE DINING, BUT NOT VICE VERSA.
PICNIC AREA	[G]		
STOCK TANKS			STOCK TANKS OF EARTHEN CONSTRUCTION.
EXPOSED WRECK			
SUNKEN WRECK			
CONTOUR LINES (EXISTING)		550	PLACE ELEVATION ON HIGH SIDE.

NAME	SYMBOL	ILLUSTRATION	REMARKS
NEW CONTOUR LINES		550	

TRANSPORTATION FEATURES

NAME	ABBREV.	LINETYPE	REMARKS
TRAIL OR UNIMPROVED DIRT ROAD	TRL UNIM DRT RD	[G]	
ROAD UNDER CONSTRUCTION	RD CONST		MAY BE IDENTIFIED WITH THE ABBREVIATION U.C.
UNPAVED ROAD	UNPV RD	[G]	
PAVED ROAD	PV		
TRAIL OR UNIMPROVED DIRT ROAD	TRL UNIM DRT RD	[G]	
ROAD UNDER CONSTRUCTION	RD CONST		MAY BE IDENTIFIED WITH THE ABBREVIATION U.C.

NAME	ABBREV.	LINETYPE	REMARKS
UNPAVED ROAD	UNPV RD	[G] 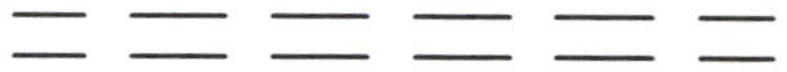	
PAVED ROAD	PV		
LEVEE WITH ROAD	LV RD	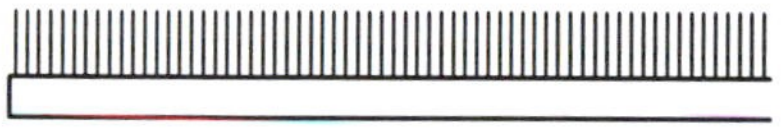 [G] 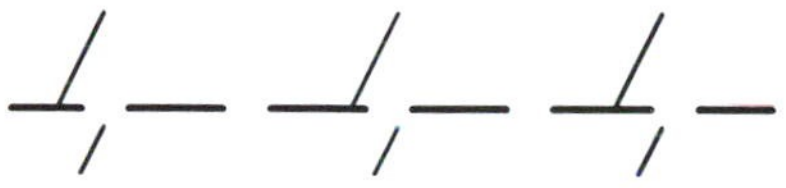	
HARD-SURFACE MEDIUM DUTY ROAD - 2 OR 3 LANES	HRD SUR MED DTY RD		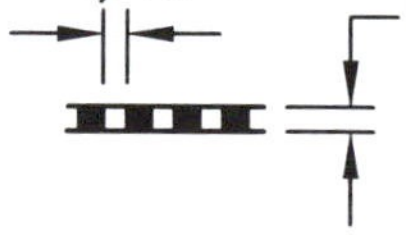 1/16" 1/16"
HARD-SURFACE MEDIUM DUTY ROAD - 4 OR MORE LANES	HRD SUR MED DTY RD		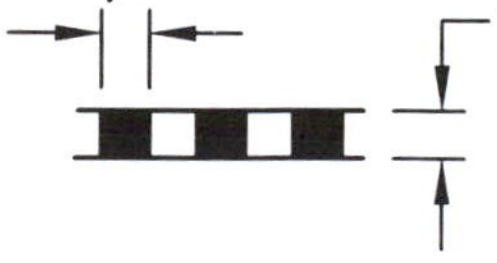 1/8" 1/8"

NAME	ABBREV.	LINETYPE	REMARKS
DUAL HIGHWAY WITH MEDIAN STRIP			
HARD-SURFACE HEAVY DUTY ROAD - 2 OR 3 LANES	HRD SUR HY DTY RD	[G] 	
HARD-SURFACE HEAVY DUTY ROAD - 4 OR MORE LANES	HRD SUR HY DTY RD		
BRIDGE OVER ROAD	BRG OV RD	[G] 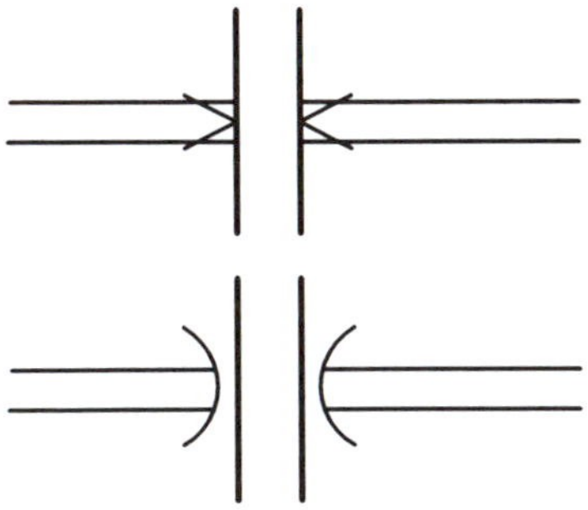	
DRAWBRIDGE		[G] 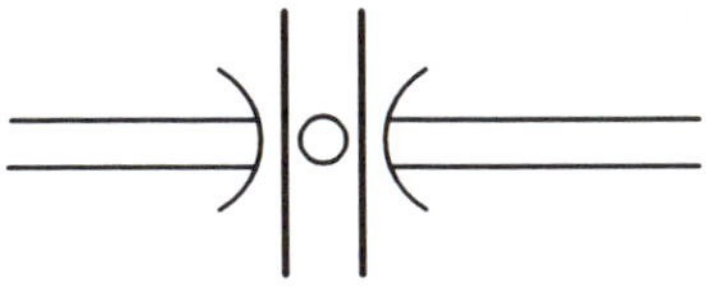	

NAME	ABBREV.	LINETYPE	REMARKS
RAILROAD TRACK	RR TRK	[G] 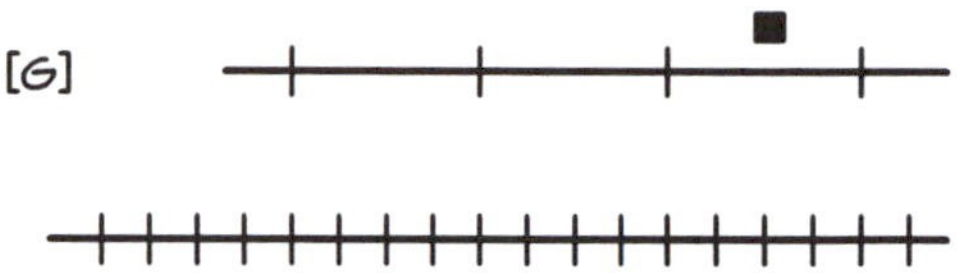	STANDARD GAUGE, SINGLE TRACK. SHOWN WITH TRAIN STATION.
RAILROAD TRACK, ABANDONED		[G] 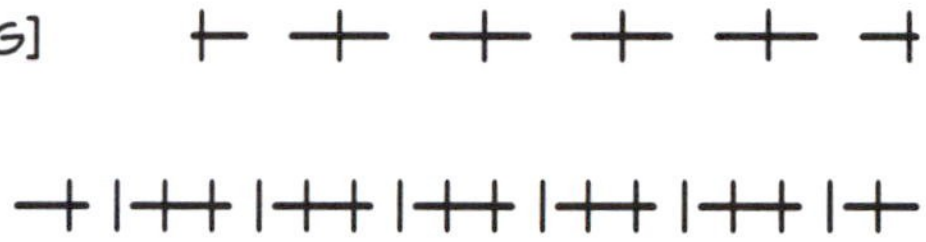	
RAILROAD TRACK, UNDER CONSTRUCTION		[G] 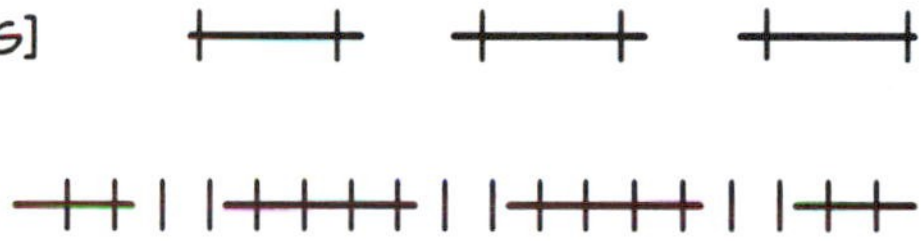	
RAILROAD, NARROW GAUGE SINGLE TRACK	RR TRK	[G] 	
RAILROAD, NARROW GAUGE SINGLE TRACK	RR TRK	[G] 	

NAME	ABBREV.	LINETYPE	REMARKS
RAILROAD TUNNEL	RR TUN	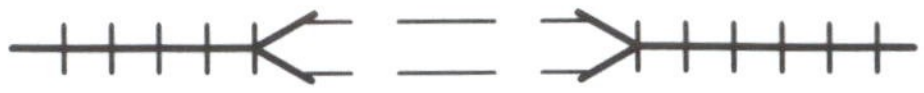	
ROUNDHOUSE AND TURNTABLE		[G] 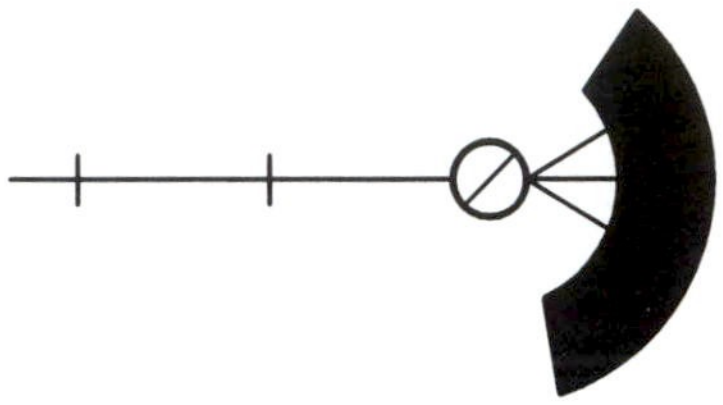	
ROAD OVERPASS	RD OVP	[G] 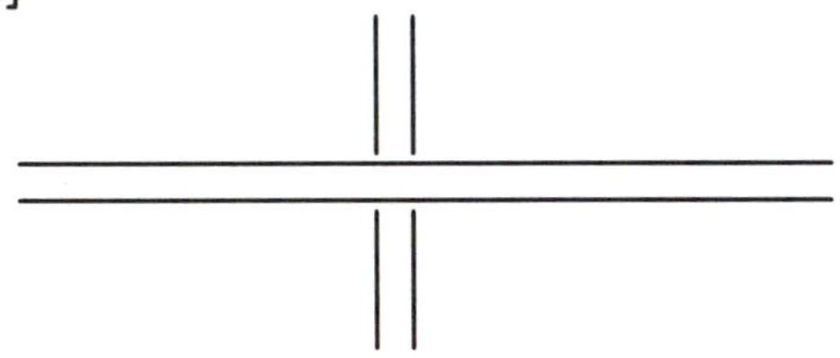	
ELEVATED AQUEDUCT	AQ TUN	[G] 	
AQUEDUCT TUNNEL	AQ TUN	[G] 	

OIL WELLS

NAME	SYMBOL	NAME	SYMBOL
LOCATION	[P]	DUAL COMPLETION - OIL	[P]
DRY HOLE	[P]	DUAL COMPLETION - GAS	[P]
DRY HOLE	[P]	DRILLED WATER-INPUT WELL	[P] W
ABANDONED OIL WELL	[P]	CONVERTED WATER-INPUT WELL	[P] W
GAS WELL	[P]	DRILLED GAS-INPUT WELL	[P] G
ABANDONED GAS WELL	[P]	CONVERTED GAS-INPUT WELL	[P] G
DISTILLATE WELL	[P]	BOTTOM-HOLE LOCATION (X = BOTTOM OF HOLE)	[P] X
ABANDONED DISTILLATE WELL	[P]	SALT-WATER DISPOSAL WELL	[P] SWD

GENERAL MINING

NAME	ABBREV.	SYMBOL	REMARKS
U.S. MINERAL OR LOCATION MONUMENT	U.S. MIN MON		
TUNNEL ENTRANCE	TUN ENT	[G] 	
MINE SHAFT	SHF	[G] 	
MINING AREA	MIN AR	[G] 	QUARRY OR OPEN PIT MINE.
BORROW PIT		[G] 	FOR GRAVEL, SAND, OR CLAY
MINE DUMP		[G] 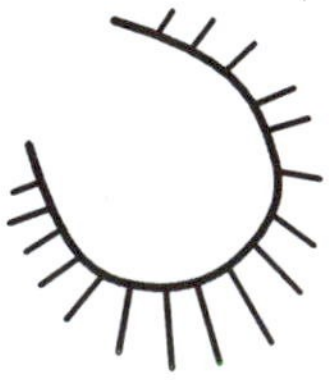	

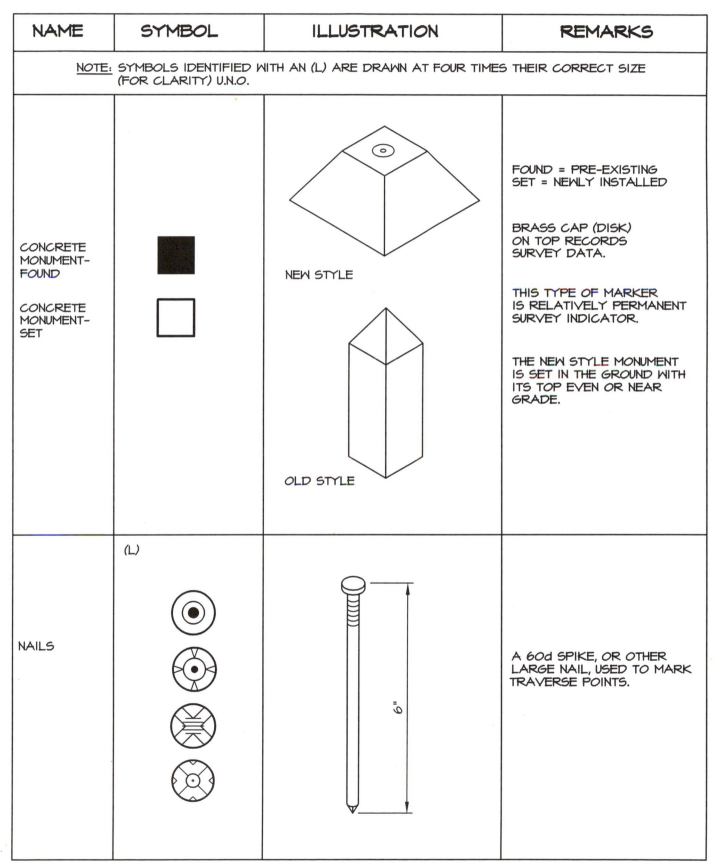

NAME	SYMBOL	ILLUSTRATION	REMARKS
NOTE: SYMBOLS IDENTIFIED WITH AN (L) ARE DRAWN AT FOUR TIMES THEIR CORRECT SIZE (FOR CLARITY) U.N.O.			
CONCRETE MONUMENT-FOUND CONCRETE MONUMENT-SET		NEW STYLE OLD STYLE	FOUND = PRE-EXISTING SET = NEWLY INSTALLED BRASS CAP (DISK) ON TOP RECORDS SURVEY DATA. THIS TYPE OF MARKER IS RELATIVELY PERMANENT SURVEY INDICATOR. THE NEW STYLE MONUMENT IS SET IN THE GROUND WITH ITS TOP EVEN OR NEAR GRADE.
NAILS	(L)	6"	A 60d SPIKE, OR OTHER LARGE NAIL, USED TO MARK TRAVERSE POINTS.

NAME	SYMBOL	ILLUSTRATION	REMARKS
PK NAIL	(L) 	 P•K	
NAIL & SHINER	(L) 		
NAIL & TAG	(L) 		
HUB & TACK	(L) 		A SQUARE WOODEN STAKE WITH THUMB TACK IN THE TOP TO IDENTIFY A TRAVERSE
PEG	(L) 		ARCHAIC. A ROUND WOODEN STAKE SERVING AS A TEMPORARY MARKER, INDICATING ELEVATION.
LEAD PLUG\ BRASS TACK	(L) 		USED TO MARK SURVEY POINT IN CONCRETE. SMALL HOLE IS FILLED WITH LEAD, AND A BRASS MARKER IS DRIVEN INTO THE CONCRETE. PERMANENT.

NAME	SYMBOL	ILLUSTRATION	REMARKS
BRASS DISK	(L) 		A DOME-SHAPED, SILVER DOLLAR SIZED MARKER SET IN CONC. MONUMENTS WITH ELEVATION OR CONTROL POINT INFORMATION.
SUB-DIVISION MONUMENT	(L) 	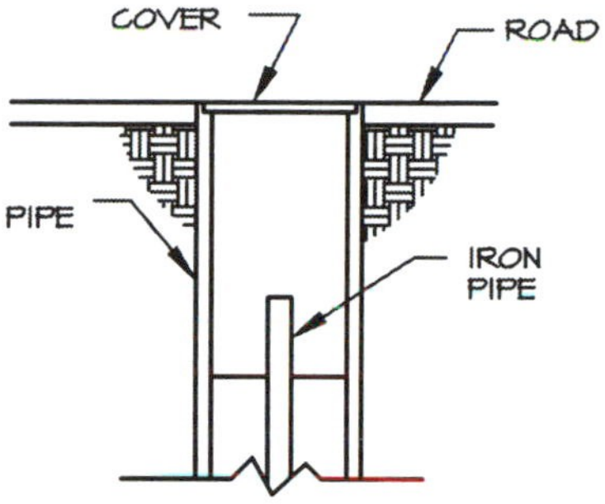 COVER ROAD PIPE IRON PIPE	TYPICALLY A 6" DIA. APPROX. STEEL PIPE SET IN A ROADWAY, CAPPED, WITH AN IRON PIPE SET IN THE GROUND WITHIN TO MARK SURVEY POINT.
TRANSIT SETUP	(L) 		
AERIAL PANEL	(L) 		AERIAL PANELS ARE LARGE CROSS SHAPED MARKERS PLACED ON THE GROUND AS REFERENCE POINTS IN AERIAL PHOTOGRAPHIC SURVEYING. (SYMBOL SHOWN AT 5:1 SCALE FOR CLARITY.)

NAME	SYMBOL	ILLUSTRATION	REMARKS
TRAVERSE POINT	(L)		
RANDOM ASPHALT SHOT	(L)		
RANDOM CONCRETE SHOT	(L)		
EXISTING ELEVATION POINT OR SPOT ELEVATION POINT	2150.0'		
NEW OR REQUIRED ELEVATION POINT	2140.0'		
TICK MARK	(L)		
GRID POINT	(L)		
COLUMNS	(L)		USE WHEN A BUILDING COLUMN IS A REFERENCE POINT IN A SURVEY. (SYMBOLS SHOWN AT 2:1 SCALE FOR CLARITY.)

NAME	SYMBOL	ILLUSTRATION	REMARKS
TOP OF CURB	(L) T/C + + T/C		
FLOW LINE	(L) + ℄		
V-MARK	(L)		V MARKS ARE USED ON THE CORNER OF STRUCTURES. A SURVEY POINT IS REFERENCED FROM THE MARK.
TICK MARK	(L)		GROUND SHOT MARKER.
FOUND IRON PIPE	(L)		(SYMBOLS SHOWN AT 2:1 SCALE.)
SET IRON PIPE	(L)		

NAME	SYMBOL	ILLUSTRATION	REMARKS
REBAR	(L)		A PIECE OF REINFORCING STEEL DRIVEN INTO THE GROUND LIKE A STAKE, TO MARK A PROPERTY CORNER.
FOUND RAILROAD SPIKE	(L)		3" TO 6½" OVERALL TYP. RAILROAD SPIKES ARE USED AS SEMI-PERMANENT SURVEY MARKERS. THE SURVEY POINT IS MARKED WITH A PUNCH IN THE TOP. IF SET IN PAVEMENT AND PAVED OVER, THE SPIKE MAY BE FOUND USING A METAL DETECTOR.
SET RAILROAD SPIKE	(L)		
BENCH MARK	BM 5280.0'		
RANDOM CONCRETE SHOT			
LEASE CORNER	(L)		INDICATES THE CORNER OF A PROPERTY WHICH HAS BEEN LEASED. FOR EXAMPLE, A SUMMER CAMP WITH A 99 YEAR LEASE.

NAME	SYMBOL	ILLUSTRATION	REMARKS
CROWS FOOT	(L)		A CROWS FOOT IS AN INCISED MARK MADE ON A BUILDING OR OTHER STRUCTURE TO INDICATE A LINE OF SURVEY, WHERE DIRECT ACCESS TO THE GROUND IS NOT AVAILABLE. FOR EXAMPLE, WHERE A A PROPERTY CORNER IS LOCATED WITHIN AN EXSTG. BUILDING, A CROWS FOOT ON THE EXTERIOR OF THE BLDG. WOULD SHOW THE LOCATION POINT FOR A FENCE.
L-CUT T-CUT	(L) (L)		L-CUTS AND T-CUTS ARE INCISED MARKERS MADE ON THE FACE OF CURBS. AN L-CUT INDICATES THE CORNER OF A SURVEYED AREA. A T-CUT INDICATES THE BORDER BETWEEN TWO SURVEYED AREAS. THEY ARE USED OFTEN IN AREAS LACKING OTHER MARKABLE FEATURES, E.G. A NEW SUBDIVISION WITH ONLY ROADS AND CURBS. (SYMBOL SHOWN AT 6:1 SCALE FOR CLARITY.)
FLAG	(L)		FLAGPOLE
WITNESS CORNER	[G] WC	WITNESS POST METAL SIGN METAL POST	A WITNESS CORNER IS A REFERENCE POINT TO LOCATE A SECTION CORNER OR TOWNSHIP CORNER WHEN THE ACTUAL SURVEY POINT CANNOT BE MARKED, E.G. WHEN THE SURVEY POINT IS UNDER WATER. (USED BY US GEOLOGICAL SURVEY.)

NAME	SYMBOL	OBJECT	REMARKS
NOTE: SYMBOLS LABLED (L) ARE SHOWN FOUR TIMES NORMAL SIZE (FOR CLARITY) U.N.O.			
THRUST BLOCK	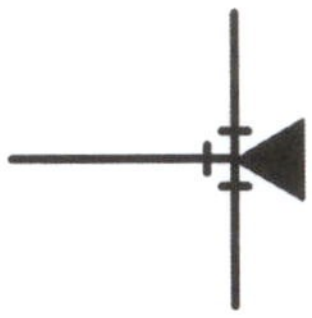	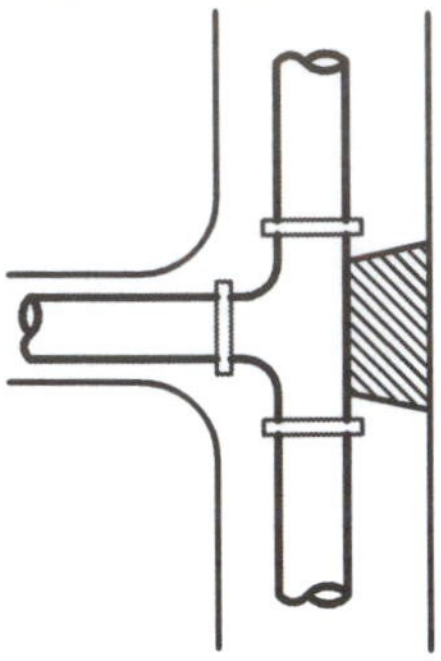 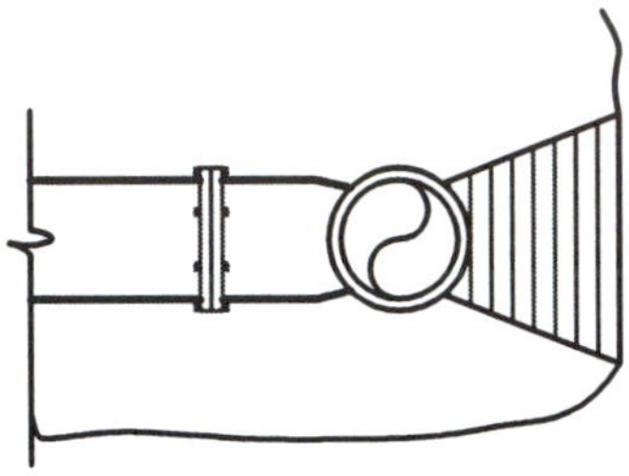	THRUST BLOCKS ARE USED IN VARIOUS POSITIONS. TYPICALLY THEY ARE UNDERNEATH AND ON THE OUTSIDE OF CONNECTIONS, & UNDERNEATH OBJECTS SUCH AS HYDRANTS. DEPICTED IN ELEVATION AND PLAN VIEWS. TYPICALLY THEY ARE CONCRETE.
MANHOLE	 KEY RIM = XXXXX INVERT = XXX OUT = XXXX  KEY RIM = XXXXX INVERT = XXX OUT = XXXX  M (L)  KEY	MANHOLE COVER 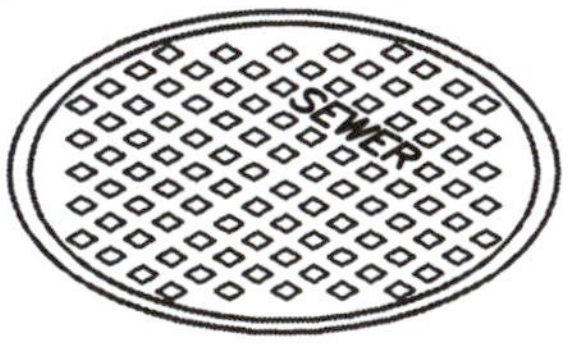 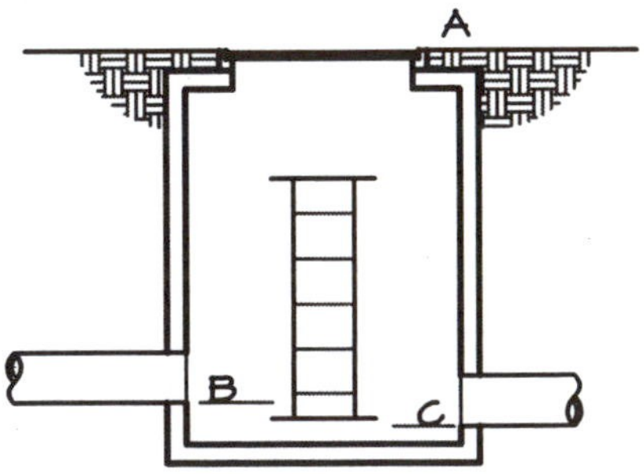 SEWER A B C MANHOLE SECTION	A = RIM ELEVATION B = INVERT ELEVATION C = OUTLET ELEVATION W = WATER MANHOLE P = POWER MANHOLE D = DRAIN MANHOLE E = ELECTRIC MANHOLE T = TELEPHONE MANHOLE S = SEWER MANHOLE U = UNKNOWN MANHOLE ST = STORM MANHOLE

NAME	SYMBOL	OBJECT	REMARKS
HANDHOLE	(L)  H	 WATER	ACCESS TO SHALLOW VALVES SUCH AS RESIDENTIAL WATER.
SEWER CLEANOUT	(L) 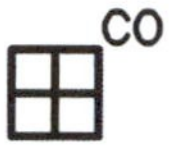 CO	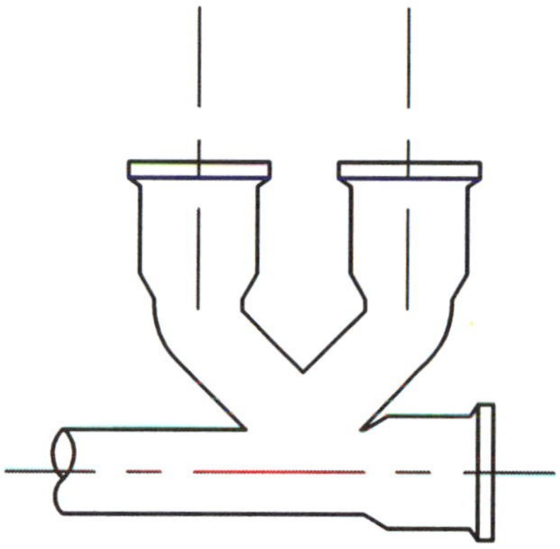	ILLUSTRATION SHOWS TWO-WAY SEWER CLEAN-OUT
BRASS DISK	(L) 		
BENCH MARK			
VALVE	(L)  KEY		FH = FIRE HYDRANT VALVE WV = WATER VALVE GV = GAS VALVE

NAME	SYMBOL	OBJECT	REMARKS
AIR VALVE	(L)  AIR		
BLOW-OFF VALVE	(L) 		
CATCH BASIN (INLET)	(L)  (L)  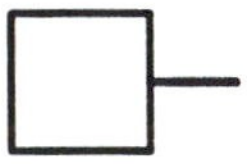 CB#	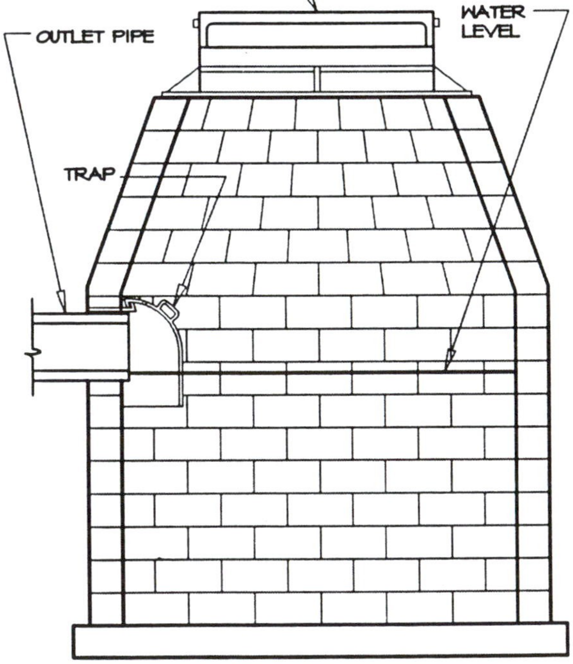 CATCH BASIN INLET OUTLET PIPE WATER LEVEL TRAP SECTION THRU CATCH BASIN	CATCH BASINS COLLECT SURFACE RUN-OFF AND CHANNEL IT INTO SEWER PIPE. CATCH BASIN TRAPS PREVENT CLOGGING OF SEWERS. ON SANITARY SEWERS THEY ALSO PROVIDE A WATERSEAL AGAINST ESCAPING GAS.

NAME	SYMBOL	OBJECT	REMARKS
STORM DRAIN	(L) 		
CONCRETE VAULT	(L) 	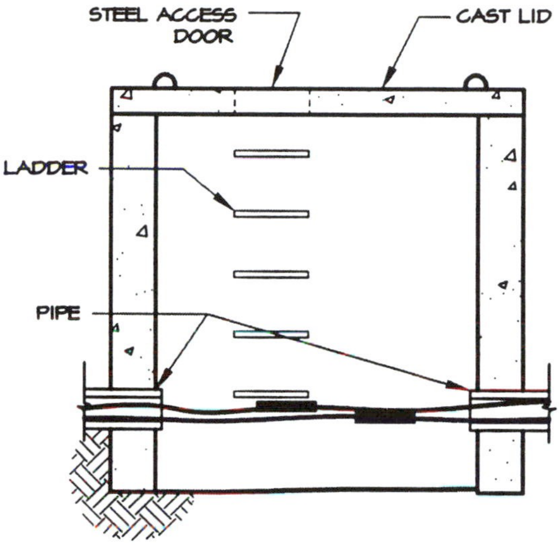 STEEL ACCESS DOOR CAST LID LADDER PIPE SECTION THRU CONC. VAULT	ILLUSTRATION SHOWS VAULT FOR UTILITY POWER LINES - ELECTRIC VAULT. E = ELECTRIC VAULT T = TELEPHONE VAULT V = UNSPECIFIED VAULT
TEST BORING	[IA] 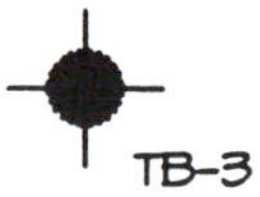 TB-3		
PULL BOX	KEY  (L) KEY 	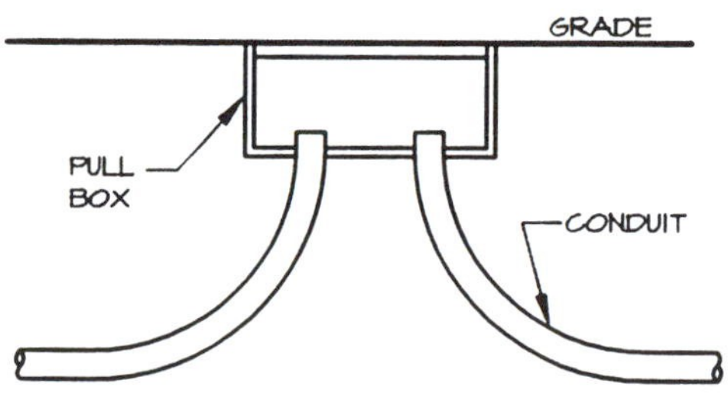 GRADE PULL BOX CONDUIT SECTION THRU PULL BOX	PULL BOXES ARE UNDERGROUND BOXES CONTAINING METERS FOR UTILITIES SUCH AS GAS ELECTRIC & WATER SUPPLY. ILLUSTRATION SHOWS BOX & CONDUIT ONLY. NOTE SWEEP OF PIPE. S = SIGNAL PULLBOX E = ELECTRIC TV = CABLE TV T = TELEPHONE U = UNKNOWN

NAME	SYMBOL	OBJECT	REMARKS
WELL	(L) 		
SUMP PIT	 (L) 	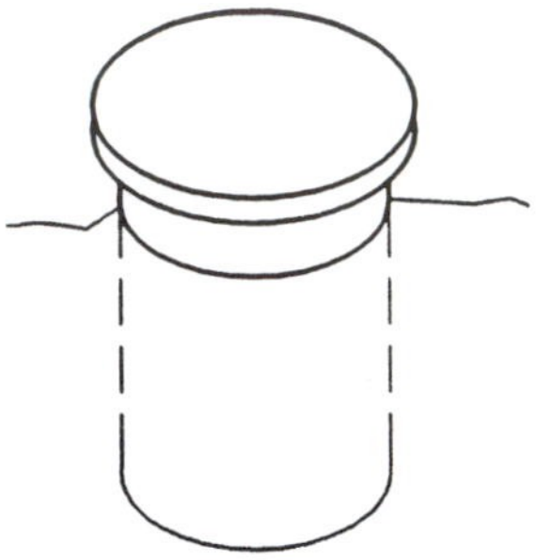	
ELECTRIC METER		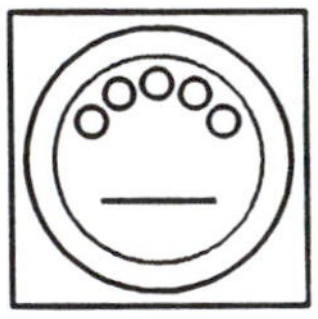	
SPRINKLER HEAD (LAWN)	(L) 	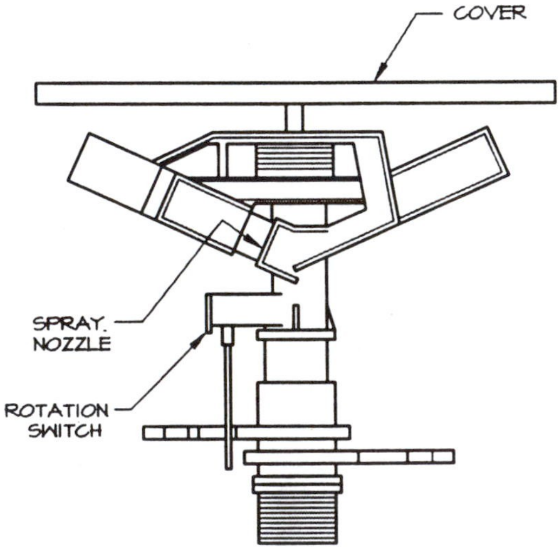 COVER SPRAY NOZZLE ROTATION SWITCH	POP-UP STYLE LAWN SPRINKLER HEAD SHOWN IN ELEVATION.

NAME	SYMBOL	OBJECT	REMARKS
UTILITY POLE	KEY KEY	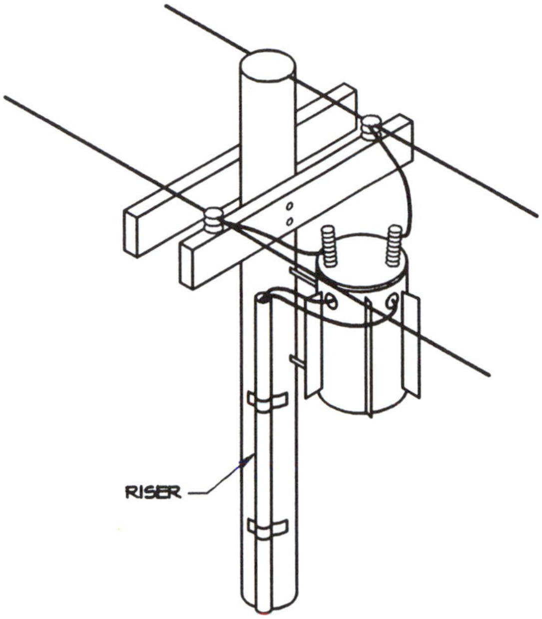 RISER	PP = POWER POLE TP = TELEPHONE E = ELECTRIC U = UNKNOWN ILLUSTRATION SHOWS TRANSFORMER ON UTILITY POLE.
TRANSFORMER	 (L) 		A RISER IS A VERTICAL CONDUIT, TYPICALY A PIPE.
RISERS: ELECTRIC	(L) 		
GAS	(L) 		
VENT			
WATER			
WATER\ ELECTRIC			
TELEPHONE			

NAME	SYMBOL	OBJECT	REMARKS
JOINT POLE	(L) 		A UTILITY POLE USED FOR MORE THAN ONE PURPOSE, E.G. POWER LINES AND TELEPHONE LINES.
LIGHTNING ROD			
LIGHTNING CONDUCTOR	 ——— L ———		
BALLARD OR GUARD POST	 GP	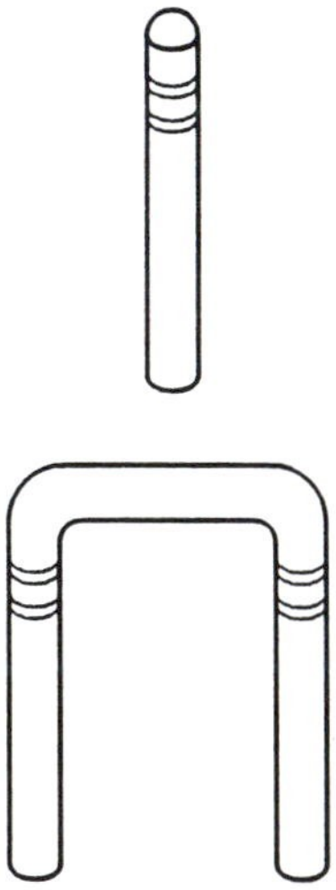	BALLARDS ARE USED TO PROTECT DEVICES FROM TRAFFIC. E.G. AT GAS PUMPS IN A GAS STATION, OR A FIRE HYDRANT IN A PARKING LOT.
GUARD RAIL	(L) 		
PARKING METER	(L) 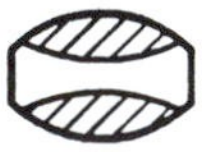		

NAME	SYMBOL	OBJECT	REMARKS
FILL CAP	(L) 		ACCESS TO UNDERGROUND FUEL TANKS, E.G. AT GAS STATIONS AND AIRPORTS.
FUEL PIT	(L)  F		UNDERGROUND FUEL STORAGE.
ANODE	(L) 		ANODES AND CATHODES ARE PROTECTIVE GROUNDING DEVICES, WHERE AN ELECTRICAL CHARGE WOULD BUILD UP IN A PIPE CARRYING MOVING WATER.
CATHODE	(L) 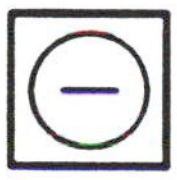		
PIEZOMETER	(L)  P		A DEVICE IN DAM STRUCTURES TO MEASURE DEPTH OR SETTLING.
GAS METER	(L)  G	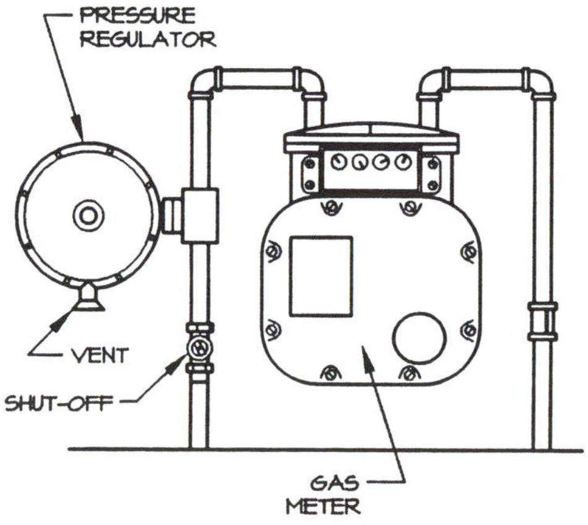 PRESSURE REGULATOR VENT SHUT-OFF GAS METER	ILLUSTRATION SHOWS TYPICAL RESIDENTIAL METER WITH PRESSURE REGULATING VALVE.

NAME	SYMBOL	OBJECT	REMARKS
WATER METER	(L)  W		
CABLE TV BOX	(L)  TV		TYPICALLY MOUNTED ON BUILDING.
SERVICE WEATHER HEAD		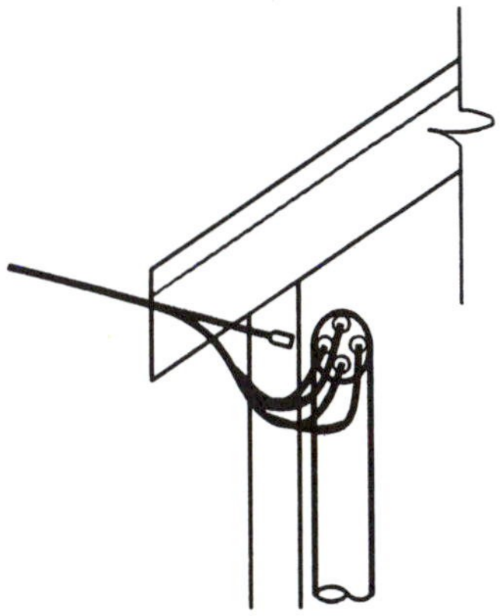	NOTE SUPPORT CABLE BOLTED TO BUILDING, IN ILLUSTRATION.
CESS POOL OR SEPTIC TANK	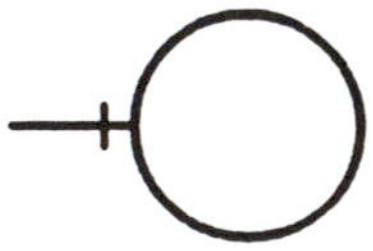		
DRY WELL	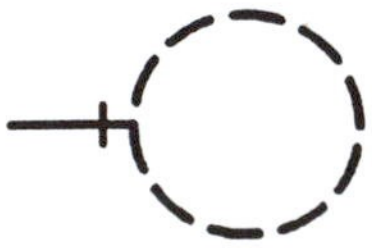		

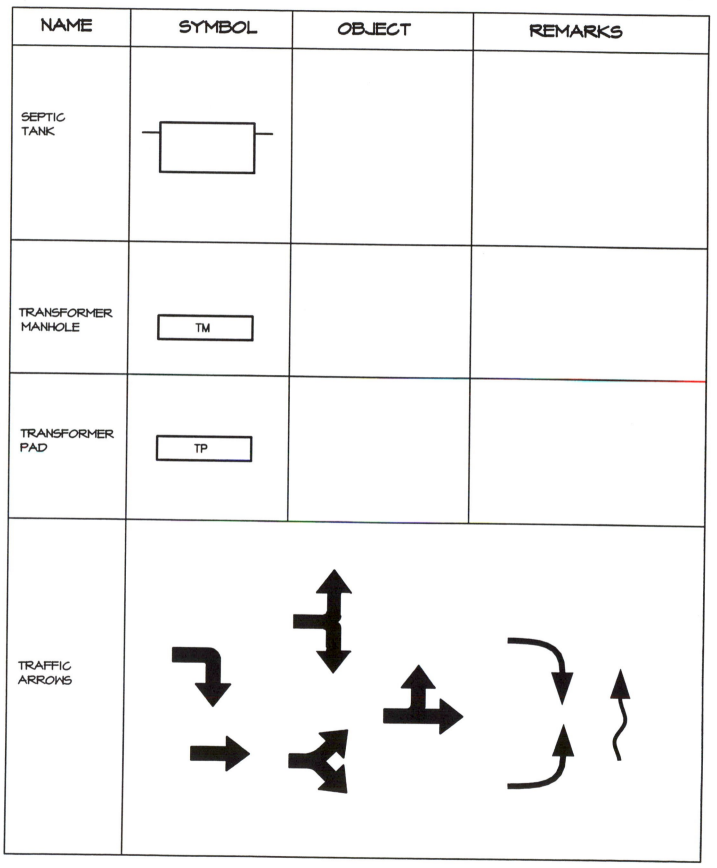

NAME	SYMBOL	OBJECT	REMARKS
SEPTIC TANK			
TRANSFORMER MANHOLE	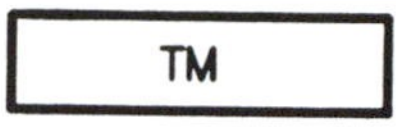 TM		
TRANSFORMER PAD	 TP		
TRAFFIC ARROWS	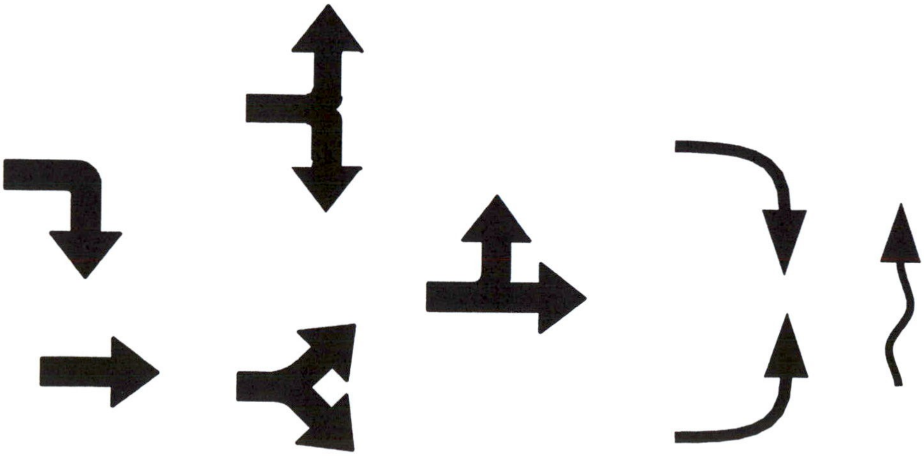		

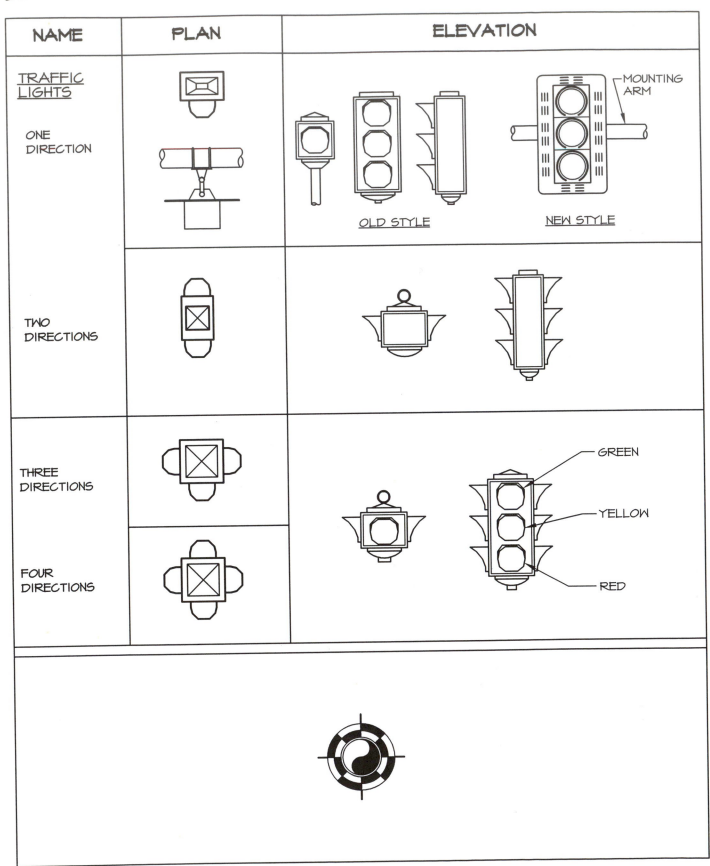

NAME	PLAN	ELEVATION
TRAFFIC LIGHTS ONE DIRECTION		MOUNTING ARM OLD STYLE NEW STYLE
TWO DIRECTIONS		
THREE DIRECTIONS		GREEN YELLOW RED
FOUR DIRECTIONS		

ELECTRICAL SYMBOLS

SYMBOL	NAME	OBJECT	REMARKS
	TELEPHONE OULET		TELEPHONE JACK. WHEN WALL MOUNTED, 16" AFF.
	FLOOR MOUNTED PRIVATE TELEPHONE OULET		
	FLOOR MOUNTED PUBLIC TELEPHONE OULET		PUBLIC PAY PHONE, COMMONLY.
D	DATA OUTLET		
D	FLOOR MOUNTED DATA OULET		
	COMBINATION TELEPHONE AND DATA OUTLET		2 TO 8 OUTLETS IN ONE PLATE.
	COMBINATION TELEPHONE AND DATA FLOOR OUTLET		

SYMBOL	NAME	OBJECT	REMARKS
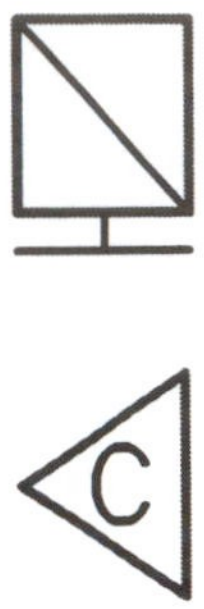	COMPUTER OUTLET		
	TELEPHONE TERMINAL CABINET OR BACKBOARD		
 TV	TELEVISION OUTLET	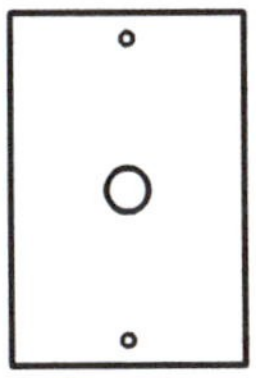	CABLE OUTLET
	CLOSED CIRCUIT TELEVISION CAMERA	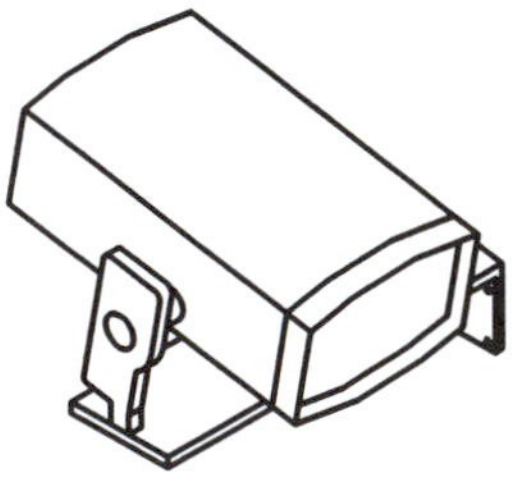	SHOWN MOUNTED ON WALL BRACKET.

SYMBOL	NAME	OBJECT	REMARKS
	SOUND SYSTEM GENERAL SYMBOL-USE WITH KEY.		FOR EXAMPLE: 1 = AMPLIFIER 2 = MICROPHONE 3 = SPEAKER ETC.
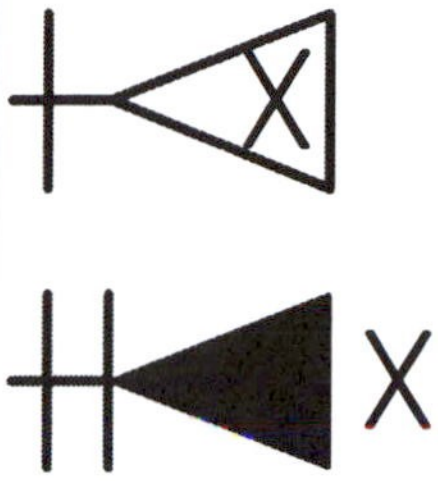	PUBLIC TELEPHONE GENERAL SYMBOL-USE WITH KEY.		FOR EXAMPLE: 1 = SWITCHBOARD 2 = WALL PHONE ETC.
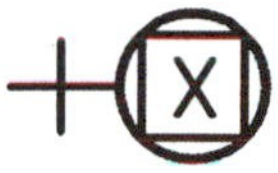	GENERIC SIGNAL SYSTEM DEVICE-USE WITH KEY.		FOR EXAMPLE: 1 = HORN 2 = SIREN ETC.
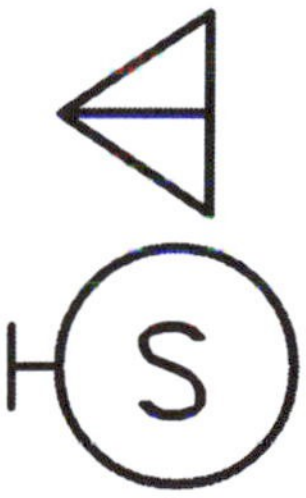	WALL MOUNTED SPEAKER		TOP SYMBOL IS PREFERABLE.
	RECESSED SPEAKER		
	MICROPHONE WALL RECEPTACLE		

SYMBOL	NAME	OBJECT	REMARKS
C	CLOCK		
G	YARD GONG		
B	BELL		
	COMBINATION BELL & BUZZER		
D	DICTATION OUTLET		USED IN SOME OFFICE FACILITIES TO CONNECT TO THE DICTATION POOL.

SYMBOL	NAME	ILLUSTRATION	REMARKS
 C	CHIME		RESIDENTIAL STYLE CHIME ILLUSTRATED.
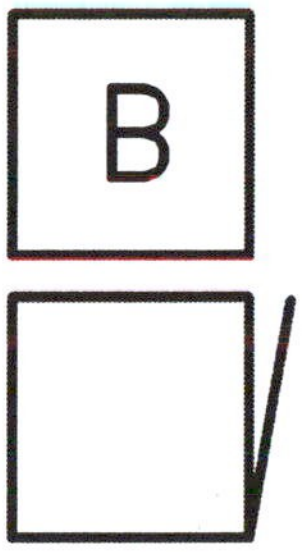 B	BUZZER	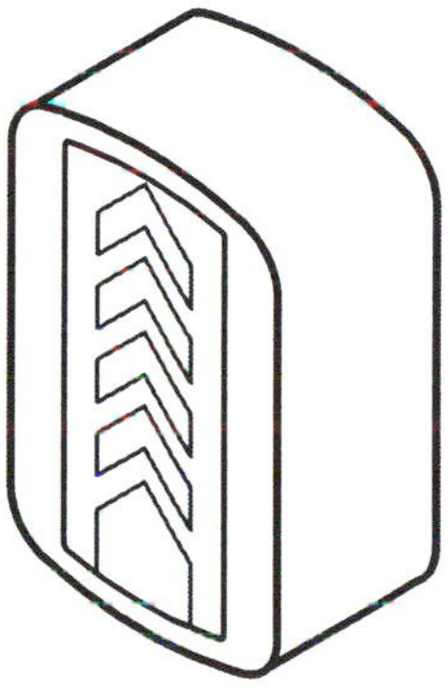	ILLUSTRATION SHOWS UNIVERSAL TYPE BUZZER USED IN HOMES, APARTMENT BUILDINGS, OEM'S AND FACTORIES.
 V	VOLUME CONTROL		
	PUSH BUTTON SWITCH	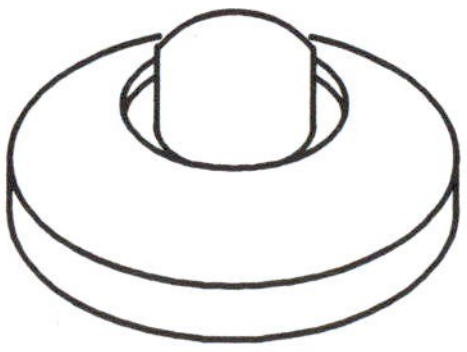	

SYMBOL	NAME	ILLUSTRATION	REMARKS
	POWER POLE WITH VARIOUS DEVICES		
	POKE-THROUGH WITH TELEPHONE OUTLETS	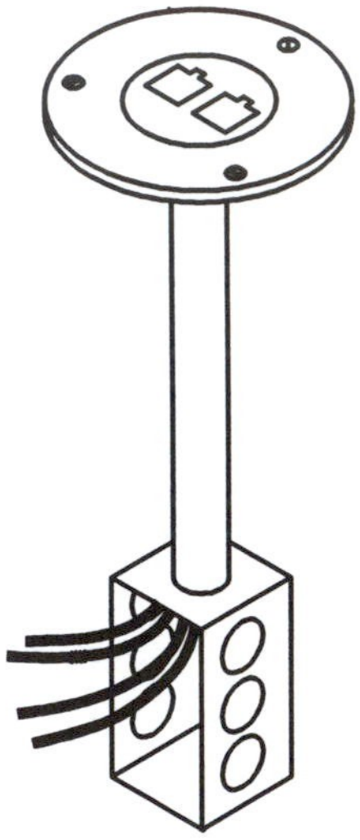	A POKE-THROUGH IS AN ELECTRICAL OUTLET IN A FLOOR CONNECTING TO A SOURCE IN THE CEILING BELOW.
 A	ABANDONED POKE-THROUGH	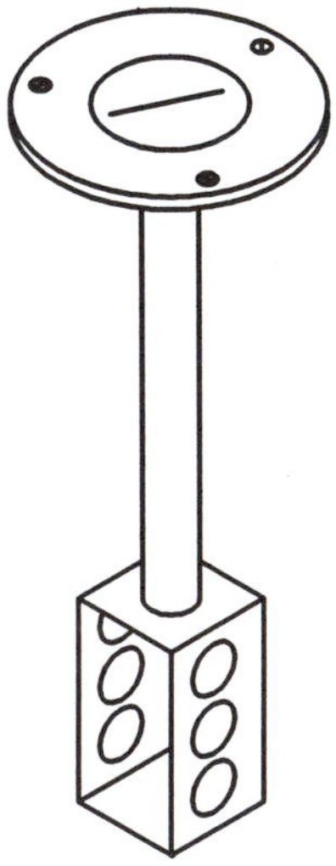	

SYMBOL	NAME	OBJECT	REMARKS
	GENERAL OUTLET-CEILING		
	GENERAL OUTLET-WALL		
	SINGLE RECEPTACLE		
 G	SINGLE GROUNDING RECEPTACLE		USE "G" SUBSCRIPT WITH ANY OUTLET TO SHOW GROUNDING.
	SINGLE RECEPTACLE - FLOOR MOUNTED	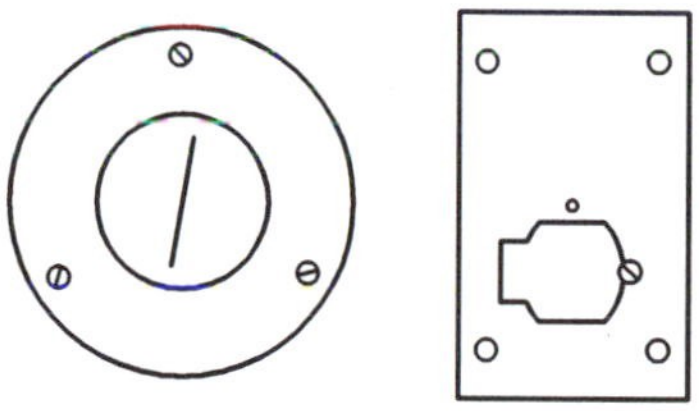	ILLUSTRATION SHOWS PROTECTIVE COVER, TYPICALLY METAL. COVER LOCKS CLOSED WITH SCREW HEADS.
	SINGLE RECEPTACLE - CEILING MOUNTED		VERY COMMON FOR OVERHEAD GARAGE DOOR OPENERS.
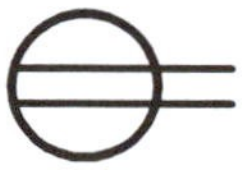	DUPLEX RECEPTACLE		

ELECTRICAL OUTLETS

SYMBOL	NAME	OBJECT	REMARKS
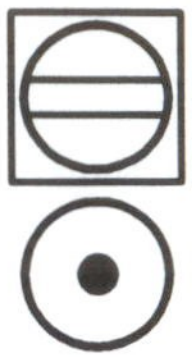	DUPLEX RECEPTACLE - FLOOR MOUNTED	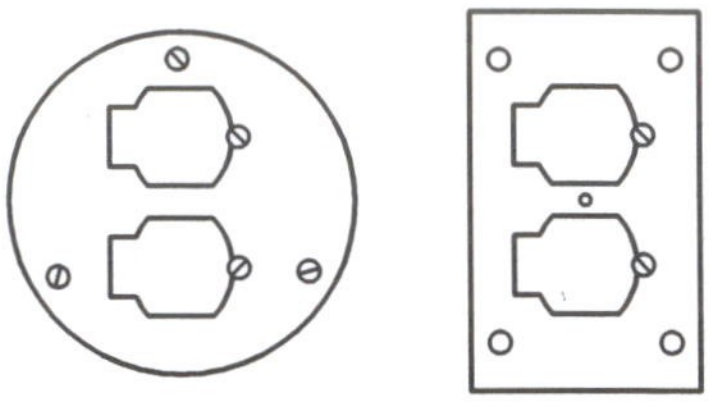	ILLUSTRATION SHOWS PROTECTIVE METAL COVERS, LOCKED CLOSED WITH SCREW HEADS.
	DUPLEX RECEPTACLE - CEILING MOUNTED		VERY COMMON POWER SOURCE FOR OVERHEAD GARAGE DOOR OPENERS.
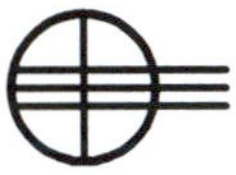	TRIPLEX RECEPTACLE		
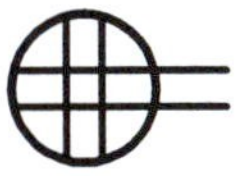	QUADRUPLEX RECEPTACLE		
	DOUBLE DUPLEX RECEPTACLE - FLOOR MOUNTED	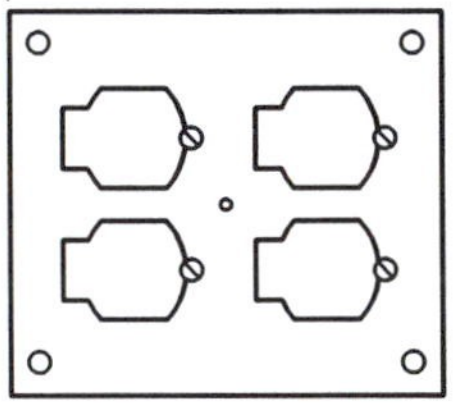	ILLUSTRATION SHOWS PROTECTIVE METAL COVERS, LOCKED CLOSED WITH SCREW HEADS.
	DOUBLE DUPLEX RECEPTACLE - CEILING MOUNTED		

SYMBOL	NAME	OBJECT	REMARKS
	DUPLEX RECEPTACLE - SPLIT WIRED		IN RESIDENTIAL WIRING, COMMONLY CONNECTED TO A WALL SWITCH, TO CONTROL A FLOOR LAMP, FOR EXAMPLE.
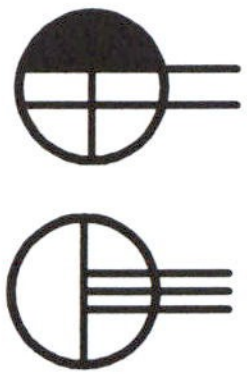	TRIPLEX RECEPTACLE - SPLIT WIRED	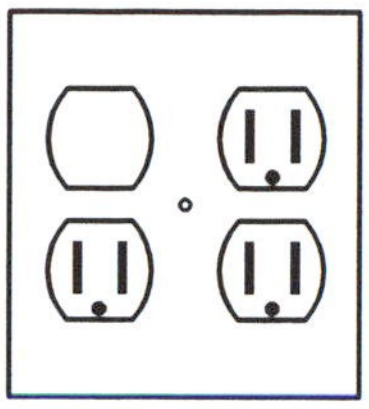	
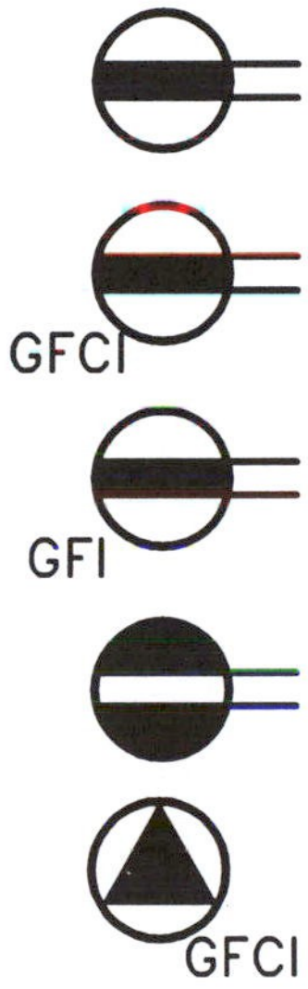 GFCI GFI GFCI	DUPLEX RECEPTACLE - WITH GROUND FAULT INTERRUPTING DEVICE	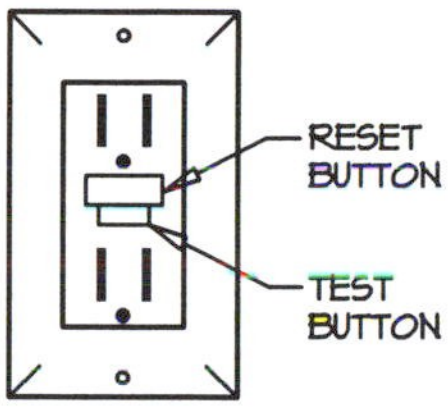 RESET BUTTON TEST BUTTON	COMMONLY USED OUT OF DOORS, AND INDOORS WHERE MOISTURE WILL BE ENCOUNTERED - FOR EXAMPLE BATHROOMS AND KITCHENS.
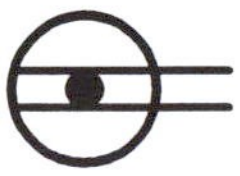	DUPLEX RECEPTACLE WITH ISOLATED GROUND	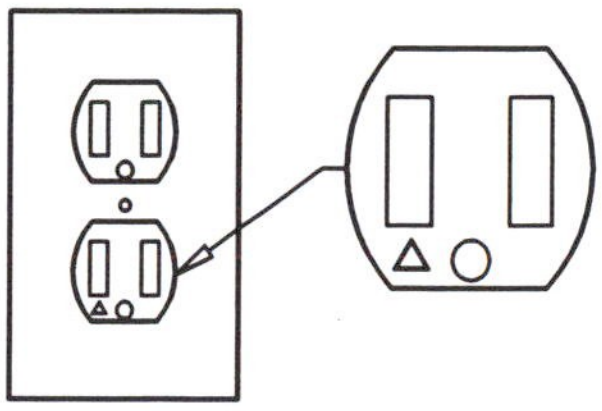	RECEPTACLE IS LABELED WITH GREEN TRIANGLE TO INDICATE ISOLATED GROUND. ISOLATING THE GROUND PROVIDES A CLEANER CURRENT. IN SENSITIVE AREAS SUCH AS COMPUTER CIRCUITS.
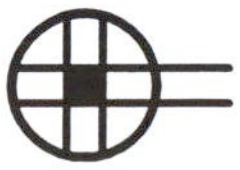	QUADRUPLEX RECEPTACLE WITH ISOLATED GROUND		

ELECTRICAL OUTLETS

SYMBOL	NAME	ELEVATION	REMARKS
D CD	DRYER OUTLET		SHOWING OLD STYLE PLUG AND BOX ON LEFT, NEW STYLE ON RIGHT.
GD GD	GARBAGE DISPOSAL		TOP SYMBOL SPECIFIES OUTLET, BOTTOM SYMBOL IS MORE AMBIGUOUS CONNECTION.
W	WASHER OUTLET		
X X	POLARIZED RECEPTACLE OUTLET		SUBSCRIPT INDICATES NEMA CONFIGURATION. ELEVATIONS SHOW 15 AMP TO THE LEFT, 20 AMP TO THE RIGHT. (PLUG SIZE EXAGERRATED FOR CLARITY.)
	220 VOLT OUTLET		SHOWING OLD STYLE TO THE LEFT, NEWER STYLE RECEPTACLE ON THE RIGHT.

SYMBOL	NAME	ELEVATION	REMARKS
 S	COMBINATION TOGGLE SWITCH AND RECEPTACLE		
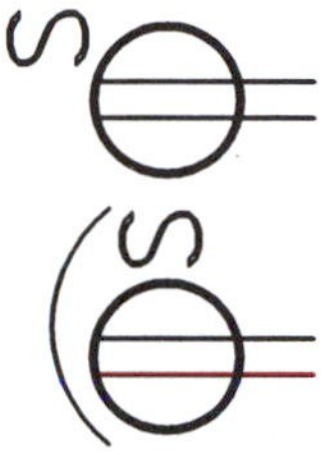 S S	COMBINATION TOGGLE SWITCH AND DOUBLE RECEPTACLE		
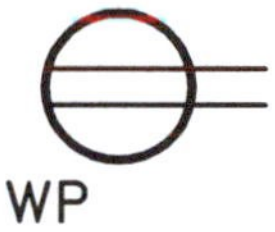 WP	WATER-PROOF DUPLEX RECEPTACLE	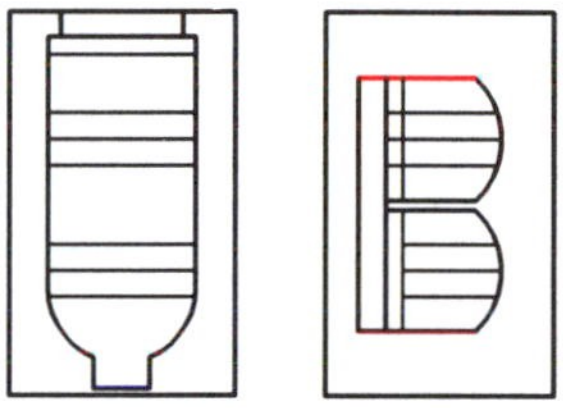	ILLUSTRATIONS SHOW SINGLE & DOUBLE STYLE COVERS.
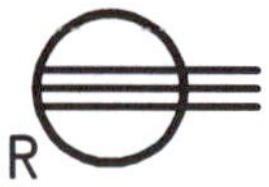 R	RANGE OUTLET	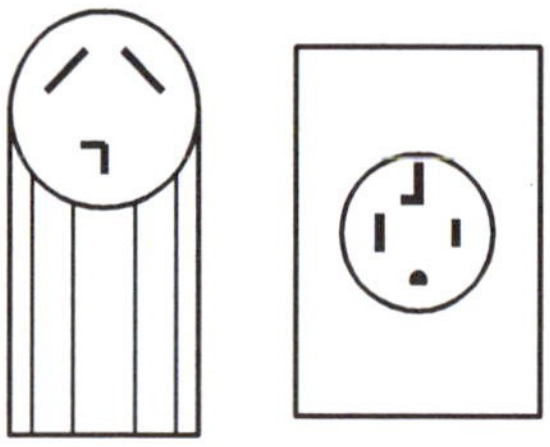	SHOWING OLD STYLE TO THE LEFT, NEWER STYLE PLUG ON THE RIGHT.
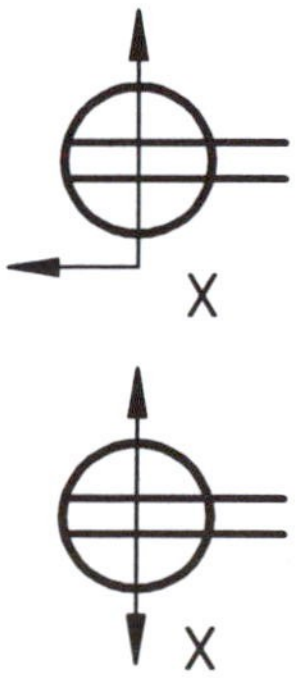 X X	MULTI-OUTLET ASSEMBLY		EXTEND ARROWS TO LIMIT OF INSTALLATION. SHOW APPROPRIATE TYPE OF SYMBOL. INDICATE SPACING OF OULETS IN INCHES, OR NUMBER OF OUTLETS.

SYMBOL	NAME	ELEVATION	REMARKS
 EM	EMERGENCY OUTLET	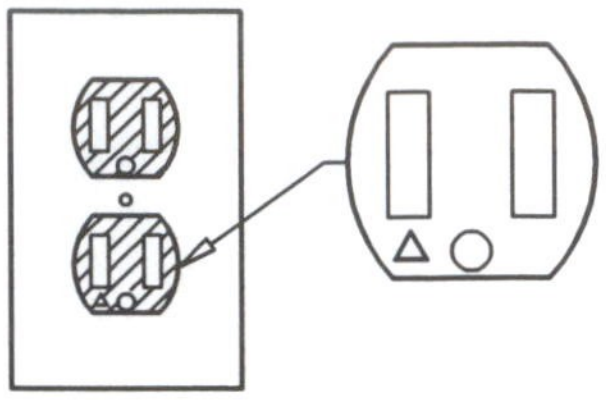	OUTLET ON EMERGENCY CIRCUIT. PLASTIC RECEPTACLES ARE RED WITH A GREEN TRIANGLE. COMMONLY USED IN HOSPITALS.
 D	DROP CORD	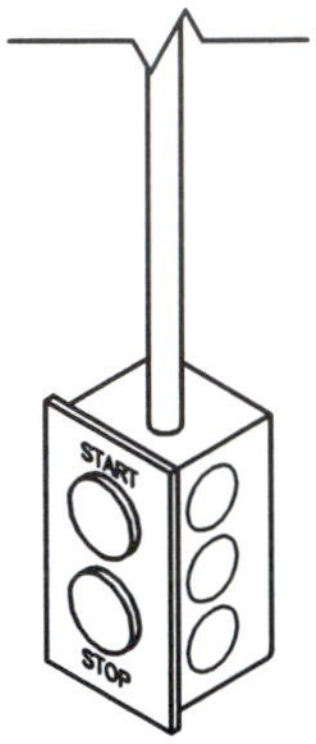 START STOP	CORD HANGS FROM CEILING. VERY COMMON IN MANUFACTURING. ILLUSTRATION SHOWS SWITCH BOX MOUNTED ON BOTTOM OF CORD.
 J	JUNCTION BOX - CEILING MOUNTED		
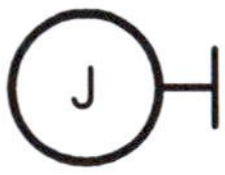 J	JUNCTION BOX - WALL MOUNTED		SHOWN WITH REMOVEABLE COVER PLATE & CONDUIT.
 L	LAMP HOLDER - CEILING MOUNTED	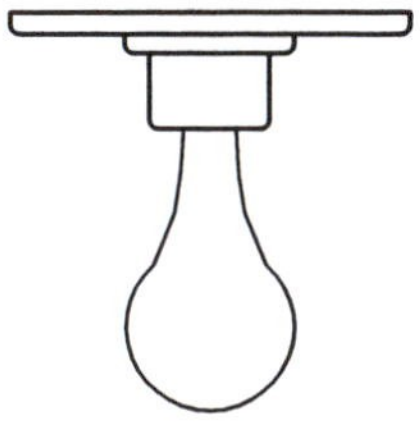	SHOWN WITH BULB

SYMBOL	NAME	OBJECT	REMARKS
 L	LAMP HOLDER - WALL MOUNTED	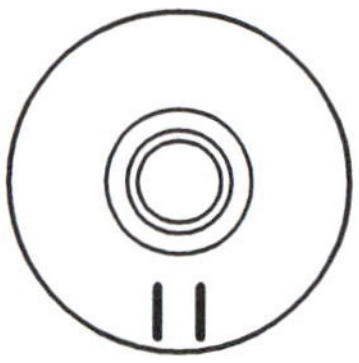	SHOWN WITH OPTIONAL OUTLET, AND NO BULB.
 L PS	LAMP HOLDER WITH PULL SWITCH	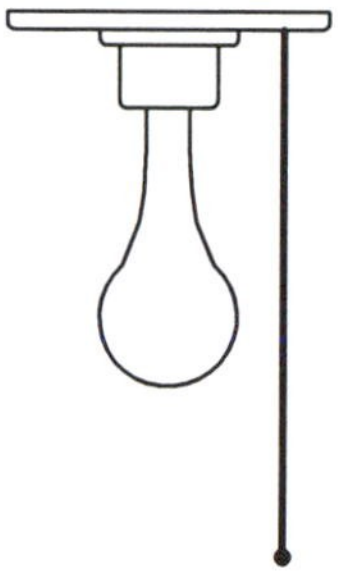	SHOWN WITH BULB
 L PS	LAMP HOLDER WITH PULL SWITCH - WALL MOUNTED		(NO BULB)
 J	JUNCTION BOX	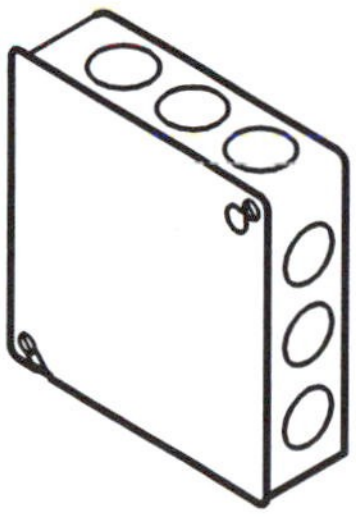	
 C	CLOCK HANGER RECEPTACLE		
 S	PULL SWITCH - CEILING MOUNTED		

SYMBOL	NAME	OBJECT	REMARKS
S	PULL SWITCH - WALL MOUNTED		
V	OUTLET FOR VAPOR DISCHARGE LAMP - CEILING MOUNTED		REFERS TO OLD STYLE MERCURY VAPOR LAMPS.
V	OUTLET FOR VAPOR DISCHARGE LAMP- WALL MOUNTED		
X	OUTLET FOR EXIT LAMP - CEILING MOUNTED		
X	OUTLET FOR EXIT LAMP- WALL MOUNTED		
B	BLANKED OUTLET - CEILING MOUNTED		OUTLET NOT IN USE, WITH COVER. COLOR OF COVER GENERALLY TO MATCH WALL PAINT.
B	BLANKED OUTLET - WALL MOUNTED		
	SPECIAL PURPOSE OUTLET		
X	SPECIAL PURPOSE OUTLET AS NOTED, FLOOR MOUNTED.		

SYMBOL	NAME	OBJECT	REMARKS
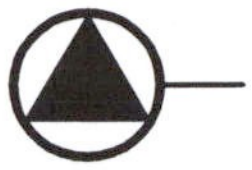	SINGLE SPECIAL PURPOSE OUTLET		
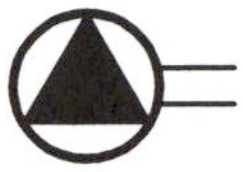	DUPLEX SPECIAL PURPOSE OUTLET		
 DW	DISHWASHER OUTLET		
 WH	WATER HEATER OUTLET		
 W	CLOTHES WASHER OUTLET		
	POKE-THROUGH WITH ELECTRICAL OUTLETS	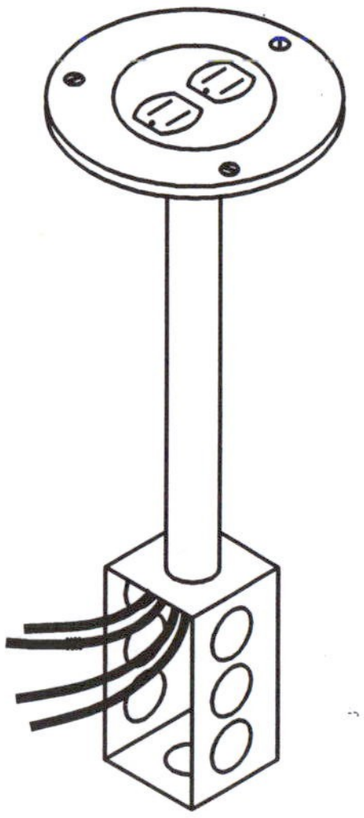	A POKE-THROUGH IS AN ELECTRICAL OUTLET IN A FLOOR CONNECTING TO A SOURCE IN THE CEILING BELOW.

SYMBOL	NAME	OBJECT	REMARKS
A	ABANDONED POKE-THROUGH		
B	BLANKED DUPLEX OUTLET		OUTLETS NOT IN USE.
B	BLANKED QUADRUPLEX OUTLET		

SYMBOL	NAME	OBJECT	REMARKS
S [A] $	SINGLE POLE SWITCH		SYMBOL COMMONLY SEEN WITH VERTICAL BAR.
S_2	DOUBLE POLE SWITCH		
S_3	THREE WAY SWITCH		ALLOWS ONE OR MORE FIXTURES TO BE OPERATED BY TWO SWITCHES.
S_4	FOUR WAY SWITCH		ALLOWS ONE OR MORE FIXTURES TO BE OPERATED BY THREE SWITCHES.
S_{CC}	CONTACTOR CONTROL SWITCH		
S_{CB}	CIRCUIT BREAKER SWITCH		

SYMBOL	NAME	OBJECT	REMARKS
S_{DM}	DIMMER SWITCH	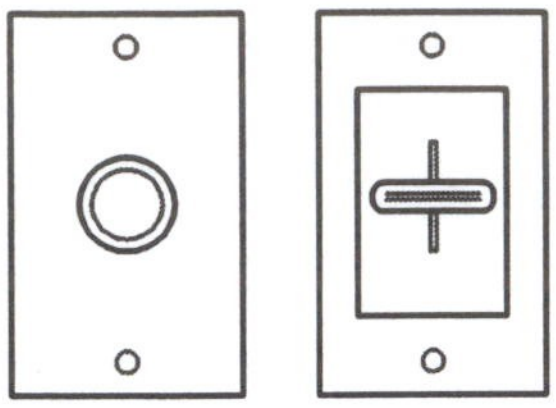	
S_{D}	DOOR SWITCH OR JAM SWITCH	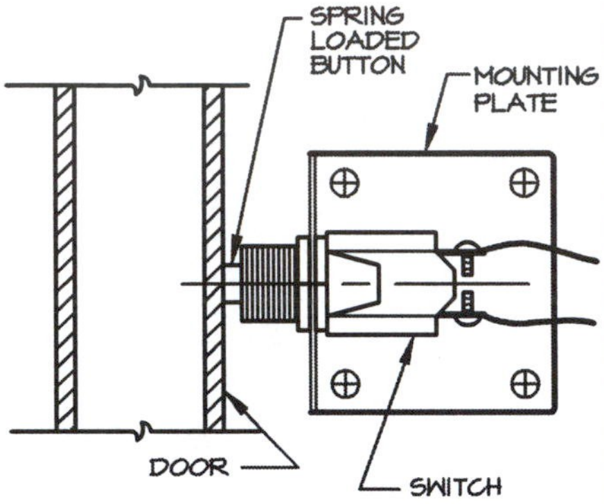 SPRING LOADED BUTTON MOUNTING PLATE DOOR SWITCH	COMMONLY USED TO CONTROL LIGHTS IN CLOSETS. SWITCH IS MOUNTED ON JAM. ILLUSTRATION SHOWS "OFF" POSITION.
S_{EC}	EVAPORATIVE COOLER SWITCH		3-WAY FAN SPEED, AND TWO SINGLE POLE STACKED SWITCHES FOR FAN & PUMP.
S_{X} S_{EP}	EXPLOSION PROOF SWITCH		
S_{F}	FUSED SWITCH		
S_{FDM}	FLUORESCENT DIMMER SWITCH		

SYMBOL	NAME	OBJECT	REMARKS
S_K	KEY CONTROLLED SWITCH		
S_L	SWITCH WITH LOCATOR LAMP		GLOW-IN-THE-DARK RESIDENTIAL LIGHT SWITCH, FOR EXAMPLE.
S_L S_{LV}	LOW VOLTAGE SWITCH		
S_{LM}	MASTER SWITCH FOR LOW VOLTAGE SWITCHING SYSTEM.		
S_M	MANUAL MOTOR STARTER		
S_{MC} [A] S_B 	MOMENTARY CONTACT SWITCH OR INTERMITTENT SWITCH OR BUTTON	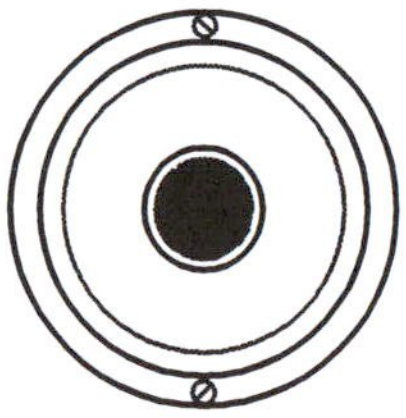	
S_{MTO}	MANUAL MOTOR STARTER WITH THERMAL OVERLOADS		

SYMBOL	NAME	OBJECT	REMARKS
S_P	PILOT LIGHT, LOAD ON		LIGHT IS ON WHEN SWITCH IS ON. LIGHT IS OFF WHEN CIRCUIT IS OPEN. COMMONLY USED WHEN FIXTURE IS LOCATED IN CLOSET, ATTIC, OR BASEMENT.
S_{PO}	PILOT LIGHT, LOAD OFF		LIGHT IS OFF WHEN SWITCH IS ON. LIGHT IS ON WHEN SWITCH IS OFF.
S_{CP}	CHAIN PULL SWITCH		
S_{RC}	REMOTE CONTROL SWITCH (RECEIVER)		

SYMBOL	NAME	OBJECT	REMARKS
S_T	TIME SWITCH OR MOTOR SWITCH WITH THERMAL OVERLOAD PROTECTION		CIRCUIT ACTIVATED BY TIMER.
S_{TH}	THERMAL RATED MOTOR SWITCH		
S_V	VARIABLE SPEED SWITCH OR VOLUME CONTROL SWITCH		
S_{WCB}	WEATHERPROOF CIRCUIT BREAKER		
S_{WP}	WEATHERPROOF SWITCH		SWITCH IS MOUNTED IN A GASEKT-SEALED BOX AND OPERATED BY A LEVER THRU THE COVER.
S_x	DEDICATED SWITCH - LOWER CASE LETTER INDICATES EQUIPMENT CONTROLLED		USE LOWER CASE LETTER TO INDICATE EQUIP.
Ⓢ	CEILING PULL SWITCH		
TC	TIME CLOCK SWITCH		CIRCUIT ACTIVATED BY TIMER.

ELECTRICAL SWITCHES

SYMBOL	NAME	OBJECT	REMARKS
 PES PE P	PHOTOELECTRIC SWITCH	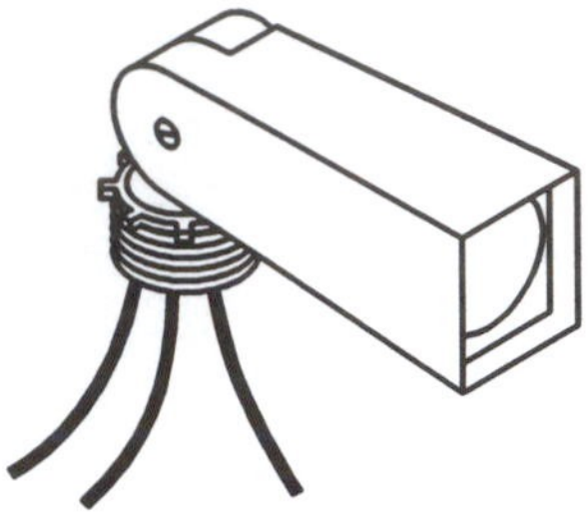	USED IN HOMES TO TURN ON OUTSIDE LIGHTS AT DUSK, FOR EXAMPLE. TYPICALLY MOUNTED FACING NORTH OR SHADED.
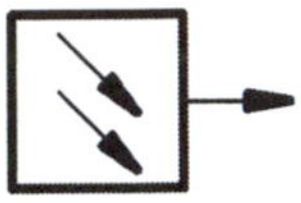	ELECTRIC EYE - BEAM SOURCE		
	ELECTRIC EYE - RELAY		
	DISCONNECT SWITCH OR SAFETY SWITCH		USED TO ISOLATE A DEVICE OR CIRCUIT, FOR PURPOES OF REMOVAL OR REPAIR. LENGTHEN SYMBOL TO SHOW LARGER BOX. NFSS: NON-FUSED SAFETY SWITCH FSS: FUSED SAFETY SWITCH ILLUSTRATION SHOWS OUT-DOORS WEATHERPROOF TYPE WITH LOCKABLE COVER AND LEVER.
	FUSIBLE SAFETY SWITCH		

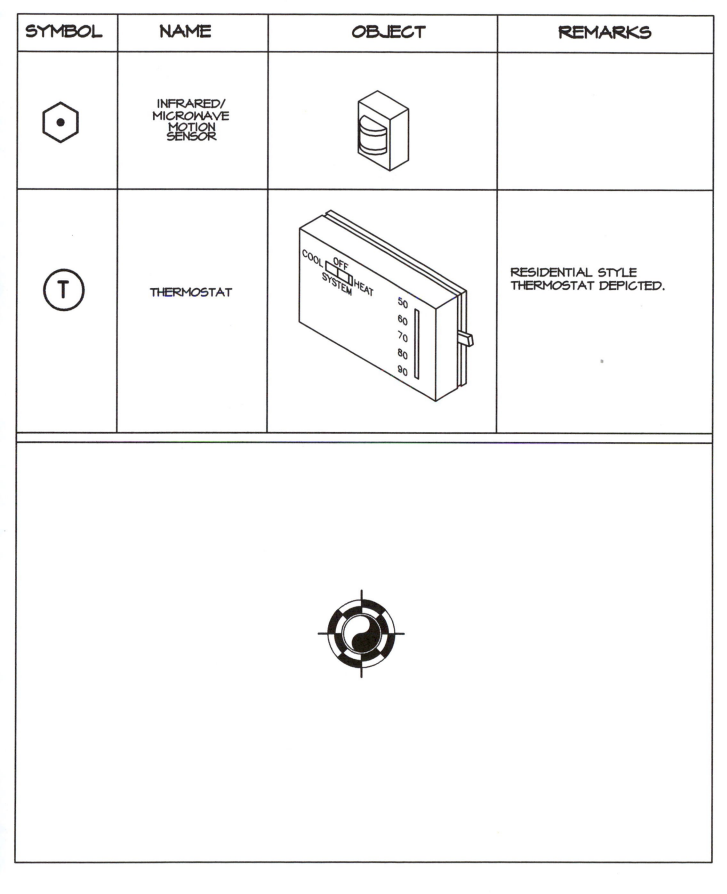

SYMBOL	NAME	OBJECT	REMARKS
	INFRARED/ MICROWAVE MOTION SENSOR		
T	THERMOSTAT	COOL OFF SYSTEM HEAT 50 60 70 80 90	RESIDENTIAL STYLE THERMOSTAT DEPICTED.

GENERAL DRAFTING NOTES:

SYMBOLS SHOULD BE DRAWN PROPORTIONATELY SO THAT THEIR CONNECTING POINTS COINCIDE WITH THE INTERSECTIONS OF A GRID:

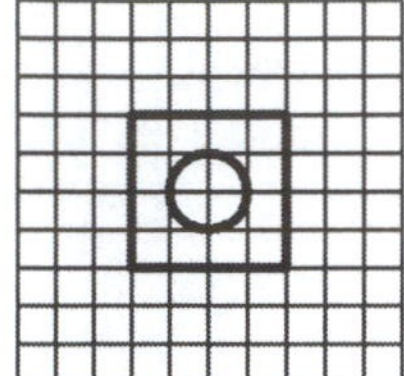

FUTURE OR ASSOCIATED COMPONENTS AND PATHS SHOULD BE SHOWN WITH HIDDEN LINE TYPE:

THE TERMINAL SYMBOL O MAY BE USED OPTIONALLY (UNLESS SHOWN AS PART OF A COMPONENT IN A STANDARD), TO SHOW POINTS OF ATTACHMENT OF WIRES TO ANY COMPONENT. THE TERMINAL SYMBOL ITSELF SHOULD BE 1/10 TO 1/16 Φ.

ø0.1 ø.0625

SYMBOL SIZE IS NOT CRITICAL, BUT ALL SYMBOLS IN A DRAWING SHOULD BE AT THE SAME SCALE.

NAME	SYMBOL	OBJECT	REMARKS
ANTENNA, GENERAL			

NAME	SYMBOL	OBJECT	REMARKS
ANTENNA, LOOP	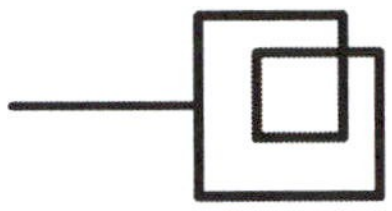	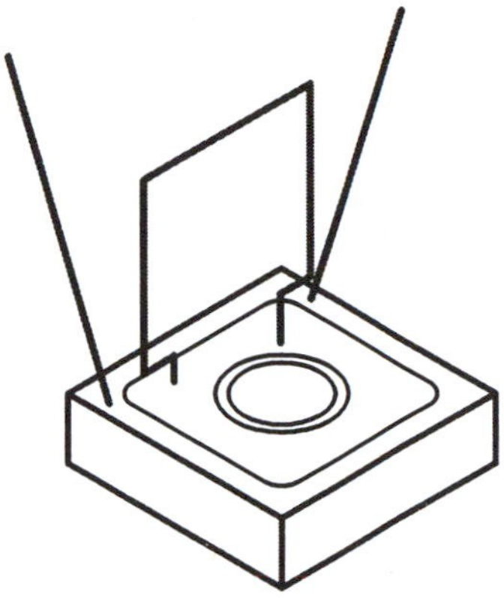	ILLUSTRATION SHOWS TELEVISION ANTENNAE; SQUARE SHAPE IS UHF LOOP, RODS ARE VHF.
ARRESTOR, GENERAL			AN ARRESTOR IS AN OVERLOAD CONTROL DEVICE.
LIGHTNING ARRESTOR			
BATTERY, SINGLE CELL	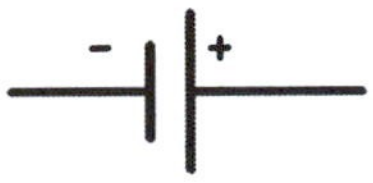 - +	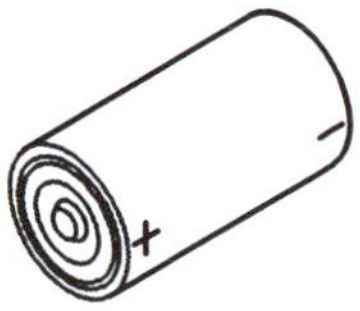 + -	ILLUSTRATION SHOWS FLASHLIGHT BATTERY WHICH IS TYP.
BATTERY, MULTIPLE CELL		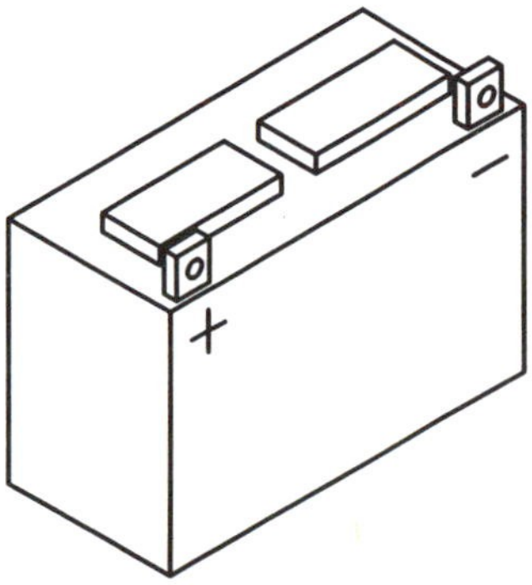 - +	IN THE SYMBOL: THE LONGER VERTICAL LINE IS THE POSITIVE POLE. PLUS AND MINUS SIGNS MAY BE ADDED. SHORTER LINE IS HALF THE LENGTH OF THE LONGER LINE.

NAME	SYMBOL	OBJECT	REMARKS
CIRCUIT BREAKER,			
GENERAL CAPACITOR OR CONDENSER	 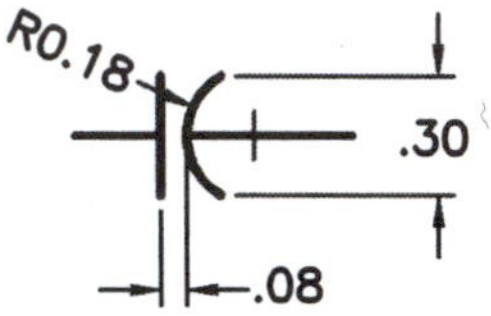 R0.18 .30 .08	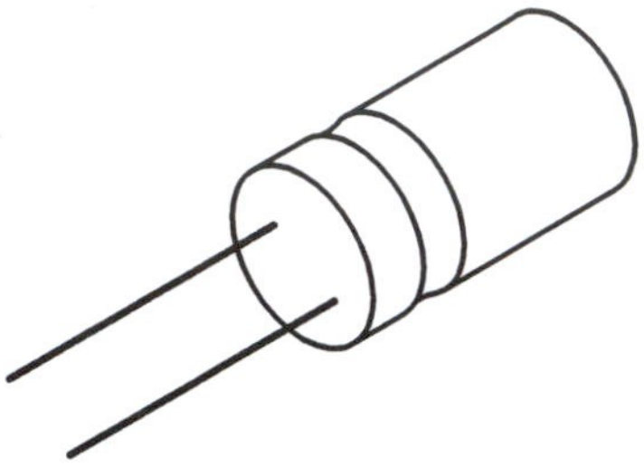	
ADJUSTABLE OR VARIABLE CAPACITOR		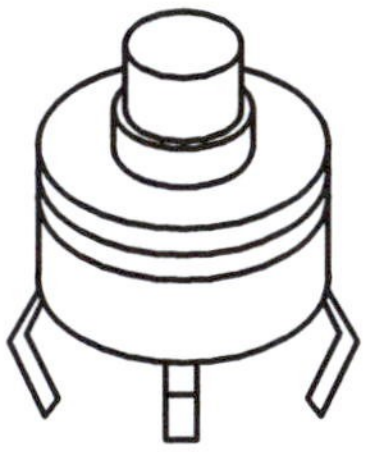	DRAW ARROW IN SYMBOL AT 45 DEGREES.
SHIELDED CONDENSER	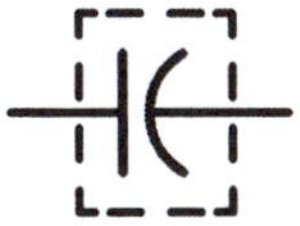		DASHED LINES SHOULD NOT TOUCH THE SIGNAL PATH.
CHASSIS, FRAME	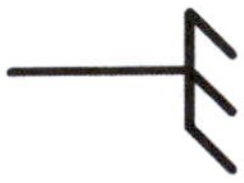		

NAME	SYMBOL	OBJECT	REMARKS
CONNECTORS		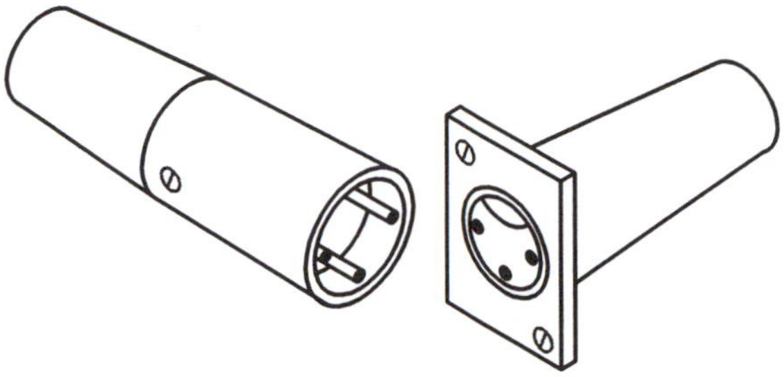	ILLUSTRATION SHOWS 3-PIN AUDIO CONNECTORS.
JACK CONNECTOR		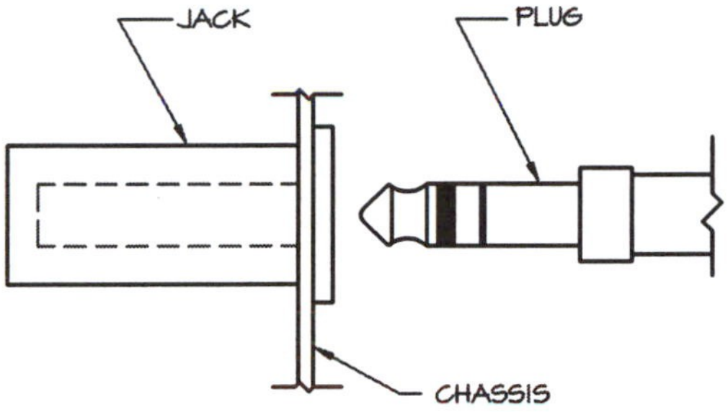 JACK PLUG CHASSIS	
CLOSED CONTACT			PARALLEL LINES SHOULD BE 1.25 TIMES THE LENGTH OF THE GAP BETWEEN THEM.
RHEOSTAT (HAND OPERATED)			SCHEMATIC SYMBOL: 
RHEOSTAT (MOTOR OPERATED)			

NAME	SYMBOL	OBJECT	DIM/REM.
OPEN CONTACT			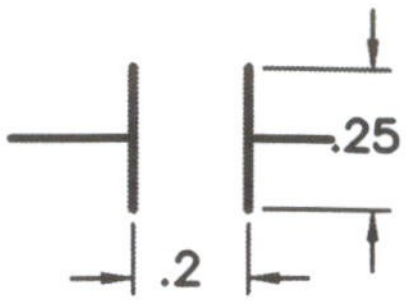 .25 .2
OPEN CONTACT WITH TIME CLOSING DEVICE	 TCD		
AUDIBLE DEVICE (SPEAKER, HORN, SIREN, ETC.)	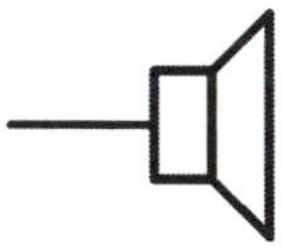	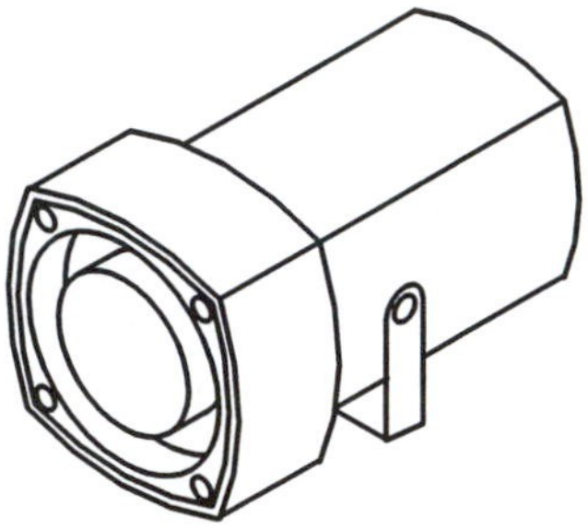	ILLUSTRATION SHOWS 'MINI SIREN'
FUSE		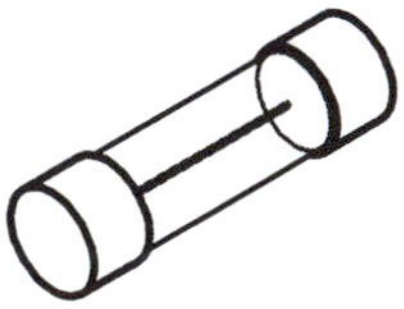	
FUSIBLE ELEMENT			USED IN FAIL-SAFE SYSTEMS SUCH AS FIRE DAMPERS.
HEADPHONES	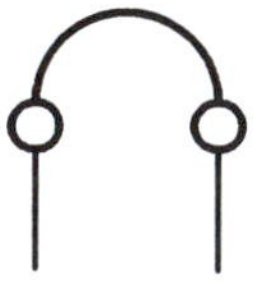		

NAME	SYMBOL	OBJECT	DIMENSIONS
GENERAL INDUCTOR			.15
ADJUSTABLE INDUCTOR			
CONTINUOUSLY ADJUSTABLE INDUCTOR			
INCANDESCENT LAMP (ILLUMINATING)			
AMMETER	A		
VOLTMETER	V		

NAME	SYMBOL	OBJECT	REMARKS
GROUND	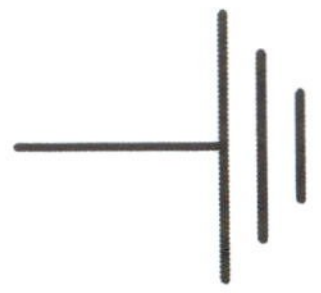	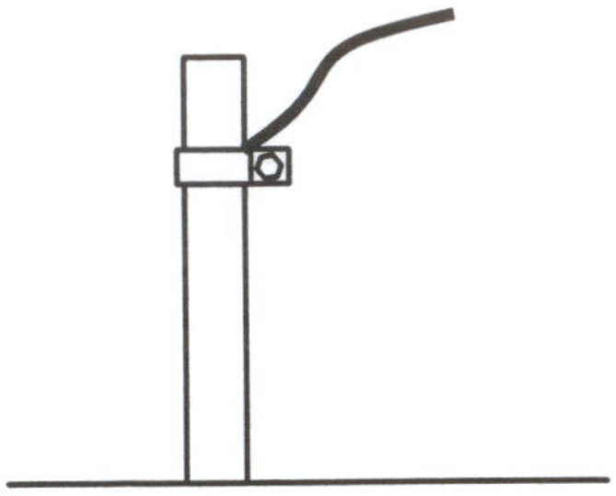	ILLUSTRATION SHOWS EARTH GROUND, USING IRON BAR.
FUSE CUT-OUT	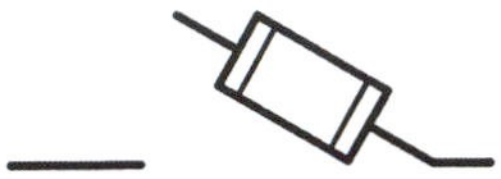		
AC MOTOR			
METALLIC RECTIFIER	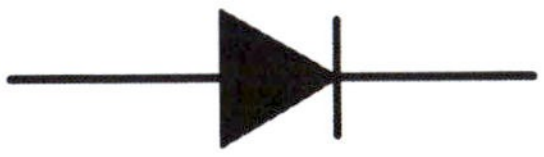	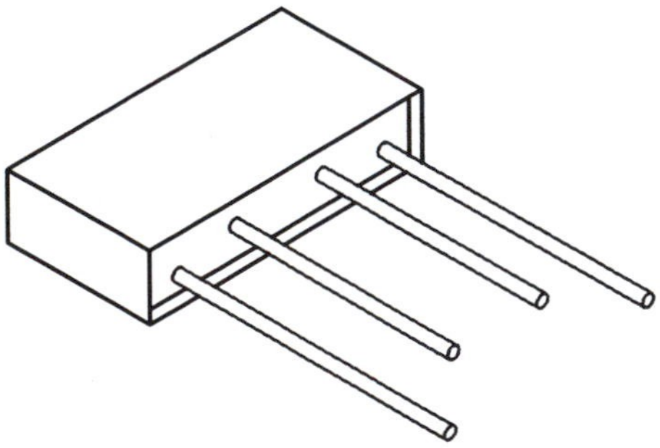	
GENERATOR	 G		
MOTOR	 M		

NAME	SYMBOL	OBJECT	REMARKS
DC MOTOR OR GENERATOR	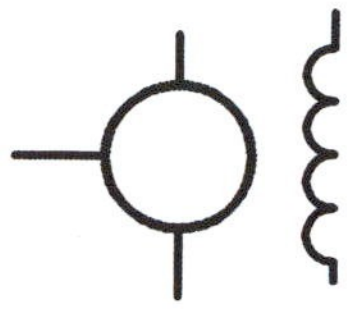		
SINGLE THROW SWITCH, GENERAL OPEN CLOSED	 		
DOUBLE THROW SWITCH, GENERAL			
DOUBLE THROW, DOUBLE POLE	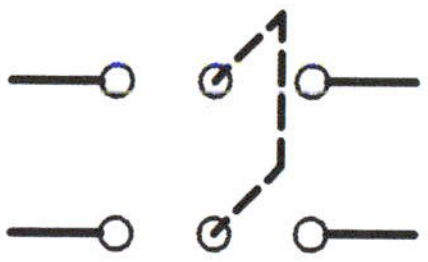		
KNIFE SWITCH			
START SWITCH	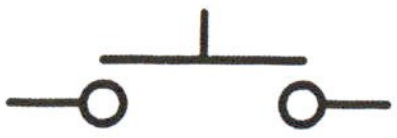	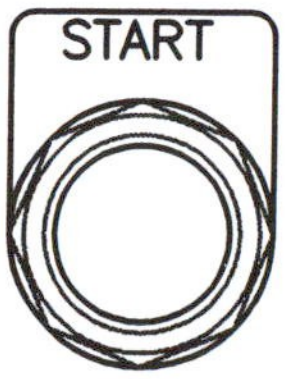 START	PUSH BUTTON OR SPRING RETURN.

NAME	SYMBOL	OBJECT	REMARKS/DIMS
STOP SWITCH		 STOP	
SWITCH, NON-LOCKING, CIRCUIT CLOSING	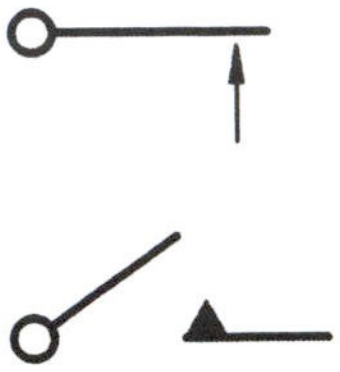		
SWITCH, LOCKING, CIRCUIT CLOSING	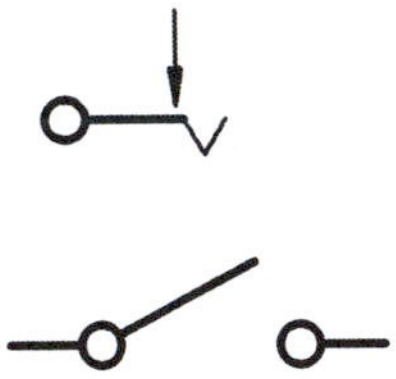		
GENERAL RESISTOR	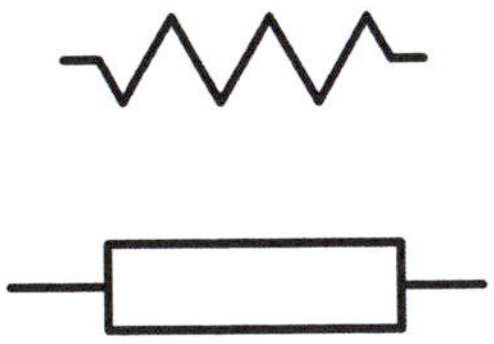	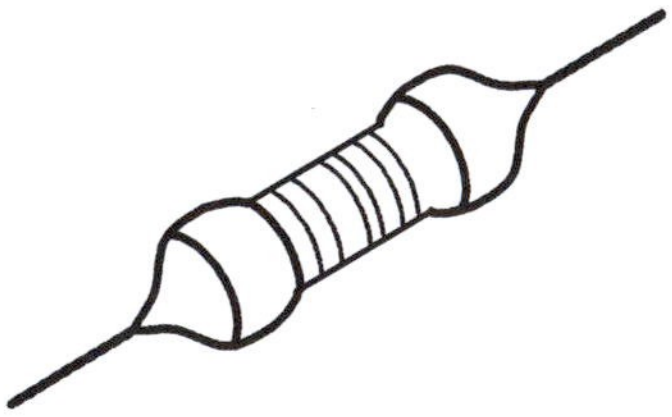	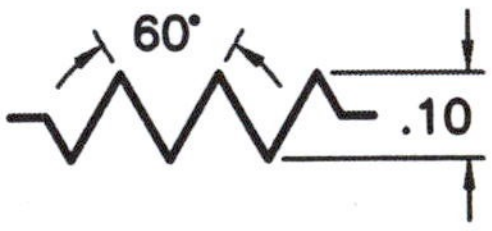 60° .10 UPPER SYMBOL TYP. FOR RADIO & TV - USE 60° ANGLE AND 3 POINTS TOP & BOTTOM. SQUARED SYMBOLS USED IN INDUSTRIAL ELECTRONICS. USE 45° ARROW.
VARIABLE RESISTOR	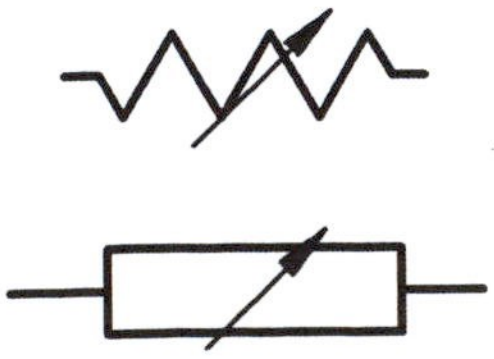		

NAME	SYMBOL	OBJECT	DIMENSIONS
COIL	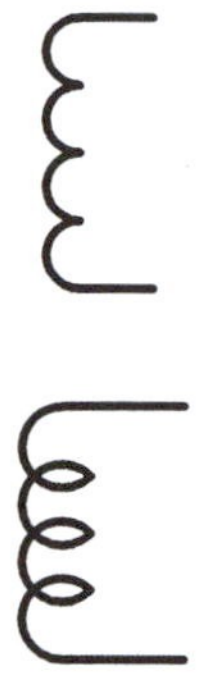		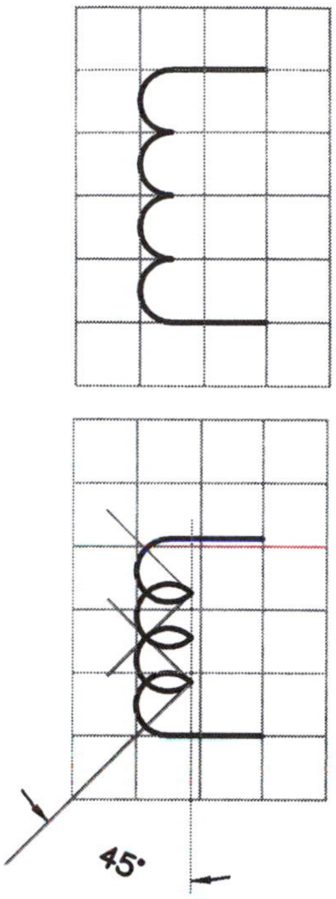 45°
TRANSFORMER	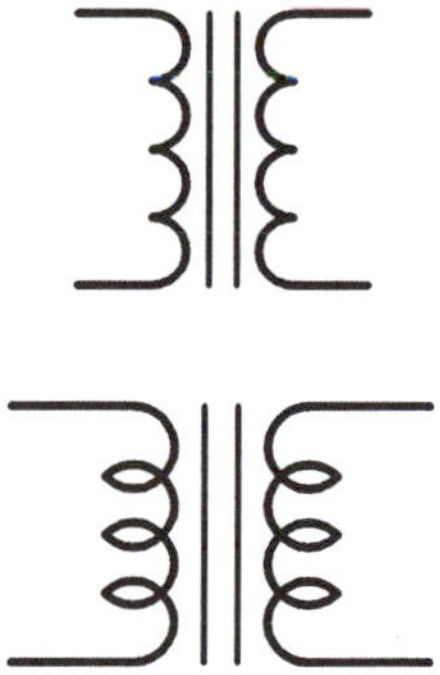	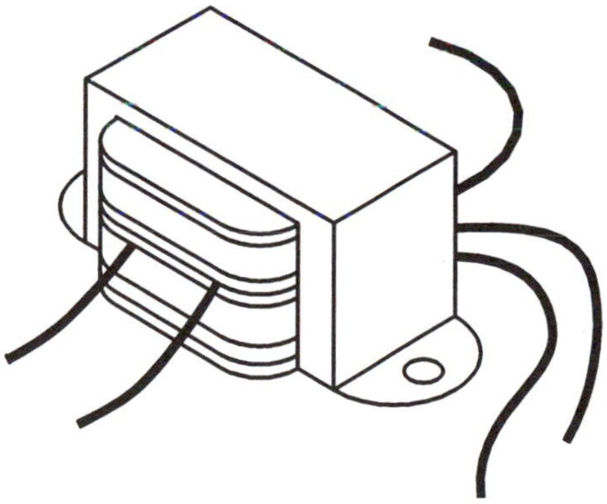	

SYMBOL	NAME	REMARKS
NOTE: DRAW INCANDESCENT AND FLUORESCENT LIGHT FIXTURES TO SCALE. INDICATE WALL MOUNTED FIXTURE HEIGHT ON SYMBOL SCHEDULE OR LIGHTING FIXTURE SCHEDULE. DIMENSION HEIGHT TO CENTER OF FIXTURE OUTLET BOX.		
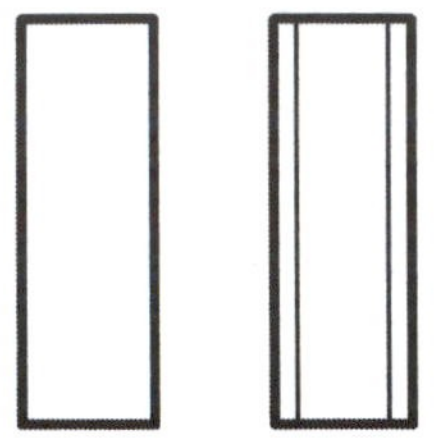	GENERAL SYMBOL FOR FLUORSCENT FIXTURE	DRAW FLUORESCENT FIXTURE TO SCALE.
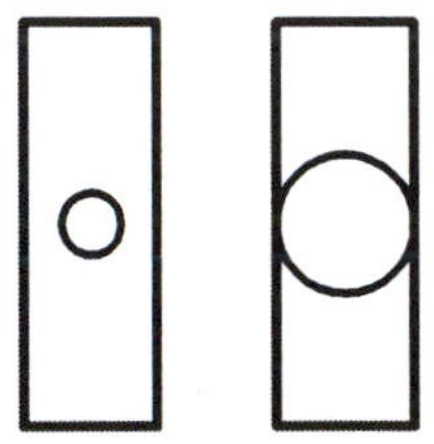	SURFACE MOUNTED FLUORSCENT FIXTURE (CEILING) WITH OUTLET BOX SHOWN	SYMBOL TO LEFT IS MORE PREVALENT.
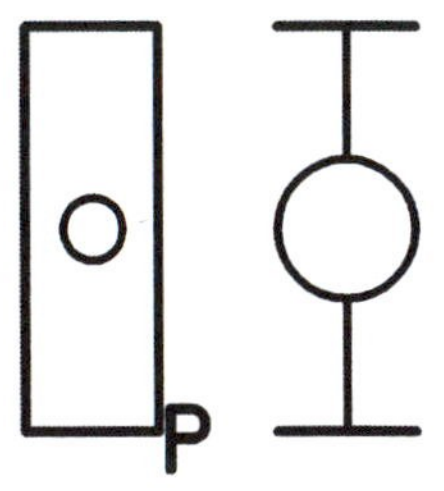 P	PENDANT STYLE FLUORSCENT FIXTURE	
 R	RECESSED CEILING FLUORSCENT FIXTURE	REFERRED TO AS "TROFFER." COMMONLY PLACED WITHIN A CEILING GRID.

SYMBOL	NAME	OBJECT	REMARKS
	WALL-MOUNTED SURFACE FLUORESCENT FIXTURE		
	CHANNEL MOUNTED FLUORESCENT CEILING FIXTURE		COMMONLY MOUNTED IN UNISTRUT.
	MODULAR WIRED RECESSED FLUORESCENT FIXTURE		FIXTURES ARE INSTALLED IN SERIES.
2	LIGHT STRIP OR FLUORESCENT STRIP		A BARE SINGLE FLUORESCENT TUBE. WHEN MULTIPLE TUBES ARE USED, CALL OUT NUMBER BY SYMBOL.
	WALL MOUNTED STRIP LIGHT		

FLUORESCENT FIXTURES

SYMBOL	NAME	OBJECT	REMARKS
	CONTINUOUS ROW FLUORESCENT CEILING FIXTURES		(ADD 'P' FOR PENDANT STYLE.)
	FLUORESCENT FIXTURE ON EMERGENCY CIRCUIT OR LIFE SAFETY BRANCH		RECESSED FIXTURE WHICH REMAINS LIGHTED WHEN ALL OTHER TROFFERS ON CIRCUIT ARE SWITCHED OFF. (AKA NIGHT TROFFER)
	FLOURESCENT FIXTURE ON CRITICAL BRANCH		FIXTURES WHICH REMAIN ON DURING GENERAL POWER LOSS. USED FOR EXAMPLE IN HOSPITAL INTENSIVE CARE UNITS.

SYMBOL	NAME	OBJECT	REMARKS
NOTE: DRAW INCANDESCENT AND FLOURESCENT LIGHT FIXTURES TO SCALE. INDICATE WALL MOUNTED FIXTURE HEIGHT ON SYMBOL SCHEDULE OR LIGHTING FIXTURE SCHEDULE. DIMENSION HEIGHT TO CENTER OF FIXTURE OUTLET BOX.			
	CEILING MOUNTED LIGHT FIXTURE		MAY ALSO BE USED FOR PENDANT STYLE FIXTURE. DRAW ALL LIGHTS TO SCALE.
X #y	STANDARD DESIGNATION FOR LIGHTS: X = UPPER CASE IS LIGHT KEY. # = CIRCUIT NO. y = LOWER CASE IS SWITCH CONTROL.		FIXTURE HEIGHT A.F.F. MAY BE INDICATED BY SYMBOL ON DRAWING. SUBSCRIPT "ALL" INDICATES ALL FIXTURES IN THE AREA ARE THE SAME.
X	CEILING LIGHT FIXTURE WITH KEY TO SCHEDULE		
	SURFACE DIRECTIONAL LIGHT FIXTURE		
	RECESSED CEILING LIGHT FIXTURE		

SYMBOL	NAME	OBJECT	REMARKS
	RECESSED DIRECTIONAL LIGHT FIXTURE		
	CEILING FIXTURE - ROUGH-IN ONLY		
B	CEILING BLANKED LIGHT FIXTURE		A CEILING MOUNTED JUNCTION BOX NOT IN USE. PLATE COLOR MATCHES WALL FINISH.
L	WALL OUTLET CONTROLLED BY LOW VOLTAGE SWITCH - RELAY INSTALLED IN OUTLET BOX.		
B	WALL BLANKED LIGHT FIXTURE		A WALL MOUNTED JUNCTION BOX NOT IN USE.

SYMBOL	NAME	OBJECT	REMARKS
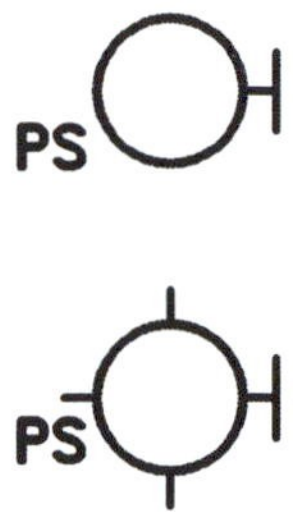 PS PS	WALL MOUNTED FIXTURE WITH PULL SWITCH	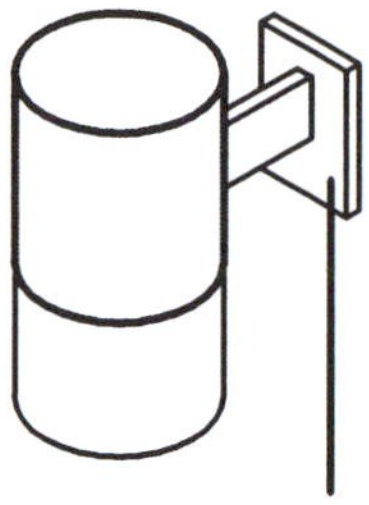	CP (CHAIN PULL) MAY ALSO BE USED INSTEAD OF PS.
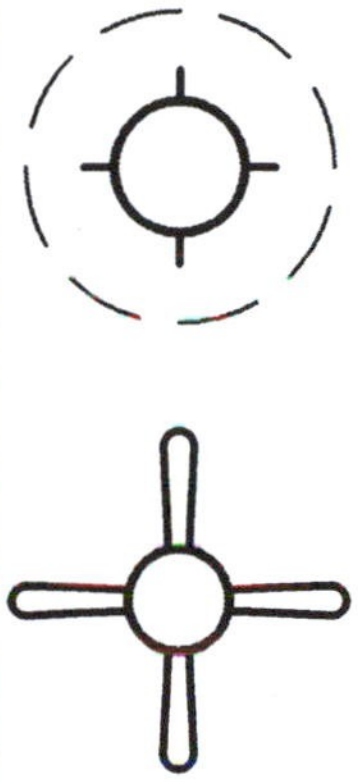	CEILING MOUNTED LIGHT WITH FAN	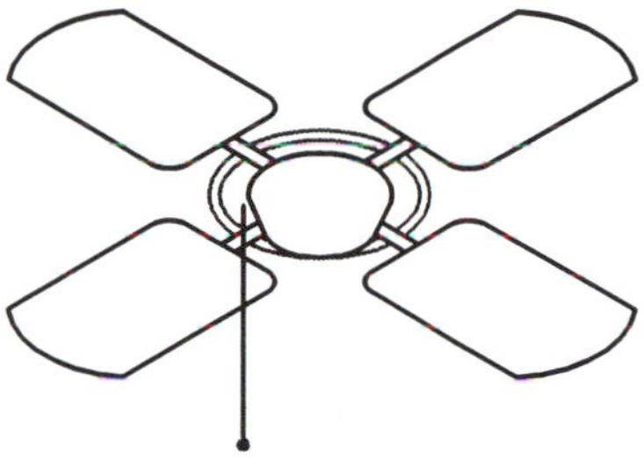	CONTROLLED BY EITHER WALL SWITCHES OR CHAIN PULLS.
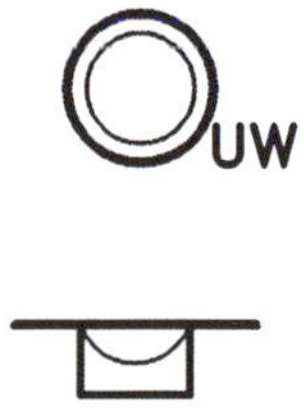 UW	UNDERWATER RECESSED LAMP		SYMBOLS FOR BOTH BOTTOM MOUNTED & SIDE WALL MOUNTED LAMPS.
 UW	UNDERWATER UP BEAM		

INCANDESCENT LIGHTS

SYMBOL	NAME	OBJECT	REMARKS
	CHANDELIER		
	TRACK LIGHTING	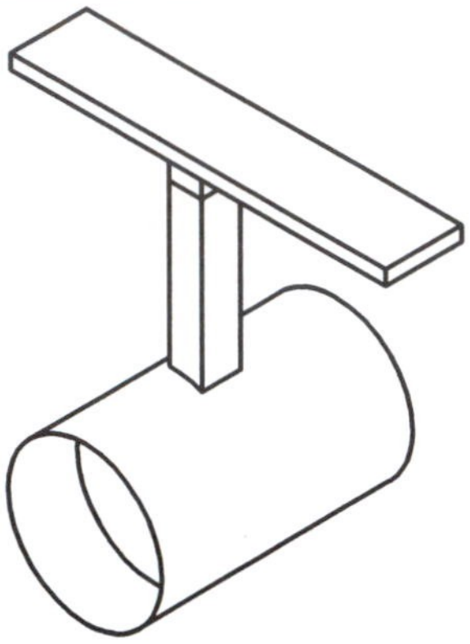	SHOW ACTUAL NUMBER OF FIXTURES ON SYMBOL.
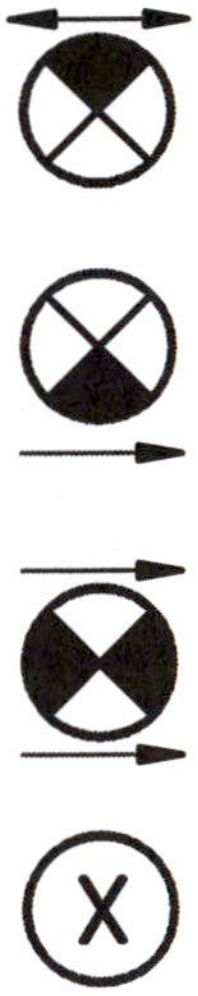	CEILING MOUNTED OR PENDANT EXIT SIGN	 	SHADING INDICATES LIGHTED FACES. ARROWS INDICATE DIRECTION OF EXIT. NO ARROW WHEN MOUNTED DIRECTLY OVER EXIT DOOR.
	WALL MOUNTED EXIT SIGN		X IN CIRCULAR SYMBOL IS LESS DESIRABLE.

SYMBOL	NAME	PERSPECTIVE	REMARKS
	EXIT LIGHT		
 XR	RECESSED EXIT LIGHT		
	BATTERY PACK WITH LUMINAIRES, AND REMOTE LUMINAIRE		DASHED LINES INDICATE POSSIBLE MULTIPLE LIGHTS AND REMOTE.
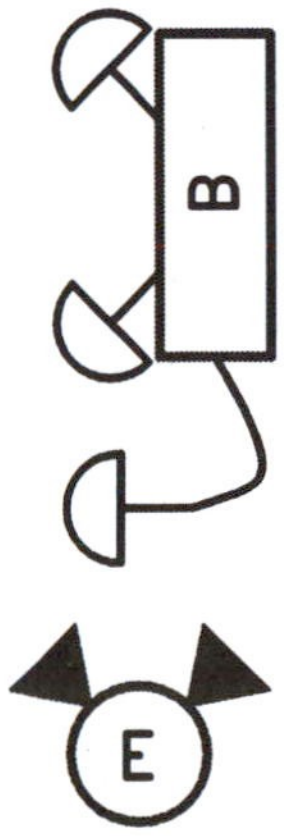 B E	EMERGENCY BATTERY PACK WITH CHARGER AND SEALED BEAM HEADS, WITH REMOTE.	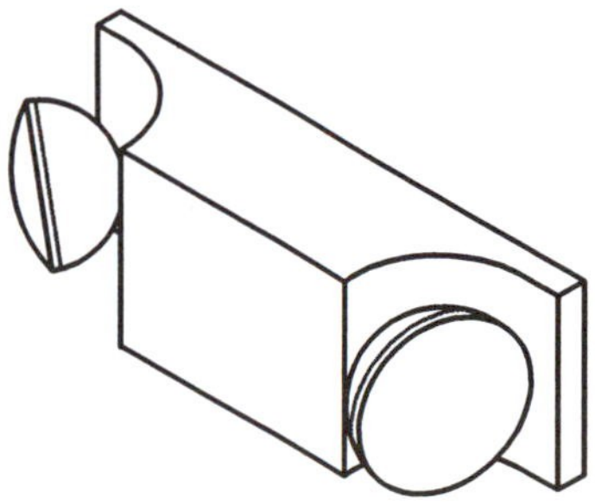	

INCANDESCENT LIGHTS

SYMBOL	NAME	OBJECT	REMARKS
	COMBINED BATTERY-POWERED & EMERGENCY LIGHT & ILLUMINATED EXIT SIGN	EXIT	INDICATE DIRECTION OF FLOW WITH ARROW.
	MULTIPLE FLOOD LIGHT ASSEMBLY		
	INCANDESCENT TRACK LIGHTING OR MIRROR LIGHTS		SIMILAR SYMBOL USED TO DESIGNATE THEATRICAL FLOOD LAMPS.
H	CEILING MOUNTED HEAT LAMP		COMMONLY USED IN BATHROOMS.
	CEILING MOUNTED SPOT LIGHT		

SYMBOL	NAME	OBJECT	REMARKS
	LIGHT FIXTURE ON CRITICAL BRANCH		CRITICAL BRANCH CIRCUIT SUPPLIES POWER TO MAXIMUM PRIORITY SITUATIONS: OPERATING ROOMS, I.C.U., ETC.
E	LIGHT FIXTURE ON EMERGENCY CIRCUIT OR LIFE SAFETY BRANCH ALSO 24-HOUR FIXTURE		THE EMERGENCY CIRCUIT USES BATTERIES AND GENERATORS IN LIFE SUSTAINING SITUATIONS. LIFE SAFETY BRANCH IS AN EMERGENCY CIRCUIT FOR LESS CRITICAL NEEDS, E.G. LIGHTING. SEE HOSPITAL CODE.
N	ILLUMINATED HOUSE NUMBER		
S	WALL BRACKET LIGHT SWITCH		

NAME	SYMBOL	OBJECT	REMARKS
PAGING SYSTEM DEVICE	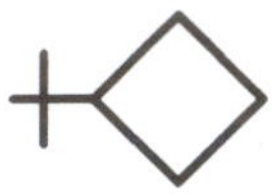		
STAFF REGISTER SYSTEM			
ELECTRICAL CLOCK SYSTEM DEVICE	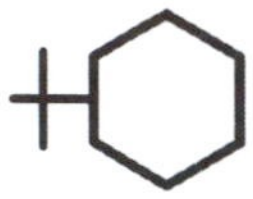		
COMPUTER DATA OR PUBLIC TELEPHONE SYSTEM DEVICES	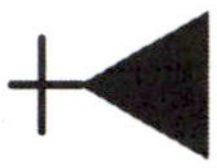		
PRIVATE SYSTEM TELEPHONE DEVICES	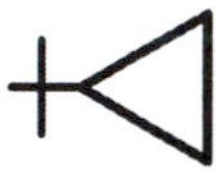		
WATCHMAN SYSTEM DEVICES			
SOUND SYSTEM DEVICES: UNSPECIFIED			
SPEAKER	 L		
VOLUME CONTROL	 V		

NAME	SYMBOL	OBJECT	REMARKS
TELEVISION ANTENNA DEVICES	CTV		
ELECTRICAL CLOCK SYSTEM DEVICES	DP		
SIGNAL CENTRAL STATION	SC		
CARD READER	CR		
SPECIAL AUXILIARY OUTLETS	X		

NURSE CALL SYMBOLS

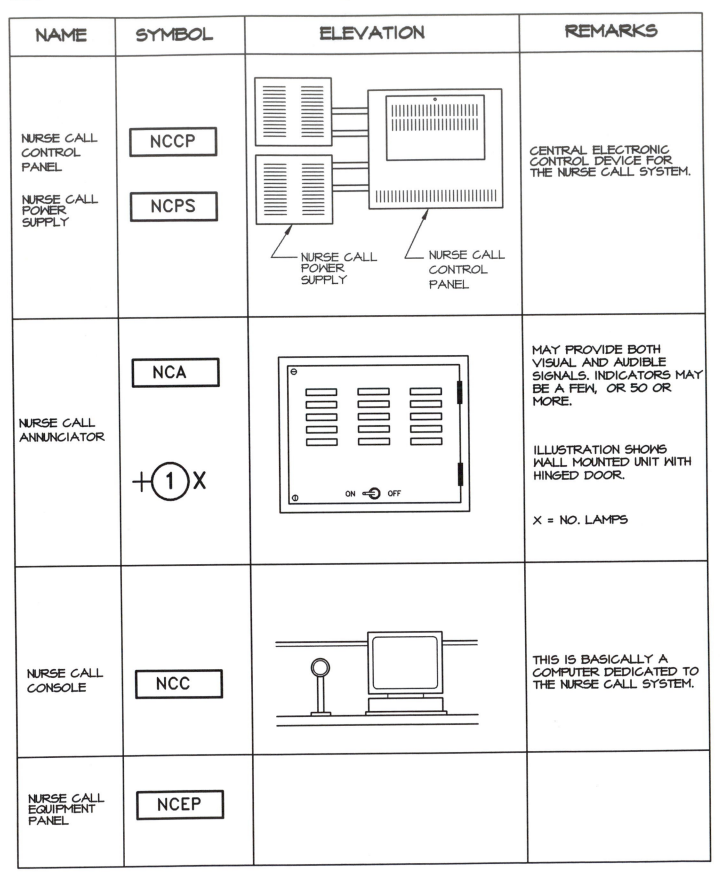

NAME	SYMBOL	ELEVATION	REMARKS
NURSE CALL CONTROL PANEL NURSE CALL POWER SUPPLY	NCCP NCPS	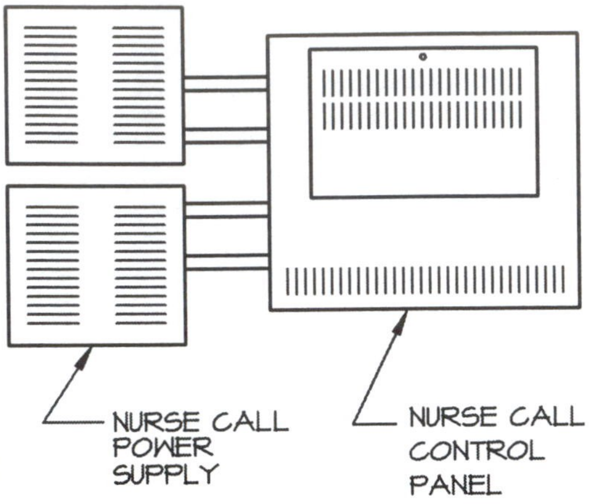 NURSE CALL POWER SUPPLY NURSE CALL CONTROL PANEL	CENTRAL ELECTRONIC CONTROL DEVICE FOR THE NURSE CALL SYSTEM.
NURSE CALL ANNUNCIATOR	NCA 1 X	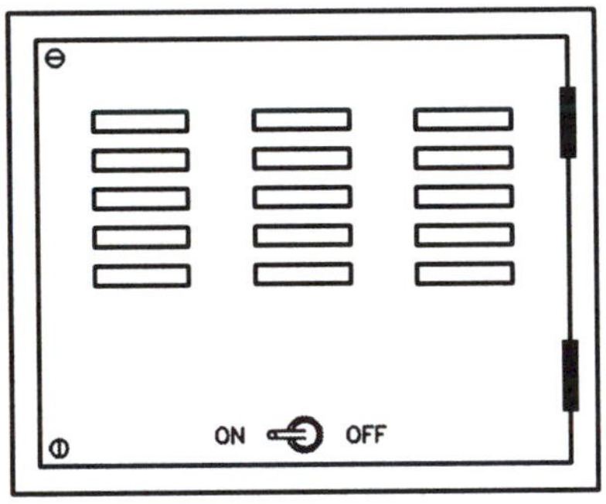 ON OFF	MAY PROVIDE BOTH VISUAL AND AUDIBLE SIGNALS. INDICATORS MAY BE A FEW, OR 50 OR MORE. ILLUSTRATION SHOWS WALL MOUNTED UNIT WITH HINGED DOOR. X = NO. LAMPS
NURSE CALL CONSOLE	NCC		THIS IS BASICALLY A COMPUTER DEDICATED TO THE NURSE CALL SYSTEM.
NURSE CALL EQUIPMENT PANEL	NCEP		

NAME	SYMBOL	ELEVATION	REMARKS
NURSE CALL DEVICES	KEY KEY N KEY		A VARIETY OF SYMBOLS USED TO DEFINE NURSE CALL DEVICES.
AUDIBLE ALARM	N A		
CORE ZERO STATION	N CZ		FOUND IN AREAS THAT DEAL WITH EXTREME EMERGENCIES, SUCH AS CARDIAC UNIT. SPEAKER AT LEFT.
CORE ZERO AUDIBLE ALARM	N CZA		

NAME	SYMBOL	ELEVATION	REMARKS
DOME LIGHT WITH NORMAL CALL LAMP	 N DL1	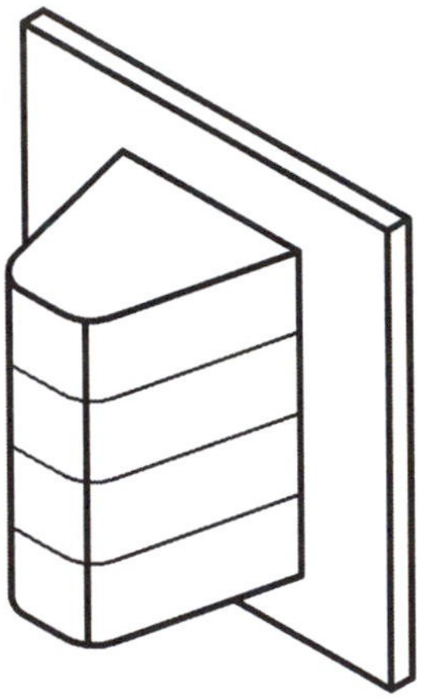	DOME LIGHT TYPICALLY MOUNTED ABOVE DOOR TO PATIENT'S ROOM. COLORS FROM TOP TO BOTTOM: WHITE, RED, ORANGE & GREEN. MAY BE ASSIGNED ACCORDING TO NEED. FOR EXAMPLE, WHITE MAY BE NORMAL CALL. RED MAY BE STAFF IN ROOM. ETC.
DOME LIGHT WITH EMERGENCY CALL LAMP	 N DL2		
DOME LIGHT WITH NORMAL EMERGENCY CALL LAMPS	 N DL3		
DOME LIGHT WITH NORMAL, EMERGENCY, CORE ZERO CALL LAMPS	 N DL4		
DOME LIGHT WITH CORE ZERO CALL LAMPS	 N DL5		
DOME LIGHT WITH EMERGENCY, CORE ZERO CALL LAMPS	 N DL6		

NURSE CALL SYMBOLS

NAME	SYMBOL	ELEVATION	REMARKS
DUTY STATION	N DS		
EMERGENCY CALL STATION WITH PUSH BUTTON	N EB	CANCEL ® PUSH FOR HELP	INSTALLED AT SHOWER. SHOWN WITH RED INDICATOR LAMP.
EMERGENCY CALL STATION WITH FOOT SWITCH	N EF		
EMERGENCY CALL STATION WITH PULL CORD	N EP	CANCEL ® PULL FOR HELP	INSTALLED AT SHOWER. SHOWN WITH RED INDICATOR LAMP.
EMERGENCY CALL STATION WITH PULL CORD, SHOWER	N EPS	CANCEL ® PULL FOR HELP	INSTALLED AT SHOWER. SHOWN WITH RED INDICATOR LAMP.

NURSE CALL SYMBOLS

NAME	SYMBOL	ELEVATION	REMARKS
PATIENT STATION	N P		MOUNTED NEAR PATIENT'S BED.
DUAL PATIENT STATION	N P2		MOUNTED BETWEEN TWO PATIENTS' BEDS.
STAFF LOCATOR STATION	N SL		SEE DOME LIGHT.
STAFF STATION	N SS		
NURSES' ANNUNCIATOR	1 X		POSTSCRIPT INDICATES NUMBER OF LAMPS.

SYMBOL	NAME	OBJECT	REMARKS
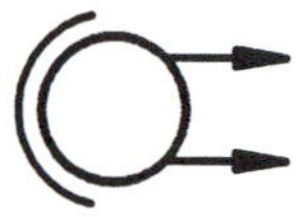	SPOT LIGHT (OUTDOORS)	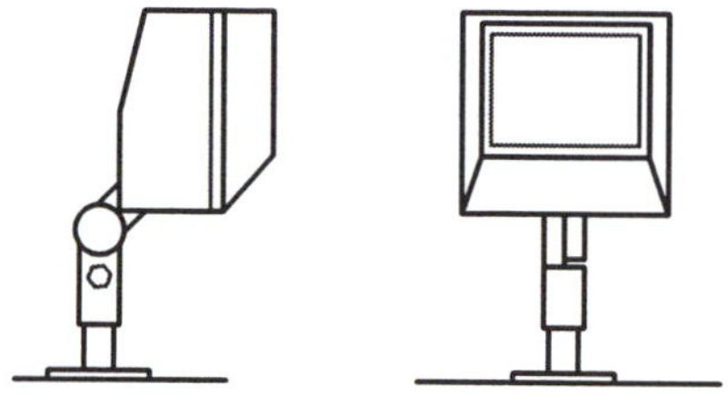	
	FLOOD LIGHT (OUTDOORS)	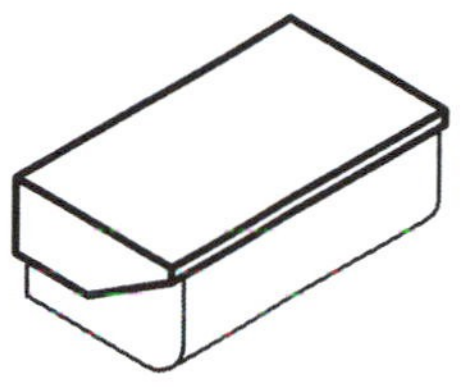	WALL MOUNTED STYLE ILLUSTRATED.
	STREET LIGHT (STANDARD FEED FROM UNDER-GROUND CIRCUIT)	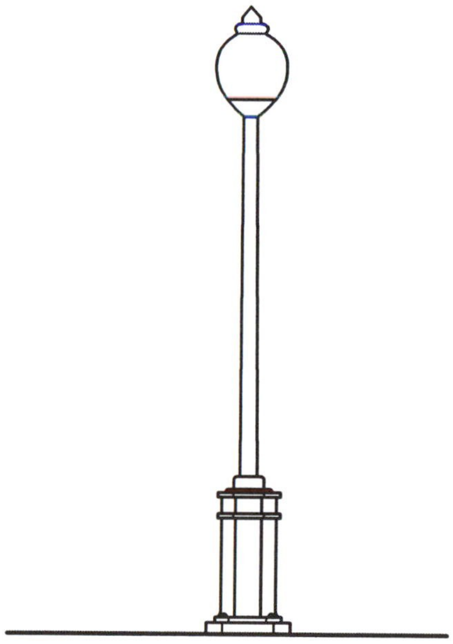	

OUTDOOR LIGHTING

SYMBOL	NAME	OBJECT	REMARKS
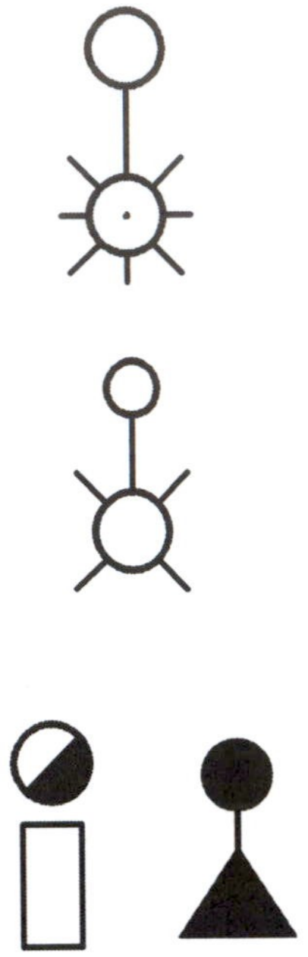	STREET LIGHT (POLE MOUNTED	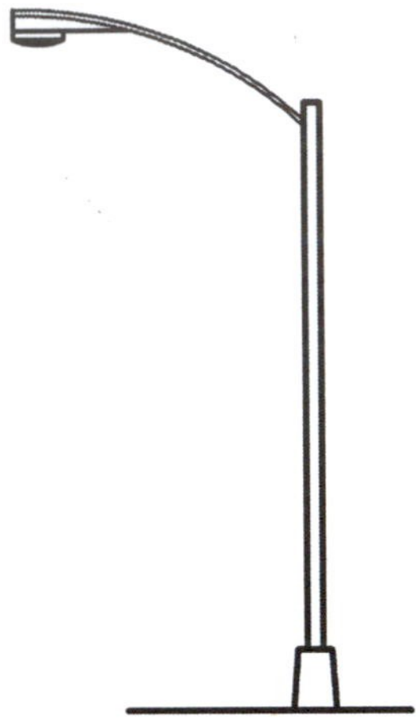	
	DUAL OUTDOOR POLE ARM FIXTURES	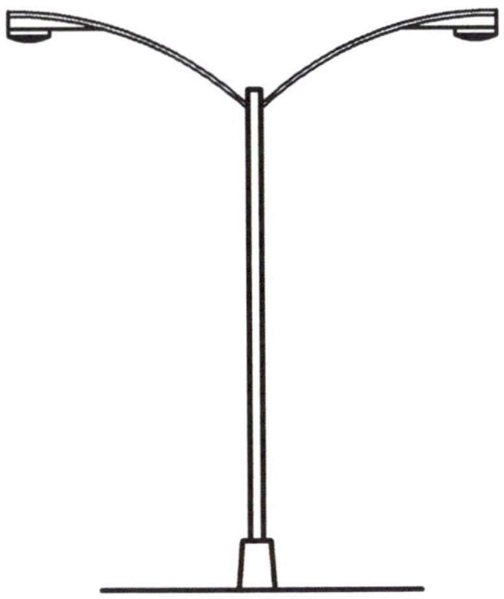	INCREASE NUMBER OF FIXTURE SYMBOLS AS REQUIRED. SQUARE HEADED LIGHTS USED COMMONLY IN PARKING LOTS.

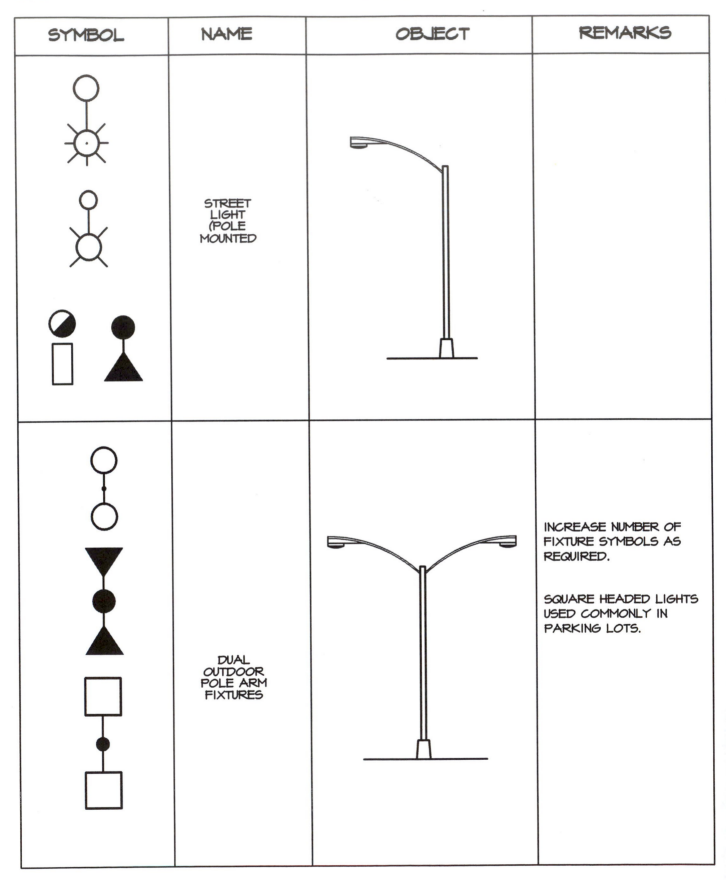

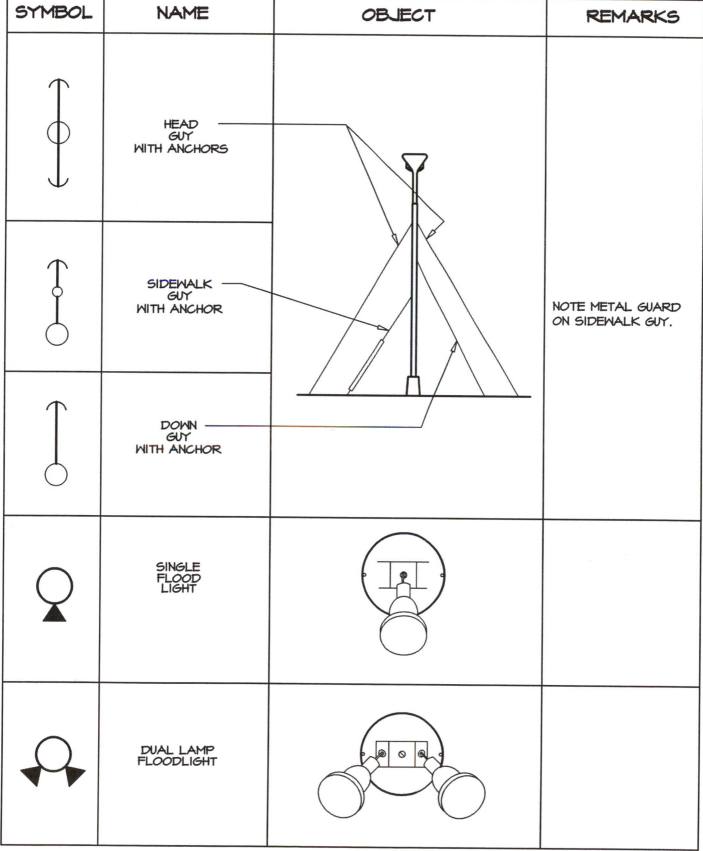

SYMBOL	NAME	OBJECT	REMARKS
	HEAD GUY WITH ANCHORS	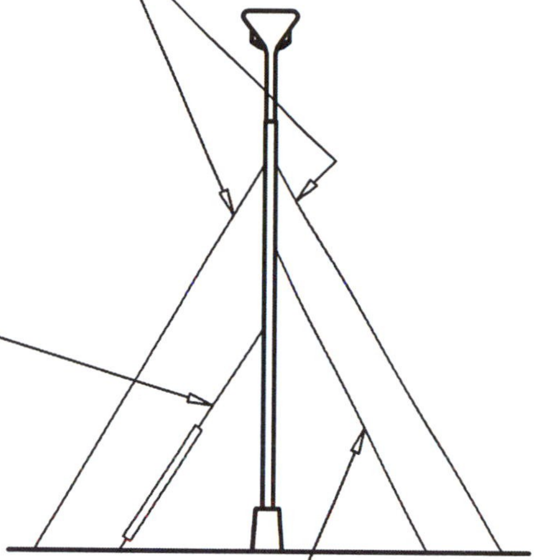	
	SIDEWALK GUY WITH ANCHOR		NOTE METAL GUARD ON SIDEWALK GUY.
	DOWN GUY WITH ANCHOR		
	SINGLE FLOOD LIGHT	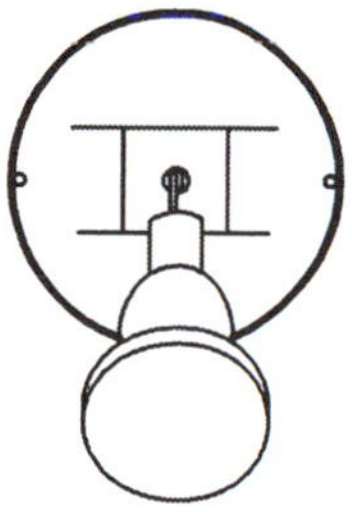	
	DUAL LAMP FLOODLIGHT	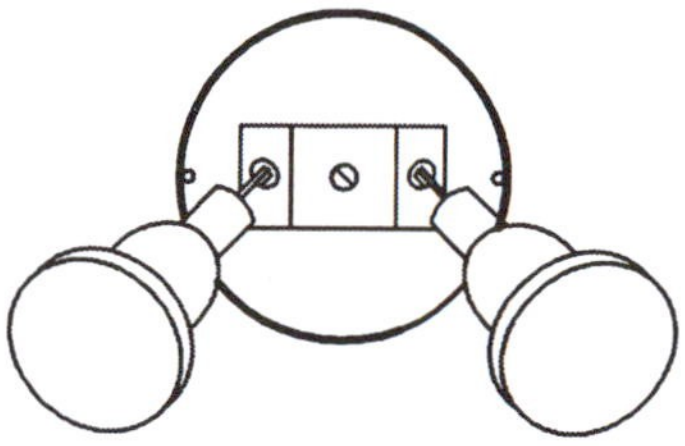	

SYMBOL	NAME	OBJECT	REMARKS
	TRIPLE LAMP FLOODLIGHT		
MD	FLOOD LIGHT WITH MOTION DETECTOR		
T	FLOOD LIGHT CONTROLLED BY TIMER		

NAME	SYMBOL	OBJECT	REMARKS
GENERAL APPLIANCES			
BOXES:			
GENERAL			
BRANCH			
CONNECTION	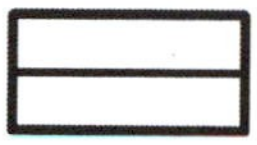	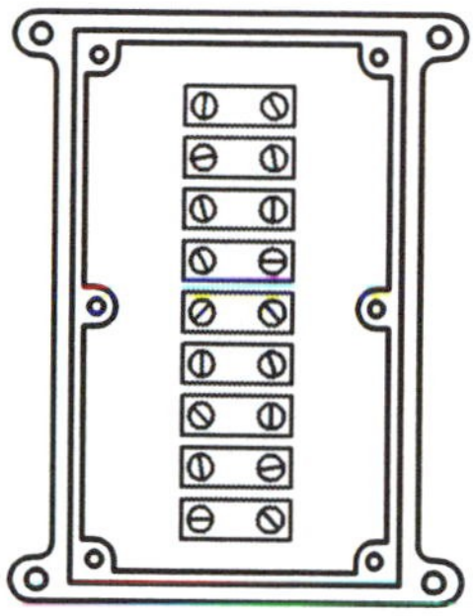	ILLUSTRATION SHOWS WATERTIGHT CONNECTION BOX, WITH GASKET AND COVER REMOVED.
DISTRIBUTION			
JUNCTION	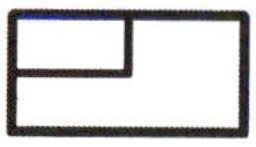	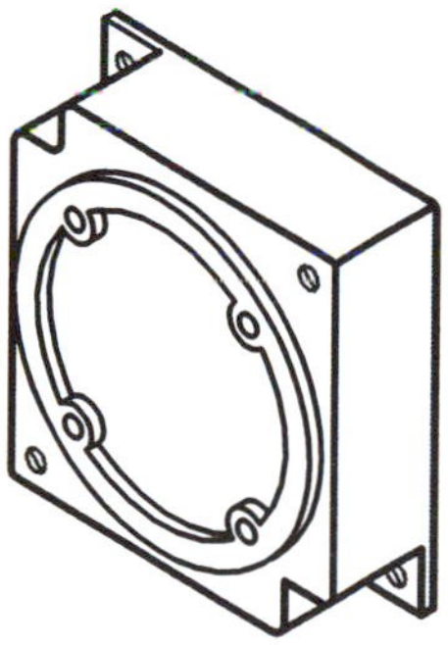	ILLUSTRATION SHOWS WATERTIGHT MARINE JUNCTION BOX (WITHOUT COVER).
TORPEDO FIRING PANEL		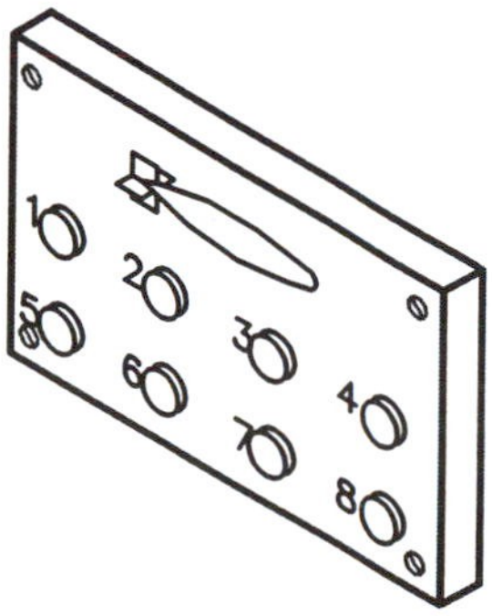	

NAME	SYMBOL	OBJECT	REMARKS
BUS TRANSFER EQUIPMENT NON-AUTOMATIC OR PUSH BUTTON AC			NON-AUTOMATIC OR PUSH BUTTON CONTROL
DC			
BUS TRANSFER EQUIPMENT AUTOMATIC:			
AC WIHTOUT PHASE FAILURE			
AC WITH PHASE FAILURE	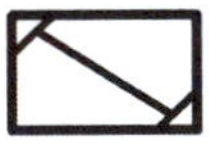		
DC			
COMMUNICATION:			
BOX, SWITCH, TELEPHONE			
JACKS			
TELEPHONE PLUGS			

NAME	SYMBOL	OBJECT	REMARKS
RECEPTACLE OR OUTLET		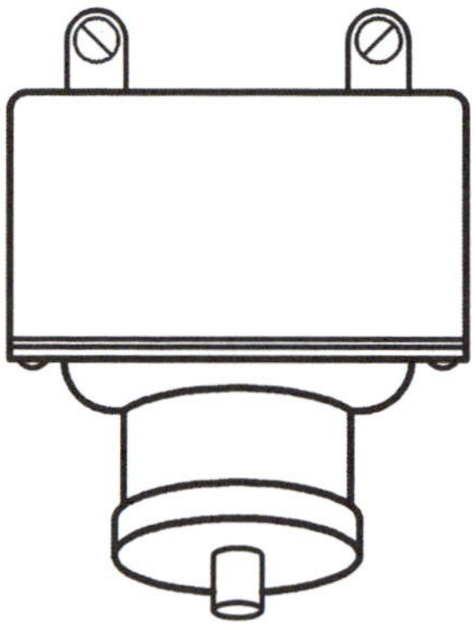	ILLUSTRATION SHOWS BULKHEAD MOUNTING PIN AND SLEEVE TYPE MARINE RECEPTACLE. (CAP HAS FITTING FOR CHAIN.)
SWITCHES PUSH BUTTON		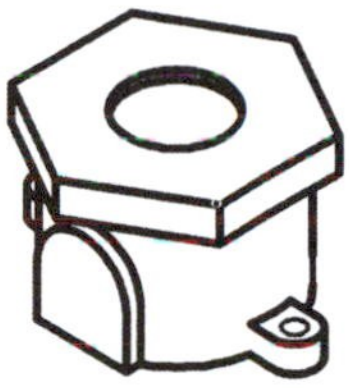	ILLUSTRATION SHOWS WATERTIGHT PUSH BUTTON SWITCH.
ON-OFF		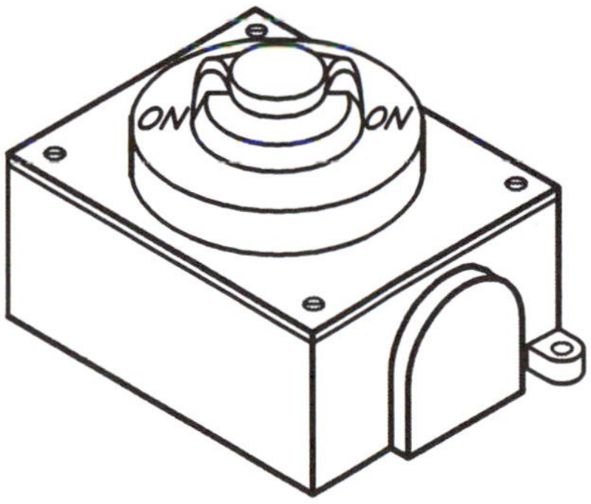 ON ON	ILLUSTRATION SHOWS SURFACE MOUNTING WATER TIGHT ON-OFF SWITCH.
SNAP			
TRANSFER			

NAME	SYMBOL	OBJECT	REMARKS
SWITCHES (CONTNUED) SNAP			
TRANSFER			
SELECTOR	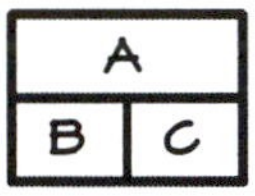 A B C		A = CIRCUIT LETTER B = PANEL OR BULKHEAD C = NUMBER OF SECTIONS
WATER SWITCH			
MOTOR CONTROLLERS: GENERAL CONTROLLER MOTOR	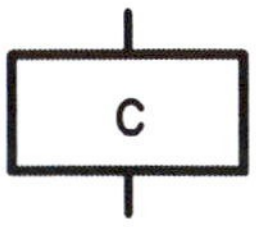 C		
CONTROLLER W/ LOW VOLTAGE RELEASE, RECLOSED UPON RETURN OF POWER.	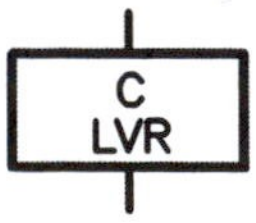 C LVR		BUILDUP EXAMPLES.

NAME	SYMBOL	OBJECT	REMARKS
MOTOR CONTROLLERS: (CONTNUED) CONTROLLER W/ LOW VOLTAGE PROTECTION, REMAINS OPEN UPON RETURN OF POWER.	C LVP		
FANS: PORTABLE BRACKET FAN	F		
OVERHEAD FAN	F		
HEATERS: GENERAL HEATER	H		
PORTABLE RADIANT HEATER	H		
LIGHTING UNITS: BULKHEAD LIGHT			

NAME	SYMBOL	OBJECT	REMARKS
LIGHTING UNITS: (CONTINUED) BULKHEAD BERTH LIGHT		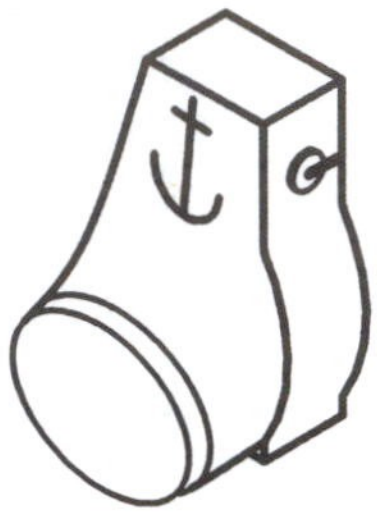	
HAND LANTERN		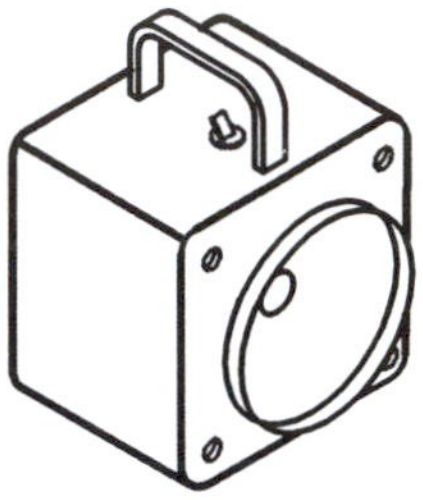	ILLUSTRATION PORTRAYS WATER TIGHT HAND LANTERN.
NAVIGATIONAL LIGHT		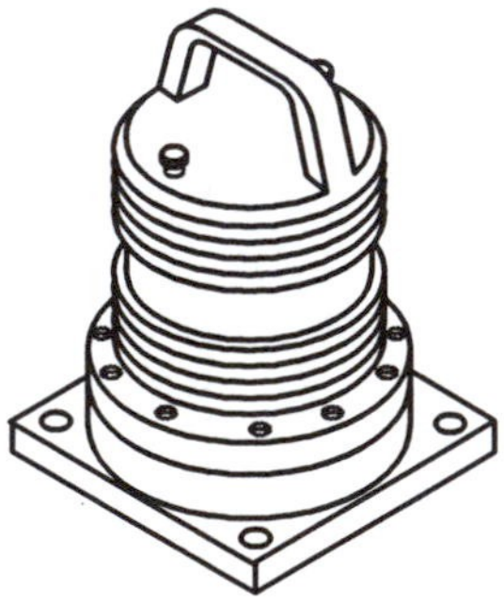	ILLUSTRATED LAMP FOR VESSELS OF 20M AND ABOVE.
NIGHT FLIGHT LIGHT		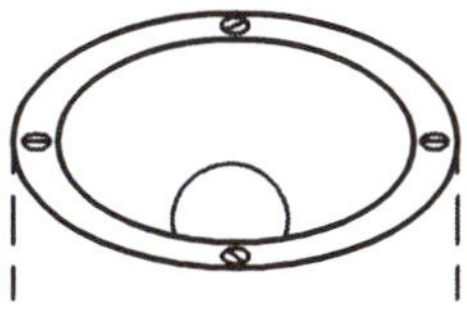	ILLUSTRATION SHOWS FLUSH DECK-MOUNTED AIRCRAFT CARRIER STYLE NIGHT FLIGHT LIGHT.

NAME	SYMBOL	OBJECT	REMARKS
LIGHTING UNITS: (CONTINUED) OVERHEAD LIGHT			
PORTABLE LIGHT		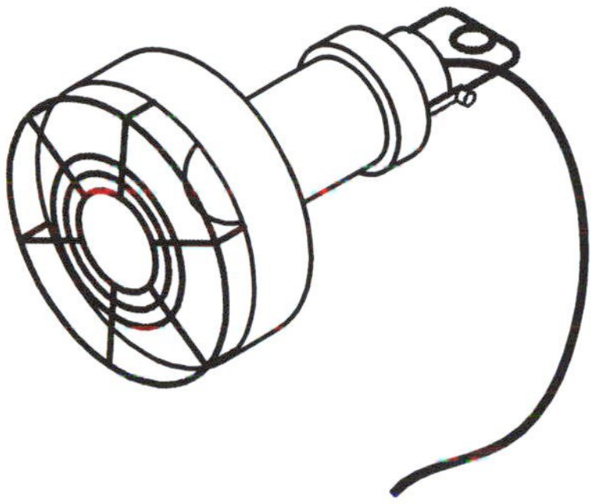	
OVERHEAD FLUORESCENT LIGHT		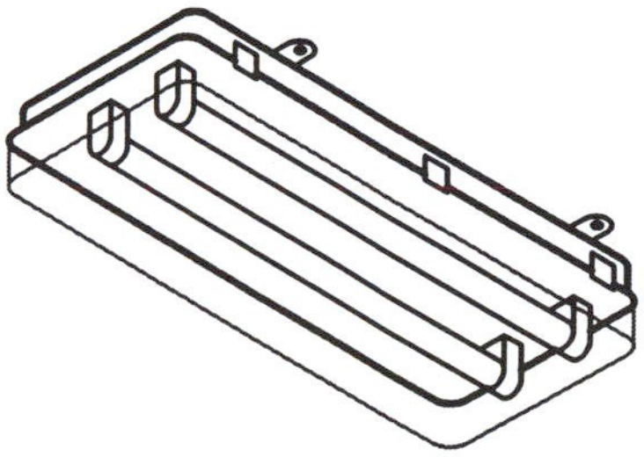	ARROWS ON SYMBOL SHOW DIRECTION OF FLUORESCENT TUBES. ILLUSTRATION SHOWS SEALED CEILING UNIT USED IN ENGINE ROOMS WORKING AREAS, ETC.

FIRE FIGHTING SYMBOLS

NAME	SYMBOL	OBJECT	REMARKS
FIRE ALARM CONTROL PANEL	FACP [F] FCP	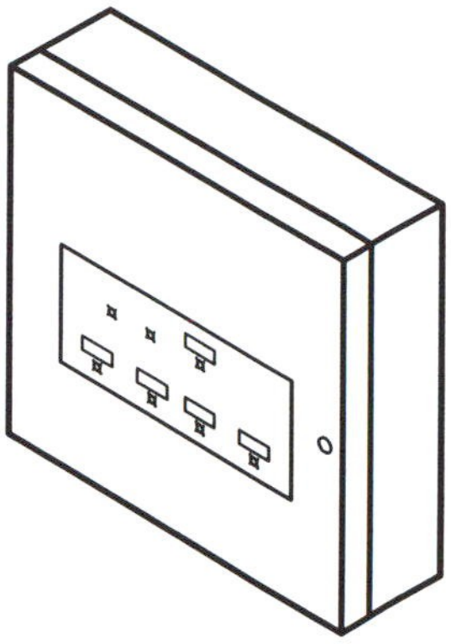	
FIRE ALARM FIELD PANEL	FAFP	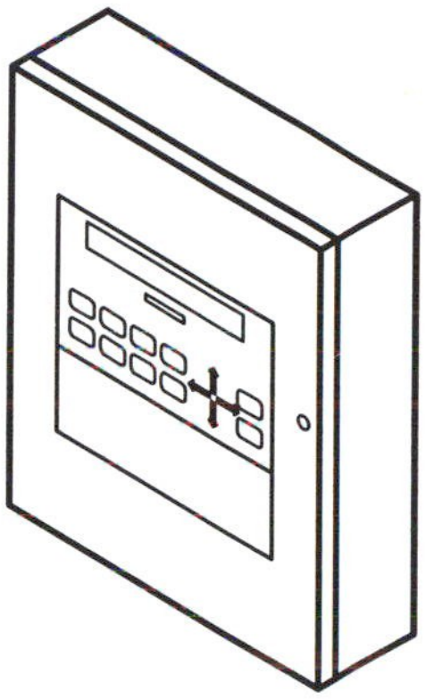	
FIRE ALARM ANNUNCIATOR	[F] FSA FAA ANN		AN ANNUNCIATOR IS A DEVICE WHICH IS USED TO ALERT PERSONNEL TO A CHANGE IN CONDITION. FIRE ALARM ANNUNCIATORS ARE DEFINED BY FLOOR, ZONE OR ROOM, COMMONLY.
FIRE ALARM REMOTE ANNUNCIATOR	FARA	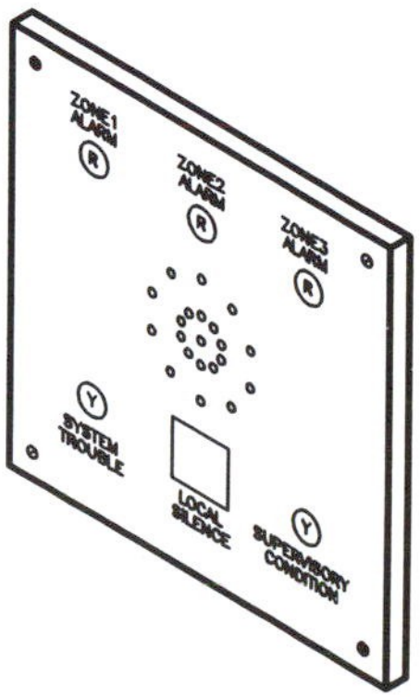 ZONE1 ALARM R ZONE2 ALARM R ZONE3 ALARM R Y SYSTEM TROUBLE LOCAL SILENCE Y SUPERVISORY CONDITION	

NAME	SYMBOL	OBJECT	REMARKS
FIRE COMMAND CENTER	FCC		
ELEVATOR STATUS/ RECALL	[F] ESR		
FIRE ALARM COMMUNICATOR	[F] FAC		
HALON CONTROL PANEL	[F] HCP		
HALON CONTROL VALVE	[N] HCV		HALON IS NO LONGER USED IN FIRE CONTROL DUE TO ITS DAMAGING EFFECT ON THE ENVIRONMENT.
HALON CONTROL PANEL	[F] HCP		

NAME	SYMBOL	OBJECT	REMARKS
REMOTE INDICATOR LIGHT		 FIRE	
REMOTE INDICATOR LIGHT W/KEY TEST/RESET SWITCH	 K	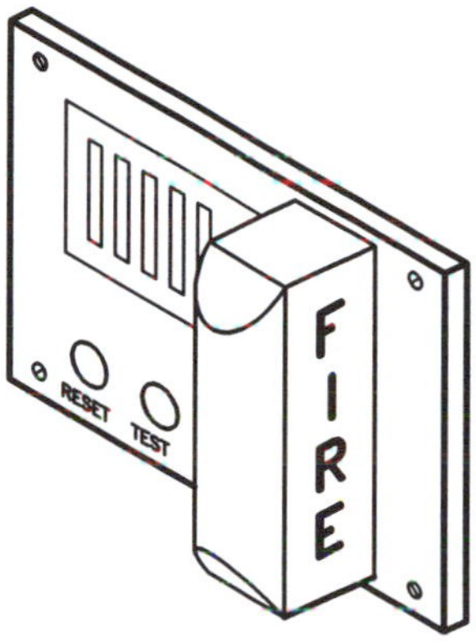 RESET TEST FIRE	
FIRE ALARM SYSTEM DEVICES	 KEY		
BELL	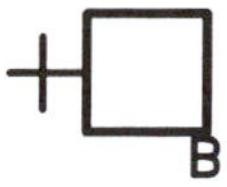 B	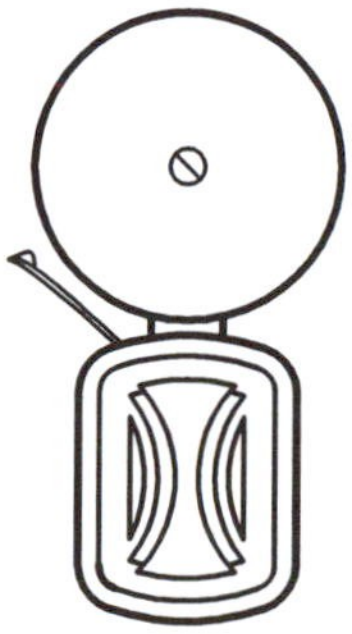	

NAME	SYMBOL	OBJECT	REMARKS
CHIME	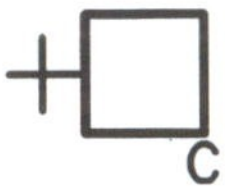 C	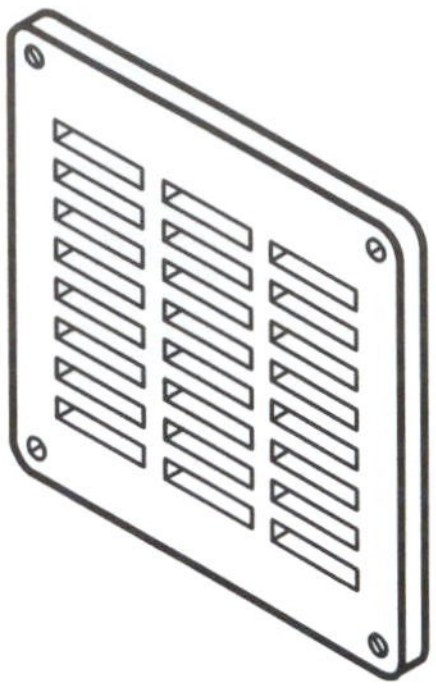	
CHIME, VISUAL INDICATOR & SPEAKER	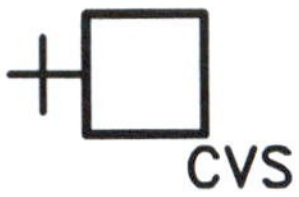 CVS		
CHIME & VISUAL INDICATOR	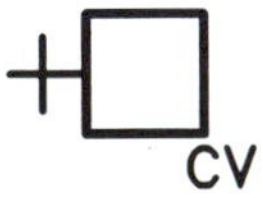 CV	 FIRE	SHOWN WITH STROBE
DUCT SMOKE DETECTOR	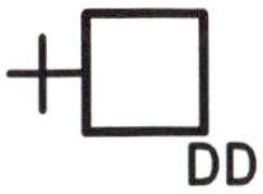 DD	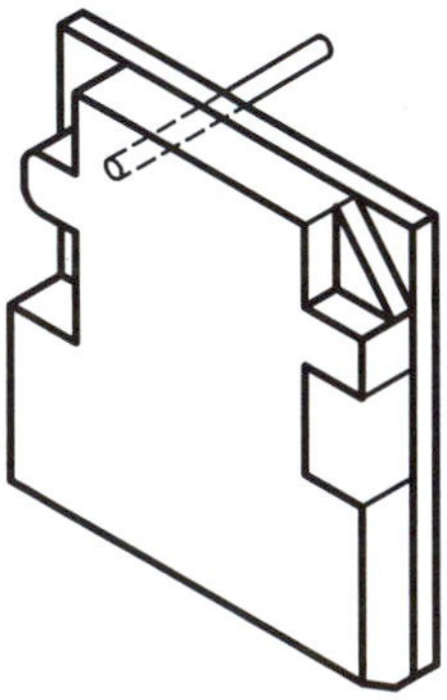	

NAME	SYMBOL	OBJECT	REMARKS
FIRE ALARM FLAME DETECTOR	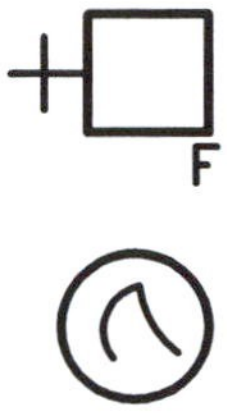 F	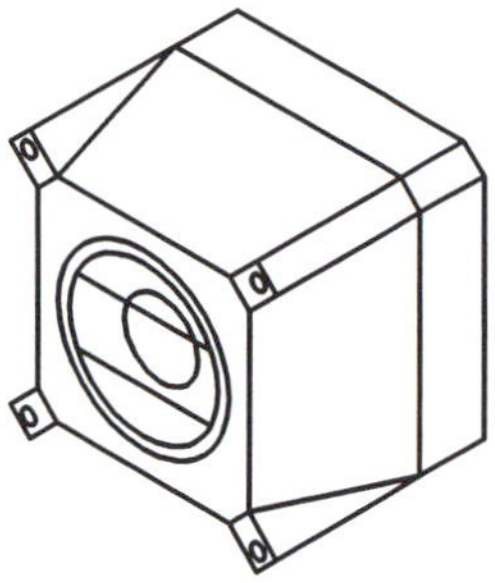	ULTRAVIOLET TYPE FLAME DETECTOR IS ILLUSTRATED.
FIRE DAMPER INTERFACE	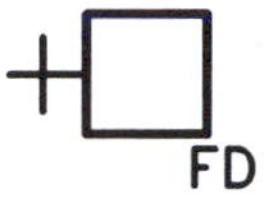 FD		
SPRINKLER WATERFLOW ALARM SWITCH	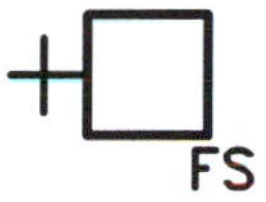 FS [F] 	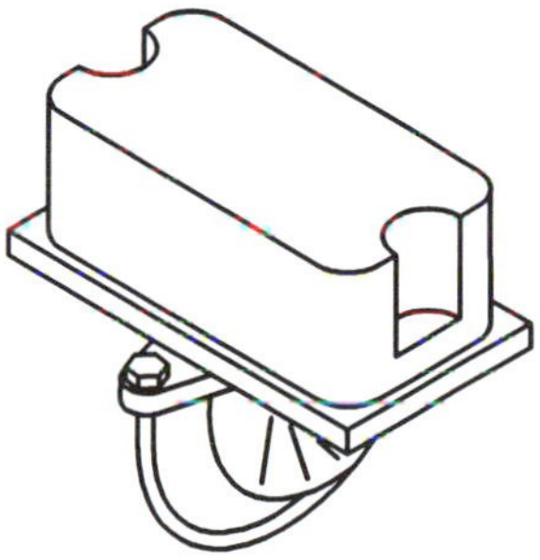	
PRESSURE DETECTOR/ SWITCH	[F] 		
LEVEL DETECTOR/ SWITCH	[F] 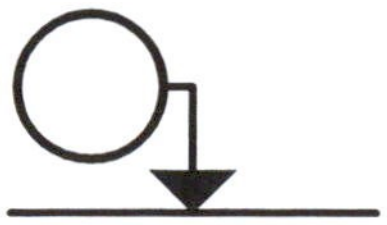		

FIRE ALARM

NAME	SYMBOL	OBJECT	REMARKS
HORN	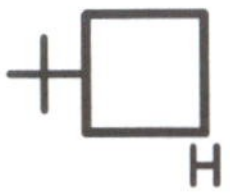 H		
HORN WITH VISUAL INDICATOR	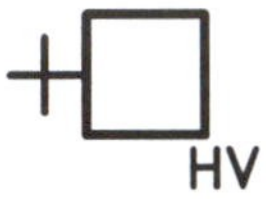 HV	 FIRE	SHOWN WITH STROBE
HORN WITH VISUAL INDICATOR & CHIME	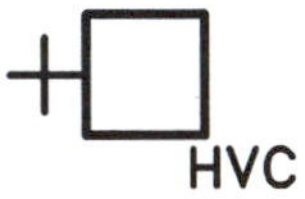 HVC		
FIRE ALARM STROBE		 FIRE	MOUNT 80" AFF OR 6" BELOW CEILING, WHICHEVER IS GREATER.
FIRE ALARM HORN & STROBE		 FIRE	

NAME	SYMBOL	OBJECT	REMARKS
MANUAL PULL STATION	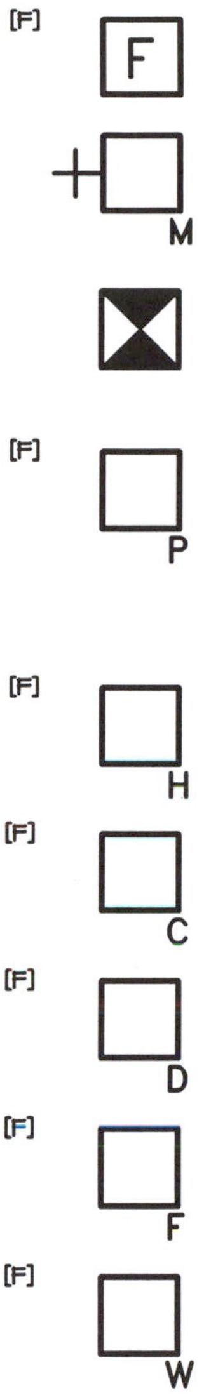 [F] F M [F] P	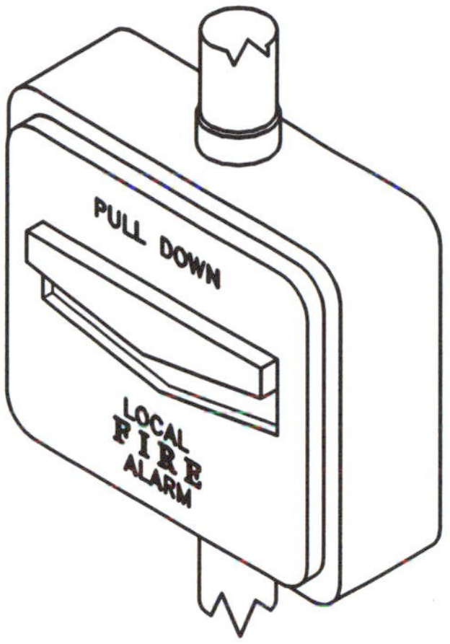	MAXIMUM HEIGHT 48" TO TO CENTERLINE. (CHECK LOCAL CODE.) RED WITH WHITE LETTERING.
HALON	[F] H		
CARBON DIOXIDE	[F] C		
DRY CHEMICAL	[F] D		
FOAM	[F] F		
WET CHEMICAL	[F] W		
PRESSURE SWITCH	 PS		
PRESSURE VALVE	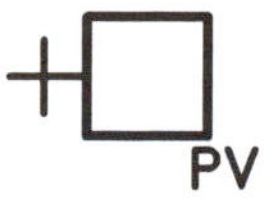 PV		SUBSCRIPT DENOTES CIRCUIT

PULL DOWN
LOCAL FIRE ALARM

NAME	SYMBOL	OBJECT	REMARKS
SPEAKER	 S	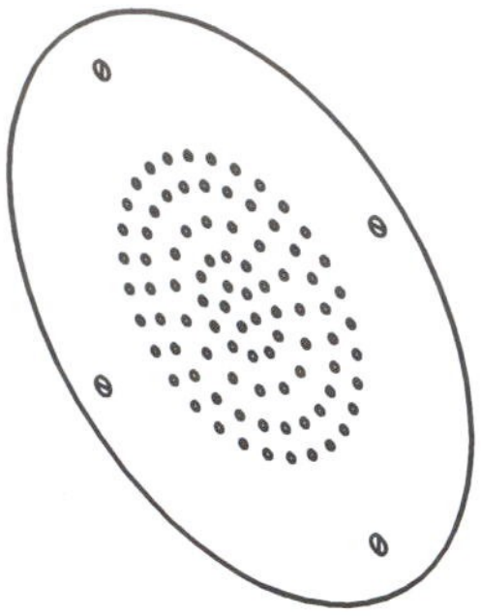	
SMOKE DETECTOR	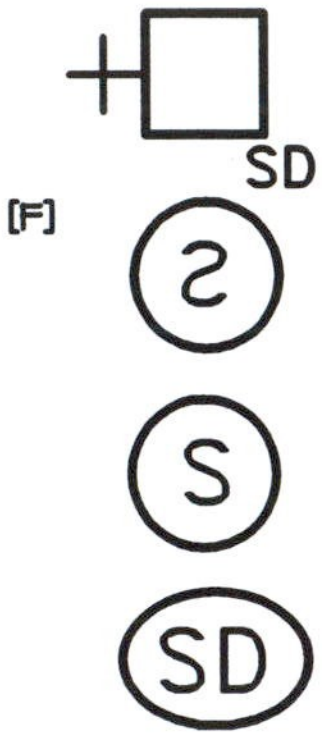 SD [F] S S SD	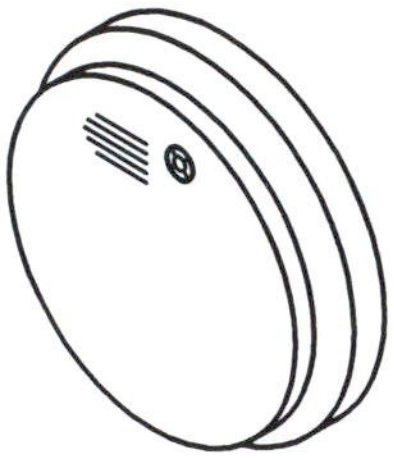	SHOWN WALL MOUNTED.
SINGLE STATION SMOKE DETECTOR	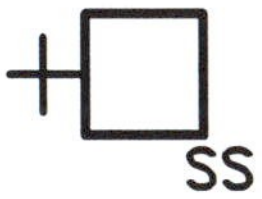 SS		
SPEAKER WITH VISUAL INDICATOR	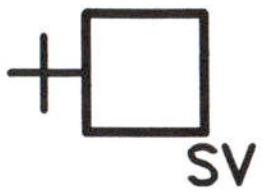 SV	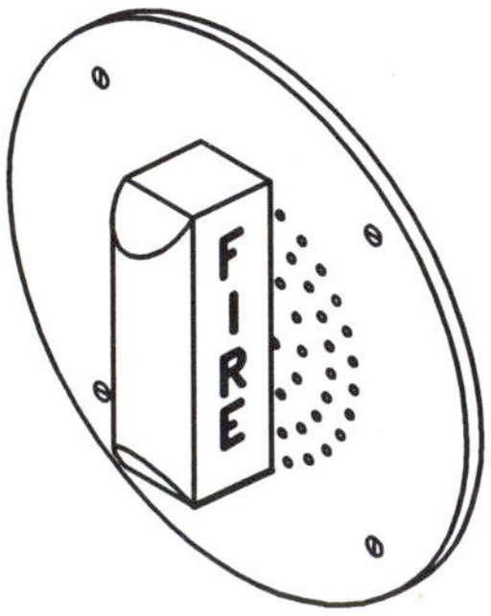 FIRE	

NAME	SYMBOL	OBJECT	REMARKS
FAULT ISOLATOR	I		
KEY OPERATED	K		
FIREMAN'S PHONE	PH [F] H		
FIREMAN'S PHONE JACK	J [F] J		
FIREMAN'S PHONE ACCESSIBLE	[F] A		

NAME	SYMBOL	OBJECT	REMARKS
SMOKE DETECTOR, BEAM TRANSMITTER	[F] BT		
SMOKE DETECTOR, BEAM RECEIVER	[F] BR		
ABORT SWITCH	AB		
THERMAL DETECTOR, RATE OF RISE	[F] R TR		
THERMAL DETECTOR, RATE OF RISE & FIXED TEMP.	[F] R/F		
THERMAL DETECTOR, RATE COMPENSATION	[F] R/C		

NAME	SYMBOL	OBJECT	REMARKS
ALARM BOX WITH TAMPER SWITCH	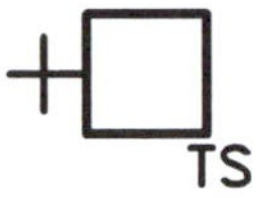 TS	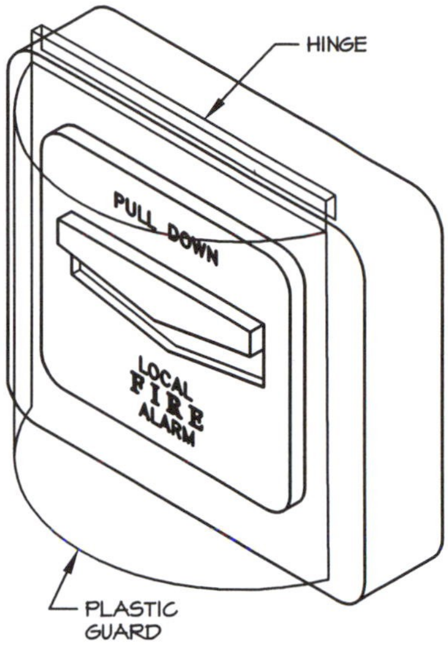 HINGE PULL DOWN LOCAL FIRE ALARM PLASTIC GUARD	PLASTIC GUARD OVER PULL STATION EMITS A LOUD NOISE WHEN LIFTED, THUS PREVENTING FALSE ALARMS AND VANDALISM.
VISUAL INDICATOR	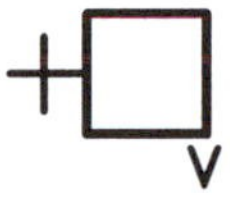 V		
FIRE ALARM HORN WITH INTEGRAL WARNING LIGHT	 F	 FIRE	RED BOX, WHITE LAMP ON TOP.
FIRE ALARM THERMO-DETECTOR, FIXED TEMPERATURE	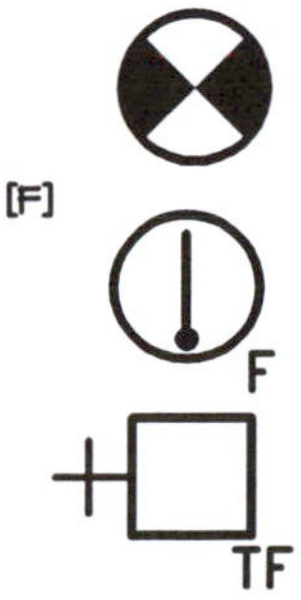 [F] F TF	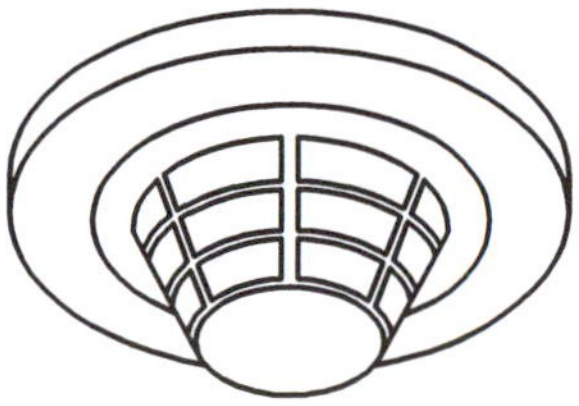	SHOWN CEILING MOUNTED.

NAME	SYMBOL	OBJECT	REMARKS
FIRE ALARM SMOKE DETECTOR, PHOTOELECTRIC TYPE	[F] P		SHOWN CEILING MOUNTED.
FIRE ALARM SMOKE DETECTOR, DUCT TYPE			SAMPLES AIR CURRENTS THROUGH A DUCT.
FIRE DETECTOR-DUCT TYPE			
FIRE DETECTOR	D		
SMOKE DETECTOR	SD		

NAME	SYMBOL	OBJECT	REMARKS
FIRE ALARM SMOKE DETECTOR, IONIZATION PRODUCTS OF COMBUSTION DETECTOR	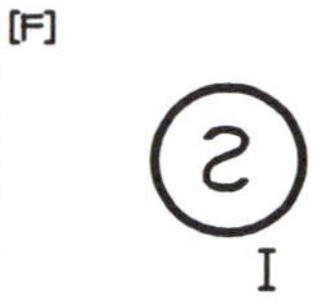 [F] S I		
FIRE ALARM MASTER BOX	 F	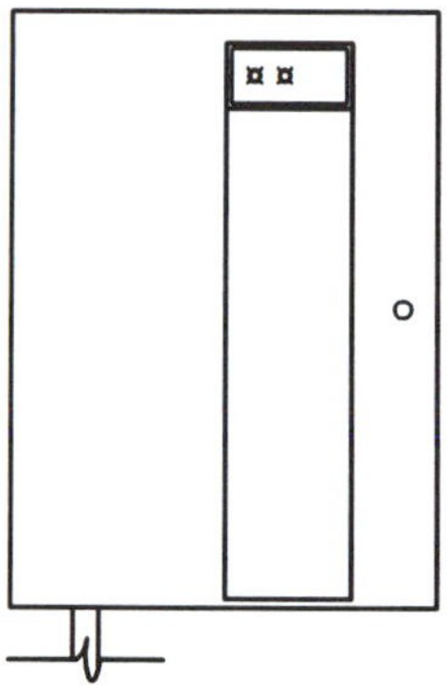	
FIRE ALARM HEAT DETECTOR		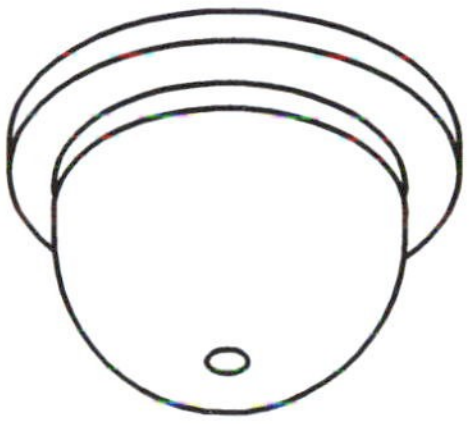	
FIRE SPRINKLER WATER FLOW TAMPER SWITCH			
FIRE ALARM HORN		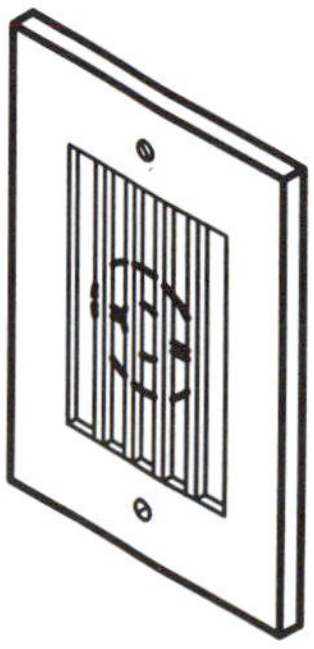	

NAME	SYMBOL	OBJECT	REMARKS
FIRE ALARM BELL	[F]		
FIRE ALARM GONG	G [F]		
WATER GONG OR WATER MOTOR ALARM	[F]	TO DRAIN PADDLE SHAFT DRIVES GONG WATER FLOW FROM VALVE/SYSTEM	SYMBOL SHOWS OPTIONAL GONG SHIELD.

NAME	SYMBOL	OBJECT	REMARKS
PRIVATE HYDRANT (ONE HOSE OUTLET)	[F] 	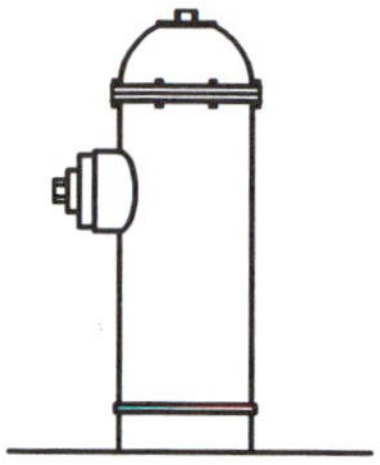	
PUBLIC HYDRANT (TWO HOSE OUTLETS)	[F] 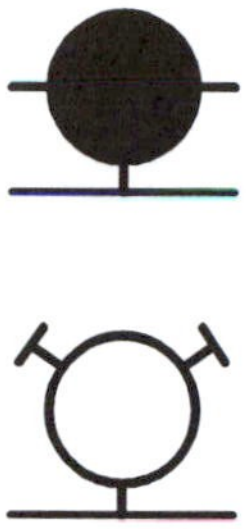	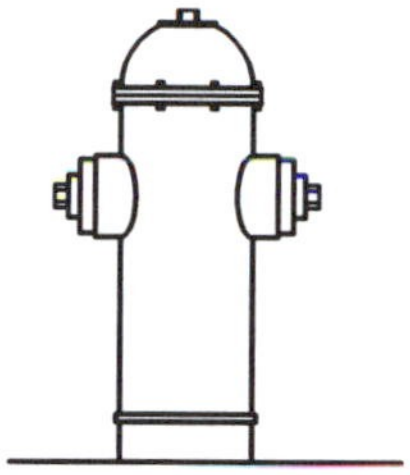	
PUBLIC HYDRANT (TWO HOSE OUTLETS & PUMPER HOSE CONNECTION)	[F] 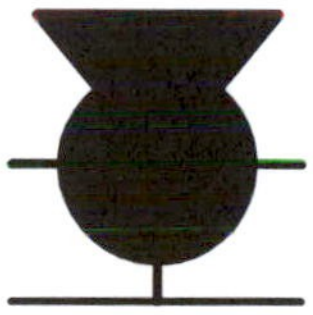	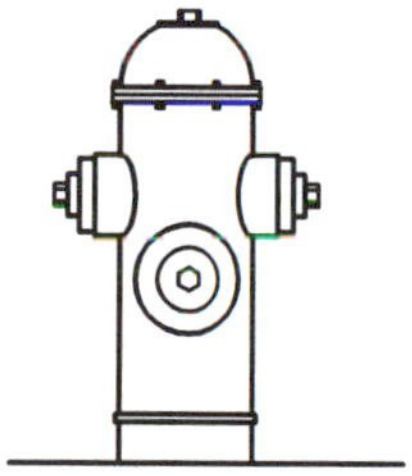	
SINGLE FIRE DEPT. CONNECTION	[F] 	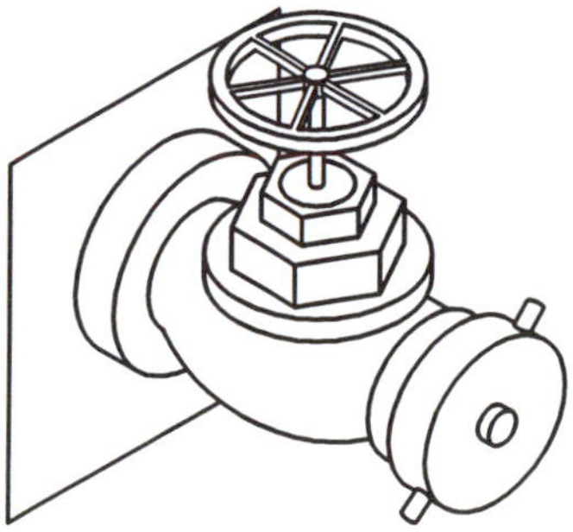	CONNECTS TO STANDPIPE OR SPRINKLER SYSTEM.

NAME	SYMBOL	OBJECT	REMARKS
MULTIPLE FIRE DEPT. CONNECTIONS	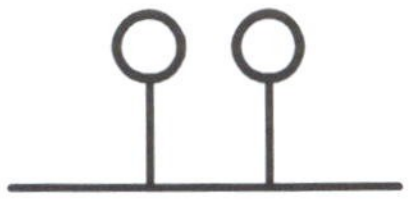	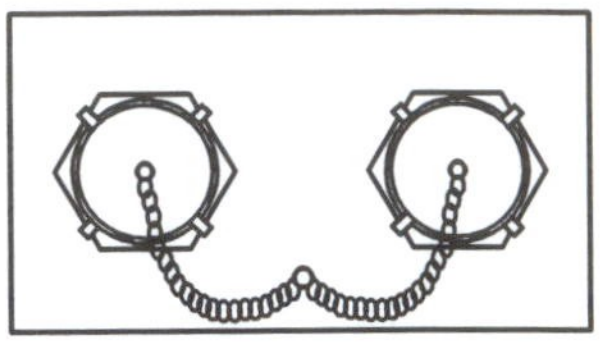	
WALL HYDRANT, TWO HOSE OUTLETS	[F] 	 HYDRANT	CONTROL VALVE FOR THESE CONNECTIONS IS WALL MOUNTED IN VICINITY, TYP.
PRIVATE HOUSED HYDRANT, TWO HOSE OUTLETS	[F] 	 FIRE HYDRANT	
HOSE CABINET		 FIRE HOSE	

NAME	SYMBOL	OBJECT	REMARKS
SIAMESE FIRE HYDRANT, DEPARTMENT CONNECTION	[F] 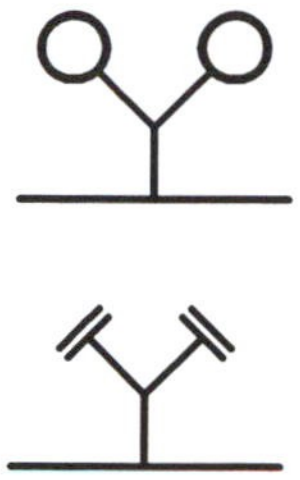	 STANDPIPE	FIRE DEPARTMENT CONNECTIONS TO PROVIDE WATER TO EITHER STANDPIPE SYSTEM OR SPRINKLER SYSTEM. WALL MOUNTED UNIT SHOWN AS STANDPIPE - SPRINKLER CONNECTION WOULD READ AUTOMATIC SPRINKLER, OR SIMILAR.
FREE-STANDING SIAMESE FIRE DEPT. CONNECTION	[F] 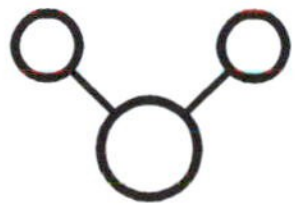	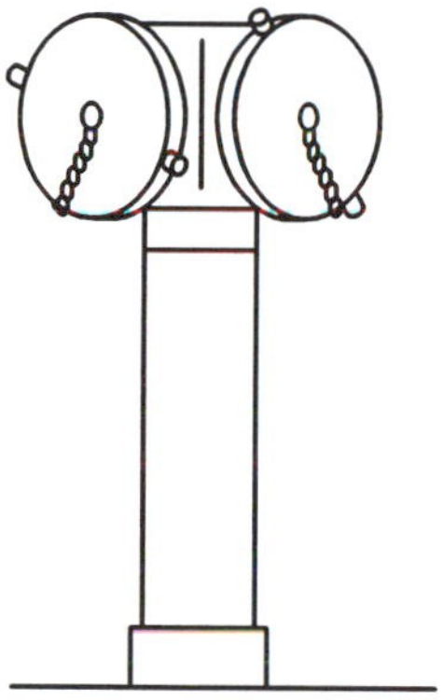	SHOWN WITH CHAINED CAPS.
UPRIGHT SPRINKLER HEAD	[F] 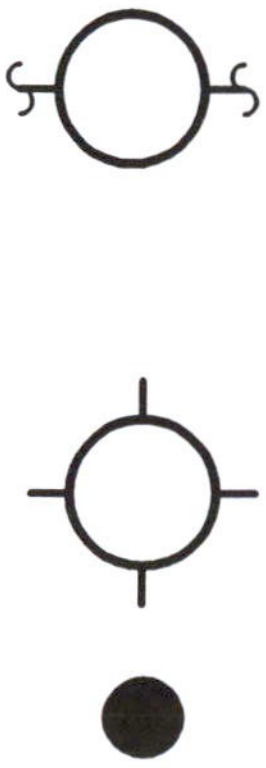	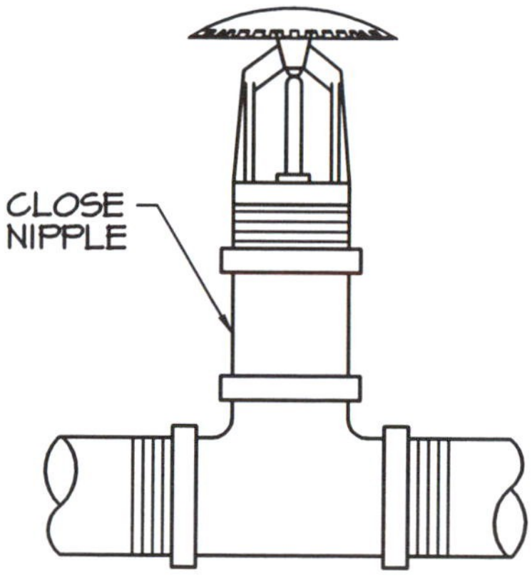 CLOSE NIPPLE	PICTURE ILLUSTRATES 'FLUID' TYPE SPRINKLER WITH 'CLOSE' NIPPLE.

NAME	SYMBOL	OBJECT	REMARKS
PENDANT SPRINKLER HEAD	[F] 	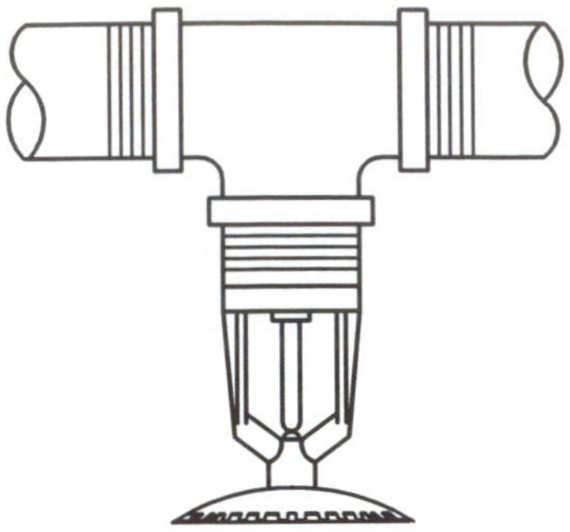	
UPRIGHT SPRINKLER, NIPPLED UP	[F] 	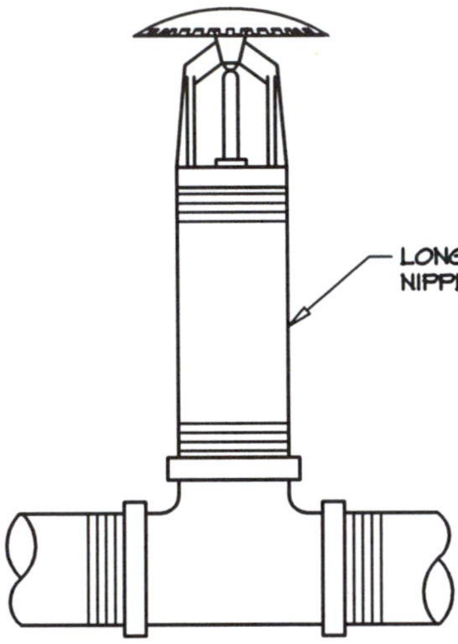 LONG NIPPLE	
PENDANT SPRINKLER, ON DROP NIPPLE	[F] 	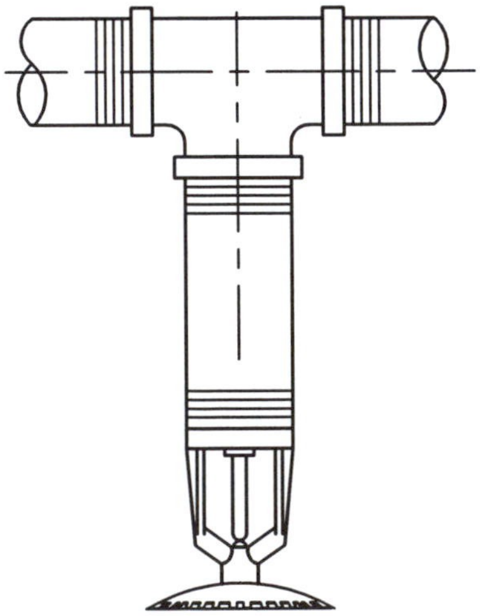	

NAME	SYMBOL	OBJECT	REMARKS
SPRINKLER WITH GUARD	[F] 	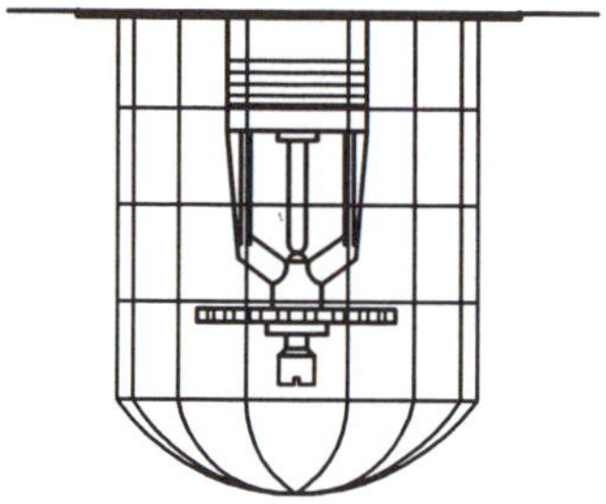	
SIDEWALL SPRINKLER	[F] 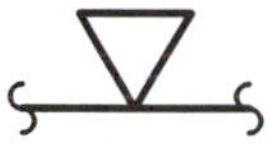	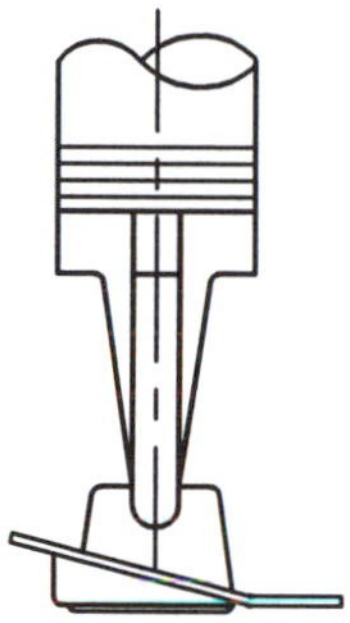	WALL MOUNTED, ANGLED HEAD DIRECTS SPRAY AWAY FROM WALL.
PIPE HANGER	[F] 		THIS PIPE HANGER SYMBOL UNIQUE TO FIRE CONTROL DRAWINGS.
OUTSIDE SPRINKLER	[N] 	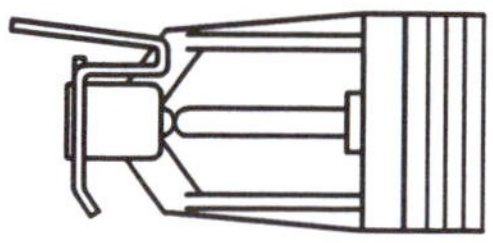	USED ON EXTERIOR WALLS TO SQUELCH PROXIMITY FIRES.

NAME	SYMBOL	OBJECT	REMARKS
FIRE PUMP WITH DRIVES	[F]	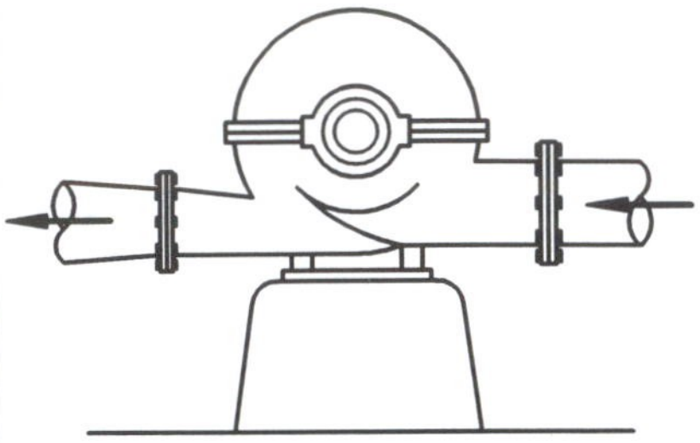	FIRE PUMPS ARE DRIVEN USUALLY BY EITHER ELECTRIC MOTOR OR DIESEL ENGINE.
FIRE PUMP-FREE STANDING	[F] 		SPECIFY NUMBER & SIZE OF OUTLETS.
TEST HEADER	[F] 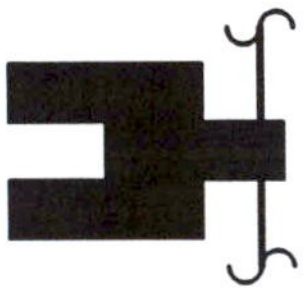		TEST HEADERS GIVE ACCESS TO FIRE CONTROL SYSTEMS TO CHECK OPERABILITY.
HALON OR CLEAN AGENT PORTABLE EXTINGUISHER	[F] 	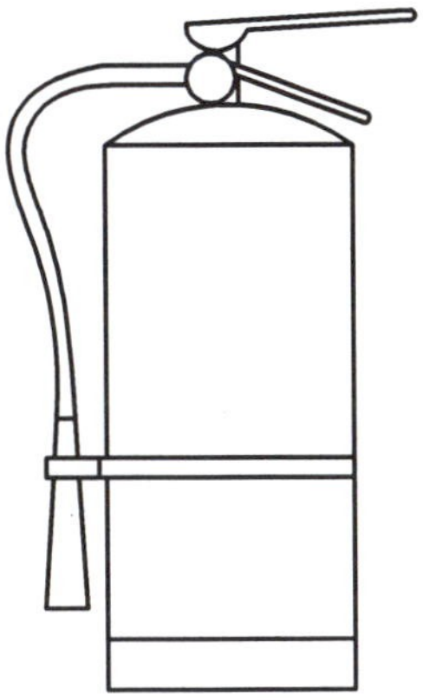	HALON IS NO LONGER IN USE, DUE TO ITS DESTRUCTIVE EFFECT ON THE OZONE LAYER. IT HAS BEEN REPLACED BY OTHER CLEAN AGENT EXTINGUISHING PRODUCTS SUCH AS 'FM200' AND 'INTERGEN' FOR EXAMPLE. CLEAN AGENT EXTINGUISHERS NOW COMMONLY USED IN PIPED SYSTEMS, ESPECIALLY WHERE ELECTRONIC EQUIP. IN USE, E.G. COMPUTERS.

NAME	SYMBOL	OBJECT	REMARKS
PORTABLE FIRE EXTINGUISHER	[F]  △		BASIC SHAPE FOR PORTABLE DEVICES CLASS A FIRE: WOOD, PAPER, ORDINARY COMBUSTIBLES. CLASS B FIRE: FIRE IN FLAM-MABLE LIQUIDS. CLASS C FIRES: FIRES INVOLVING ENERGIED ELECTRIC EQUIPMENT.
PORTABLE WATER EXTINGUISHER	[F]  △	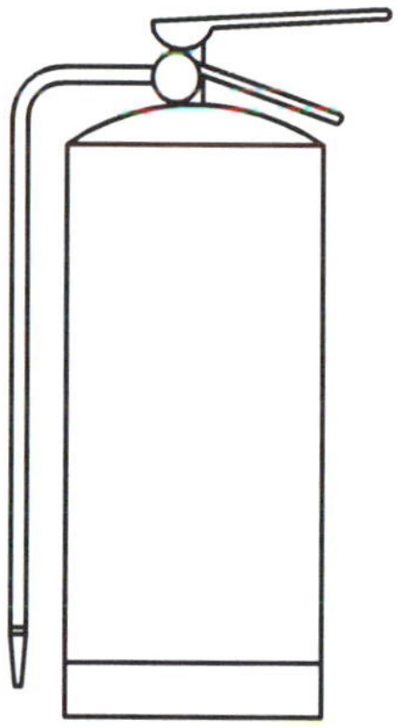	WATER UNDER PRESSURE.
PORTABLE FOAM EXTINGUISHER	[F] 	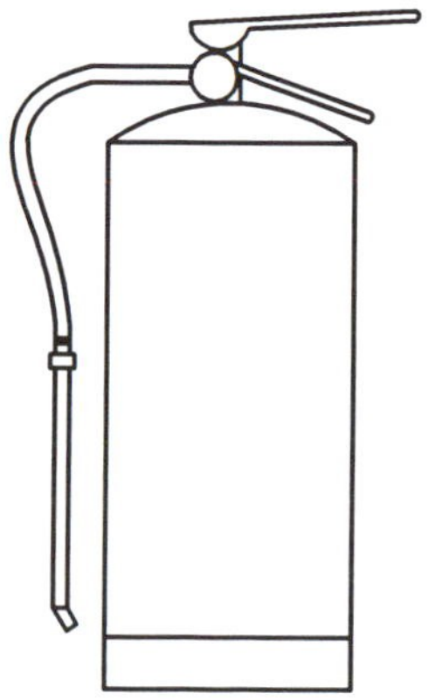	

NAME	SYMBOL	OBJECT	REMARKS
BC TYPE PORTABLE CHEMICAL FIRE EXTINGUISHER	[F] 	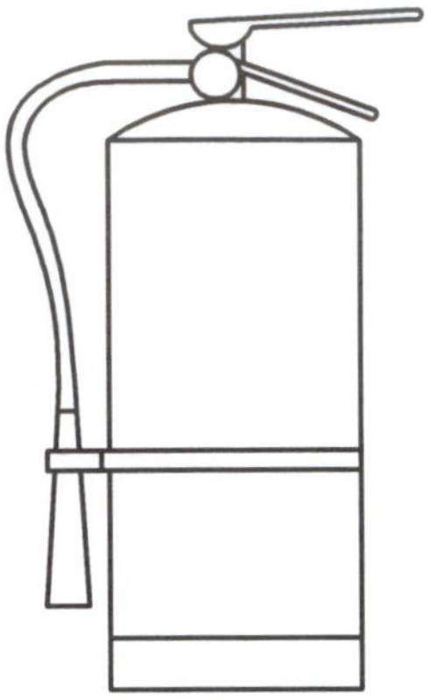	FOR FIRES OF LIQUID, GAS OR ELECTRICAL TYPE
ABC TYPE PORTABLE CHEMICAL FIRE EXTINGUISHER	[F] 	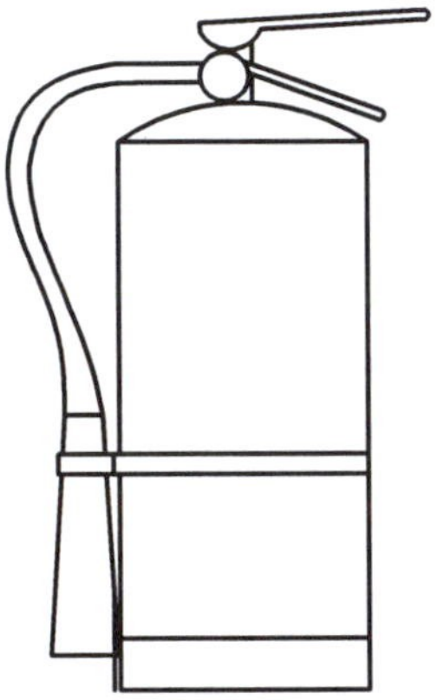	
PORTABLE CARBON DIOXIDE EXTINGUISHER	[F] 	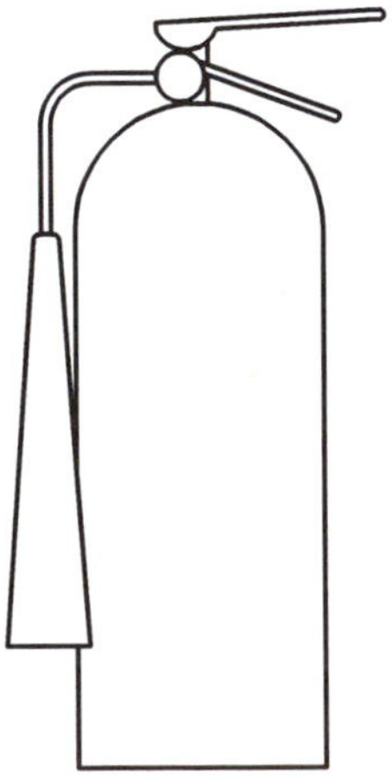	

NAME	SYMBOL	OBJECT	REMARKS
PORTABLE EXTINGUISHER FOR METAL FIRES	[F] 	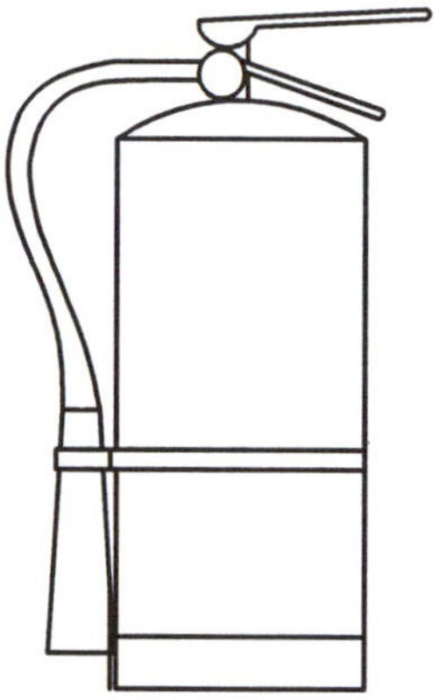	
FIRE FIGHTING EQUIPMENT	[F] 		BASIC SHAPE
CO_2 REEL STATION	[F] 	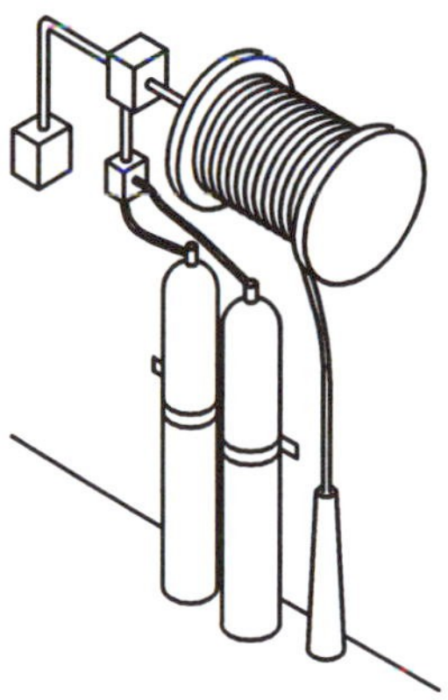	
DRY CHEMICAL REEL STATION	[F] 		
FOAM REEL STATION	[F] 		RARE

NAME	SYMBOL	OBJECT	REMARKS
HOSE STATION DRY STANDPIPE	[F]		SHOWN IN WALL CABINET
HOSE STATION CHARGED STANDPIPE	[F]		
MONITOR NOZZLE, DRY	[F]		A FIRE NOZZLE WHICH MAY BE LEVER-OPERATED, GEAR-OPERATED, MOUNTED ON A TRUCK, HYDRANT OR BLDG. SPECIFY ORIFICE SIZE.
MONITOR NOZZLE, CHARGED	[F]		
FULLY SPRINKLERED SPACE	[F] AS		

NAME	SYMBOL	OBJECT	REMARKS
PARTIALLY SPRINKLERED SPACE	[F]  AS		
NON-SPRINKLERED SPACE	[F]  NS		
ALARM CHECK VALVE	[F] 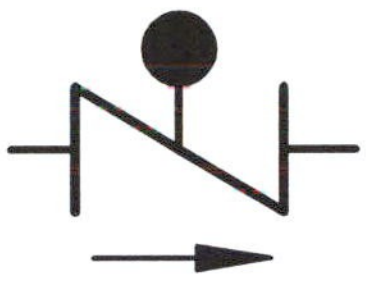		SPECIFY SIZE & DIRECTION OF FLOW
DRY PIPE VALVE	[F] 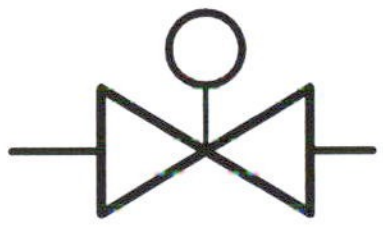	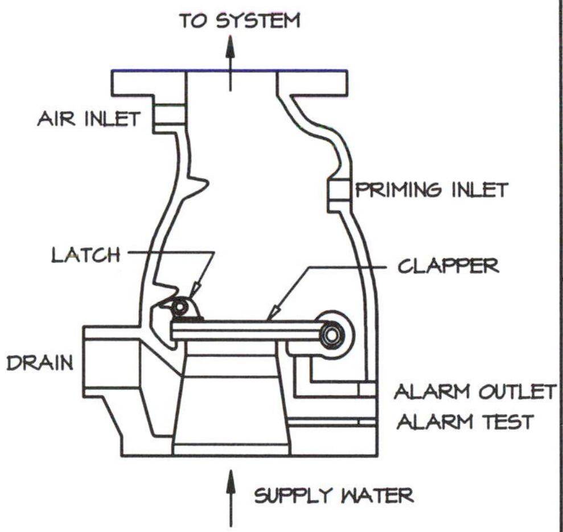 TO SYSTEM AIR INLET PRIMING INLET LATCH CLAPPER DRAIN ALARM OUTLET ALARM TEST SUPPLY WATER	SPECIFY SIZE. ACTIVATED BY DROP IN PRESSURE TO SPRINKLER HEAD LINES. WATER FLOW ACTIVATES ALARM SYSTEM.CLAPPER SHOWN IN CLOSED POSITION.
DRY PIPE VALVE WITH QUICK OPENING DEVICE	[F] 		SPECIFY SIZE & TYPE. SIMILAR TO DRY PIPE VALVE, BUT FLOW IS AUGMENTED BY AIR RELEASE MECHANISM. RECOMMENDED FOR SYSTEMS EXCEEDING 500 GAL. CAPACITY.
PREACTION VALVE	[F] 		SPECIFY SIZE & TYPE. MAY BE CONTROLLED BY SMOKE DETECTOR OR SMOKE DETECTOR & FLOW SENSOR, OR NEITHER.

NAME	SYMBOL	OBJECT	REMARKS
DELUGE VALVE	[F] 	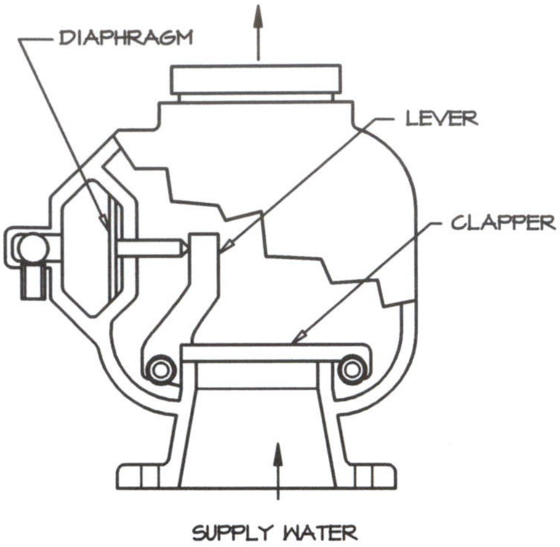 DIAPHRAGM LEVER CLAPPER SUPPLY WATER	SPECIFY SIZE & TYPE. USED WITH OPEN PIPE SPRINKLER SYSTEM. ALL HEADS SPRAY WHEN VALVE OPENS (LIKE IN THE MOVIES). SPRING LOADED CLAPPER LOCKED IN PLACE BY LEVER. DIAPHRAGM RELEASES LEVER.
FIRE DAMPER	[F] 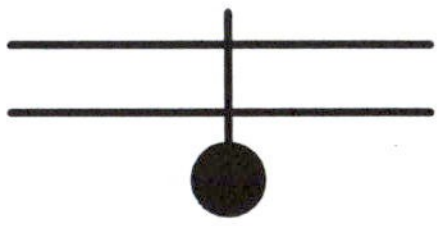	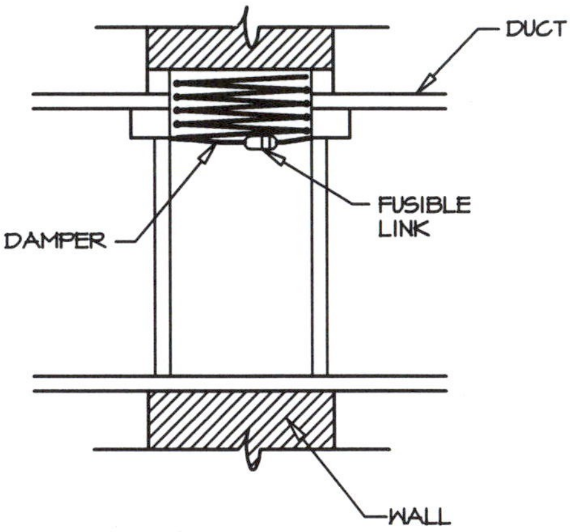 DUCT DAMPER FUSIBLE LINK WALL	
SMOKE DAMPER	[F] 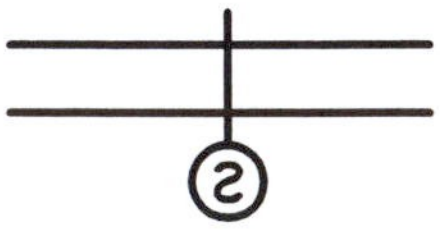		
FIRE/SMOKE DAMPER	[F] 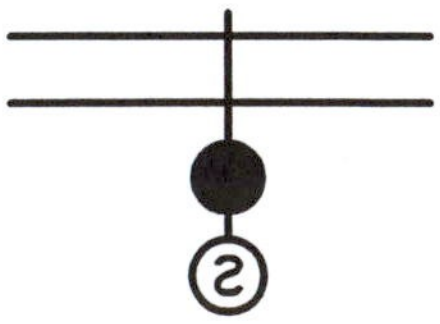		

NAME	SYMBOL	OBJECT	REMARKS
BAROMETRIC DAMPER	[F]		
DOOR HOLDER	DH [N]		MAGNETIC DOOR HOLDERS HOLD FIRE & SMOKE BARRIER DOORS OPEN UNTIL RELEASED BY REMOTE SIGNAL. THE MAGNET IS MOUNTED ON THE WALL, THE PIVOTING PLATE IS MOUNTED ON THE DOOR. THE UNIT PICTURED IS A WALL MOUNTING STYLE. OTHER STYLES ARE FLOOR MOUNT, LOW-PROFILE, DOUBLE DOOR & HAZARDOUS LOCATION STYLE.

NAME	AUTOMATICALLY ACTUATED	MANUALLY ACTUATED
WATER-BASED SYSTEMS WET CHARGED SYSTEM	[F]	[F]
DRY SYSTEM	[F]	[F]
FOAM SYSTEM	[F]	[F]
DRY-CHEMICAL SYSTEMS FOR LIQUID, GAS AND ELECTRICAL TYPE FIRES	[F]	[F]
FOR ALL TYPES OF FIRES EXCEPT METALS	[F]	[F]

NAME	AUTOMATICALLY ACTUATED	MANUALLY ACTUATED
GASEOUS MEDIUM SYSTEMS CARBON DIOXIDE SYSTEM	[F]	[F]
HALON SYSTEM OR CLEAN AGENT EXTINGUISHING SYSTEM	[F]	[F]

MANUFACTURING & MACHINERY SYMBOLS

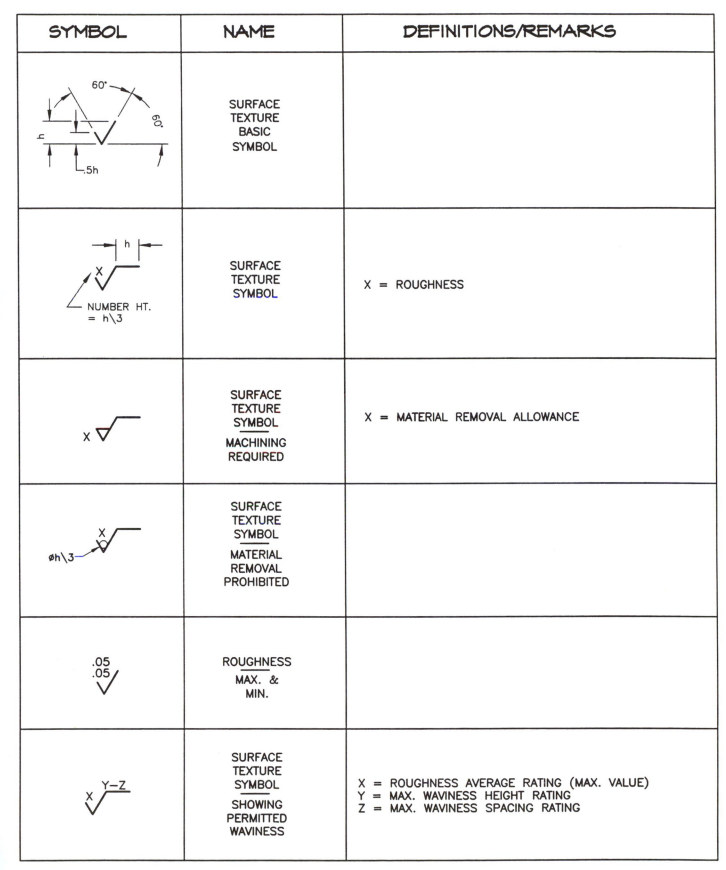

SYMBOL	NAME	DEFINITIONS/REMARKS
60° 60° h .5h	SURFACE TEXTURE BASIC SYMBOL	
h X NUMBER HT. = h\3	SURFACE TEXTURE SYMBOL	X = ROUGHNESS
X	SURFACE TEXTURE SYMBOL — MACHINING REQUIRED	X = MATERIAL REMOVAL ALLOWANCE
X øh\3	SURFACE TEXTURE SYMBOL — MATERIAL REMOVAL PROHIBITED	
.05 .05	ROUGHNESS — MAX. & MIN.	
X Y–Z	SURFACE TEXTURE SYMBOL — SHOWING PERMITTED WAVINESS	X = ROUGHNESS AVERAGE RATING (MAX. VALUE) Y = MAX. WAVINESS HEIGHT RATING Z = MAX. WAVINESS SPACING RATING

SYMBOL	NAME	DEFINITIONS/REMARKS
X √ Y–Z	SURFACE TEXTURE SYMBOL — SHOWING PERMITTED WAVINESS	X = ROUGHNESS AVERAGE RATING (MAX. VALUE) Y = MAX. WAVINESS HEIGHT RATING Z = MAX. WAVINESS SPACING RATING
√ = h/3	LAY SYMBOL DESIGNATION	
√ X	ROUGHNESS SAMPLING LENGTH *OR* CUTOFF RATING	
√⊥ X h/3 2X	MAXIMUM ROUGHNESS SPACING	

NAME	SYMBOL	OBJECT	REMARKS
POSITIVE DISPLACEMENT METER			
VORTEX METER	 V		
ROTAMETER			
TURBINE METER	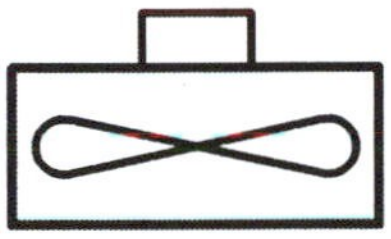		
FLOWMETER, ORIFICE	[A]  IN FITTING: 	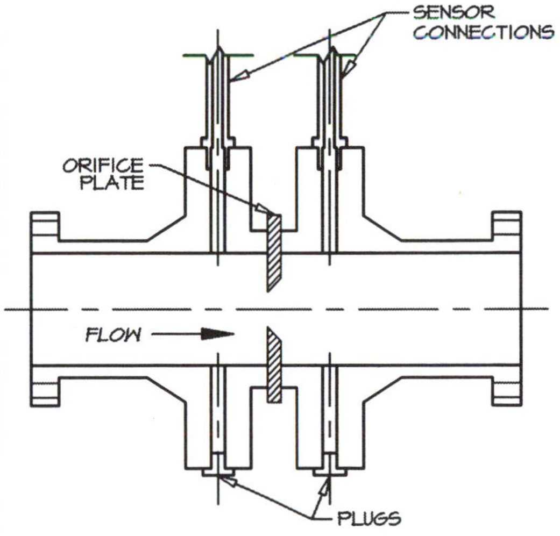 SENSOR CONNECTIONS ORIFICE PLATE FLOW PLUGS	
FLEXIBLE VENTURI	 VFM-1		

NAME	SYMBOL	OBJECT	REMARKS
WORKING LINE			
PILOT LINE			LENGTH OF DASH SHOULD BE AT LEAST 20 LINE WIDTHS WITH SPACE APPROX. 5 LINE WIDTHS.
LIQUID DRAIN OR AIR EXHAUST LINE			LENGTH OF DASH AND SPACE ARE EQUAL, EACH LESS THAN 5 LINE WIDTHS.
CROSSING LINES			TWO WORKING LINES CROSSING.
	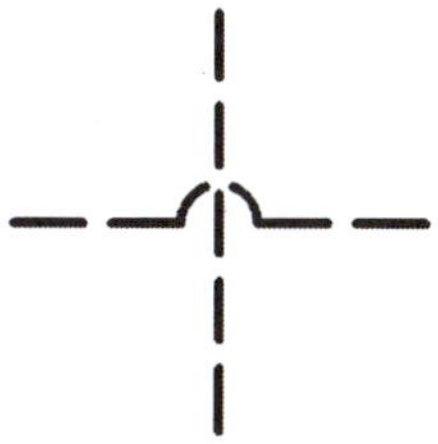		TWO PILOT LINES CROSSING.
	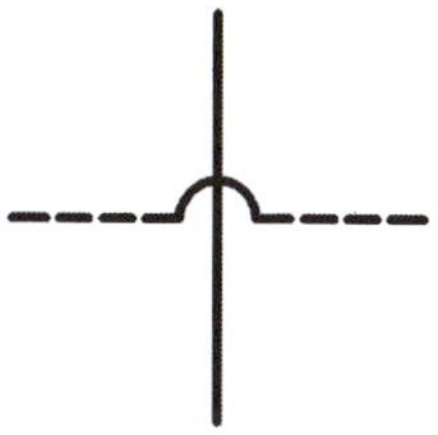		DRAIN OR EXHAUST PASSING OVER WORKING LINE.

NAME	SYMBOL	OBJECT	REMARKS
JOINING LINES	 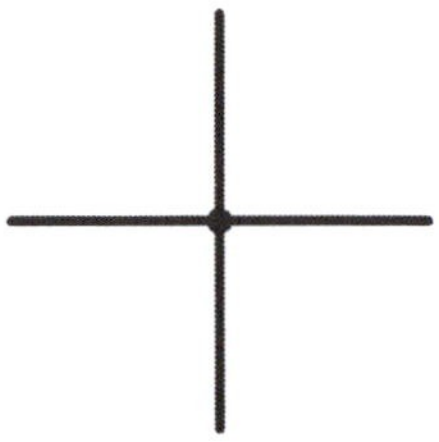 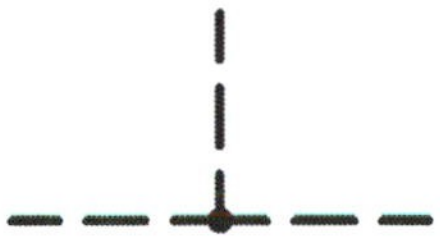		DOT IS 5 WIDTHS OF ADJACENT LINES. TWO CONNECTING PILOT LINES.
FLEXIBLE LINES		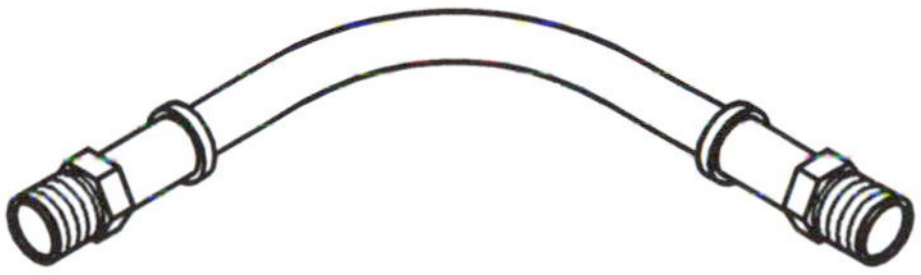	TYPICALLY A SYNTHETIC RUBBER TUBE SURROUNDED BY STEEL WIRE BRAID AND AN OIL RESISTANT RUBBER COVER.
LINES TO RESERVOIR BELOW FLUID LEVEL ABOVE FLUID LEVEL	 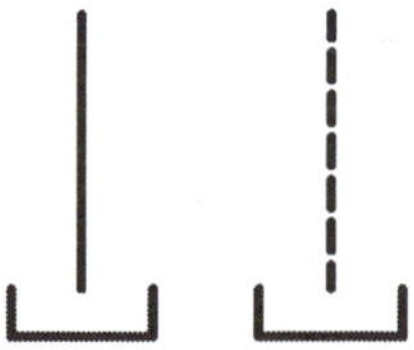		

NAME	SYMBOL	OBJECT	REMARKS
PLUG OR PLUGGED CONNECTION			
TESTING STATION	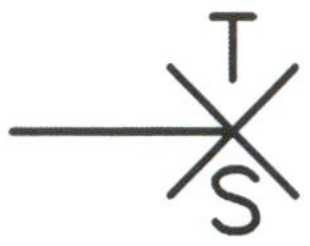 T S		
FLUID POWER TAKE-OFF STATION	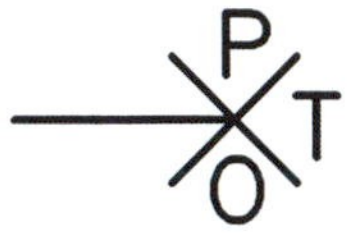 P T O		
FIXED RESTRICTION			
ROTATING SHAFTS	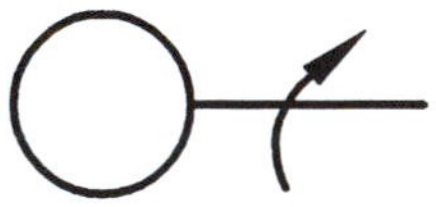 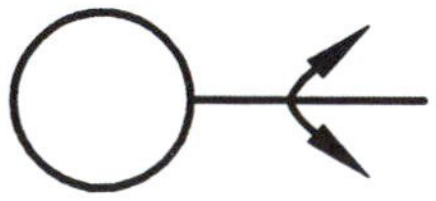		INTERPRET ARROW AS BEING ON NEAR SIDE OF SHAFT.

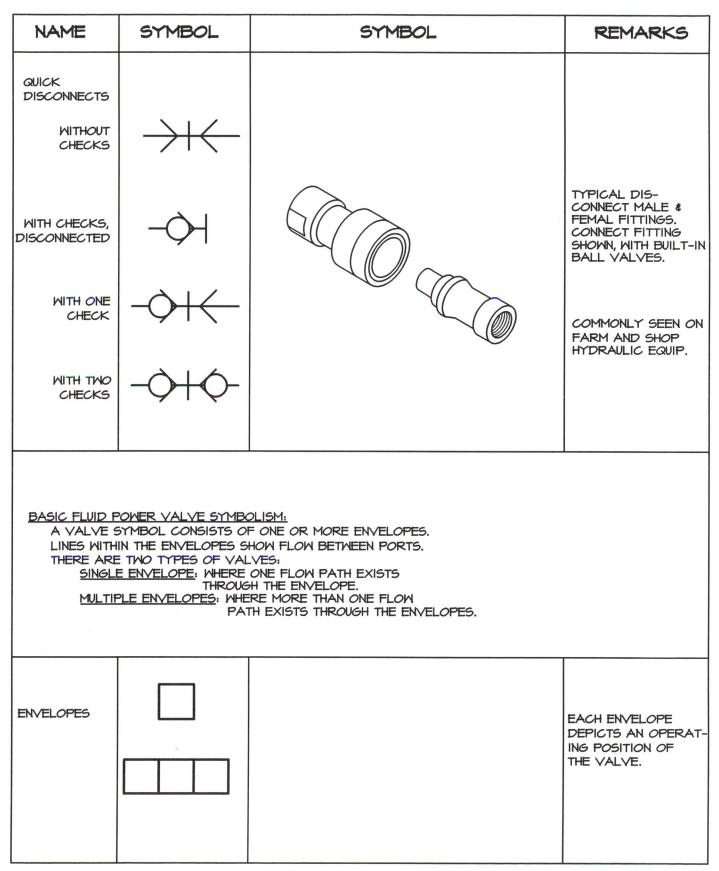

NAME	SYMBOL	SYMBOL	REMARKS
QUICK DISCONNECTS		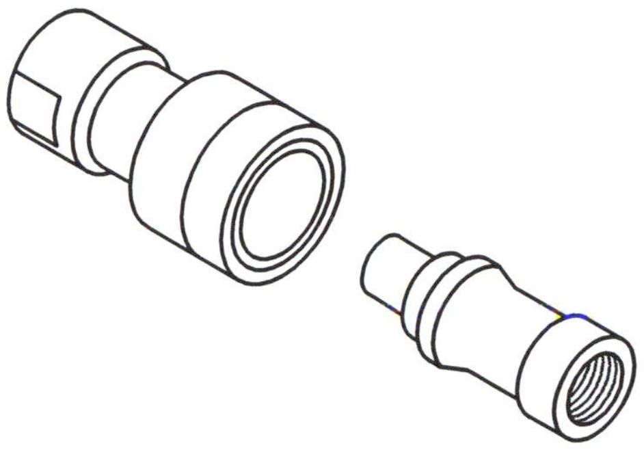	
WITHOUT CHECKS	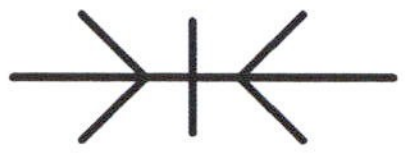		
WITH CHECKS, DISCONNECTED			TYPICAL DIS-CONNECT MALE & FEMAL FITTINGS. CONNECT FITTING SHOWN, WITH BUILT-IN BALL VALVES.
WITH ONE CHECK			COMMONLY SEEN ON FARM AND SHOP HYDRAULIC EQUIP.
WITH TWO CHECKS	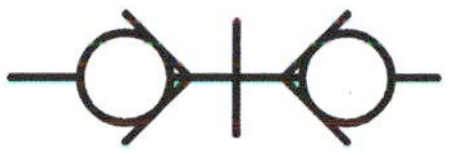		

BASIC FLUID POWER VALVE SYMBOLISM:

A VALVE SYMBOL CONSISTS OF ONE OR MORE ENVELOPES.
LINES WITHIN THE ENVELOPES SHOW FLOW BETWEEN PORTS.
THERE ARE TWO TYPES OF VALVES:

SINGLE ENVELOPE: WHERE ONE FLOW PATH EXISTS THROUGH THE ENVELOPE.
MULTIPLE ENVELOPES: WHERE MORE THAN ONE FLOW PATH EXISTS THROUGH THE ENVELOPES.

NAME	SYMBOL	SYMBOL	REMARKS
ENVELOPES	 		EACH ENVELOPE DEPICTS AN OPERATING POSITION OF THE VALVE.

NAME	SYMBOL	OBJECT	REMARKS
ENVELOPES WITH PORTS	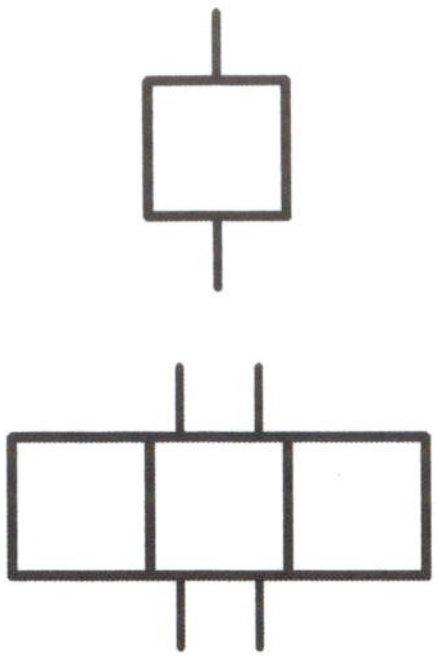		
PORTS WHICH ARE BLOCKED INTERNALLY	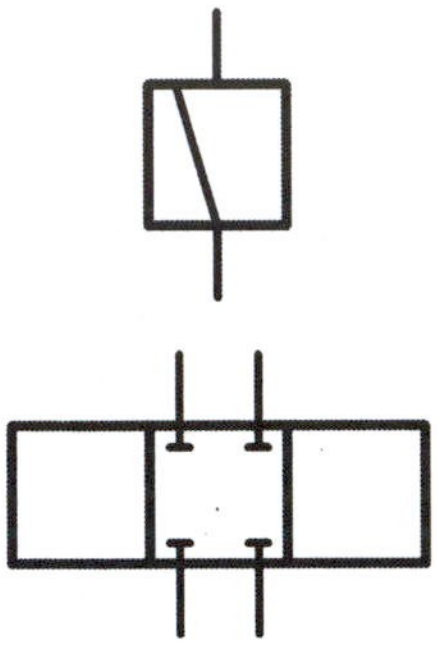		
INTERNAL FLOW PATHS	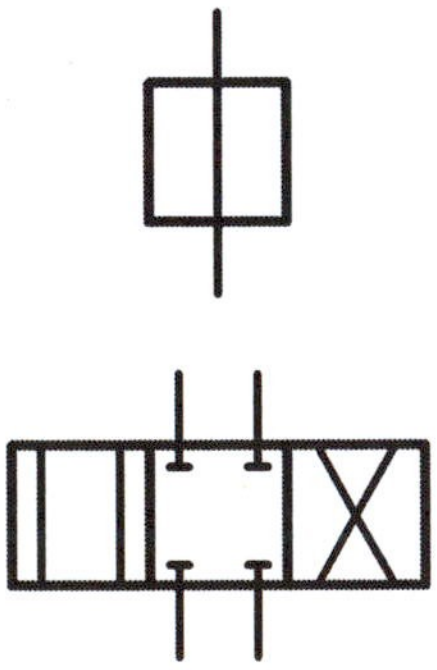		
INTERNAL FLOW PATHS SHOWING DIRECTION OF FLOW			
	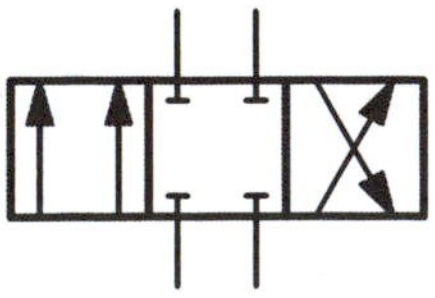		SOMETIMES REFERRED TO AS A 4/3-WAY VALVE: 4 PORTS & 3 POSITIONS.

NAME	SYMBOL	OBJECT	REMARKS
SINGLE ENVELOPE VALVE, NORMALLY OPEN			2/2 VALVE
SINGLE ENVELOPE VALVE, NORMALLY OPEN, ACTUATED (CLOSED)			2/2 VALVE
SINGLE ENVELOPE VALVE, NORMALLY CLOSED			2/2 VALVE
SINGLE ENVELOPE VALVE, NORMALLY CLOSED, ACTUATED (OPEN)			2/2 VALVE
2 PORT DOUBLE POSITION VALVE, NORMALLY CLOSED	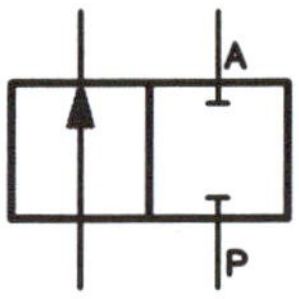 A P		2/2-WAY VALVE A = ACTION P = PRESSURIZED

NAME	SYMBOL	OBJECT	REMARKS
2 PORT DOUBLE POSITION VALVE, NORMALLY OPEN	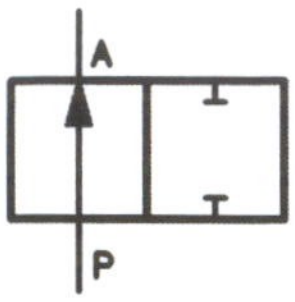 A P		2/2-WAY VALVE A = ACTIVE P = PRESSURIZED
3 PORT DOUBLE POSITION VALVE, 1 PORT EXHAUSTED AT REST	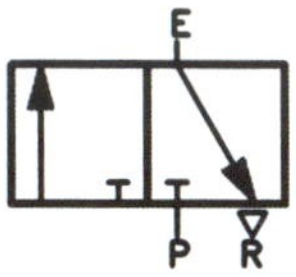 E P R		E = EXHAUSTED PORT P = PRESSURIZED R = REST POSITION
3 PORT DOUBLE POSITION VALVE, 1 PORT PRESSURIZED AT REST	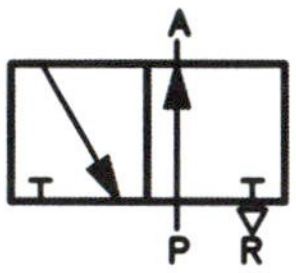 A P R		A = ACTIVE P = PRESSURIZED R = REST POSITION
4 PORT DOUBLE POSITION VALVE, 1 PORT ACTIVE, 1 PORT EXHAUSTED AT REST.	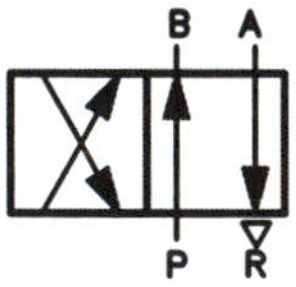 B A P R		WHEN VALVE ACTUATED FROM REST POSITION, ACTIVE/EXHAUST PORTS REVERSED.
RELIEF VALVE CLOSED OPEN	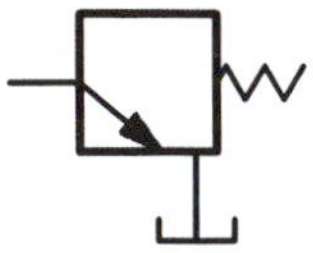 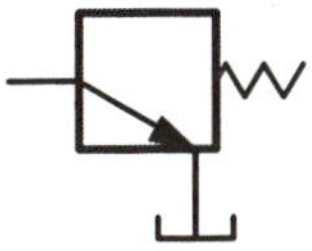		ZIGZAG LINE REPRESENTS SPRING CONTROL ON ENVELOPE. SEE "TYPES OF CONTROLS" FOR OTHER SYMBOLS.

NAME	SYMBOL	OBJECT	REMARKS
RELIEF VALVE, REMOTELY CONTROLLED	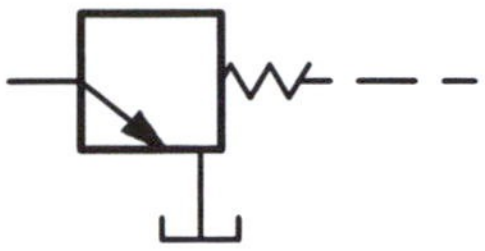		DASHED LINE REPRESENTS PILOT CONTROL ON ENVELOPE. SEE "TYPES OF CONTROLS" FOR OTHER SYMBOLS.
UNLOADING VALVE, REMOTELY CONTROLLED	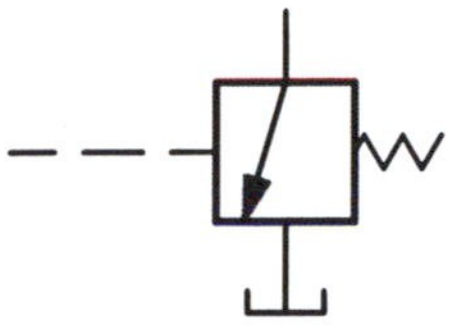		
DECELERATION VALVE, NORMALLY OPEN	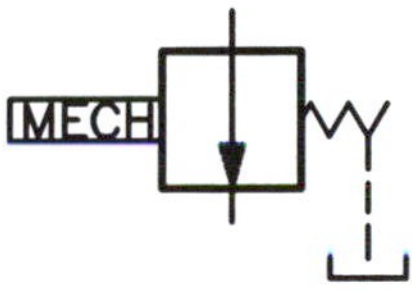 MECH		
REDUCING VALVE PRESSURE REGULATOR (RELIEVING)	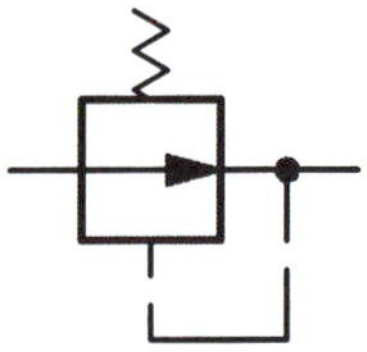 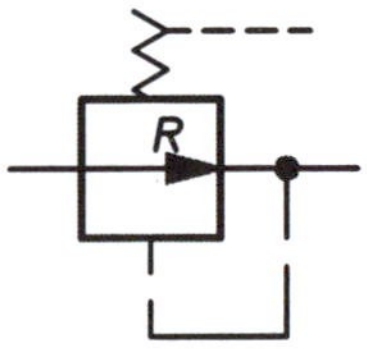 R		
SINGLE ENVELOPE VALVES (GENERAL)	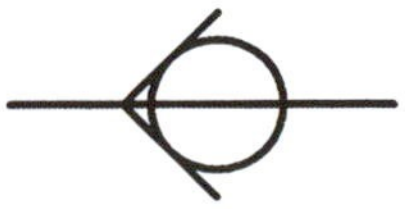		

<table>
<tr><th>NAME</th><th>SYMBOL</th><th>OBJECT</th><th>REMARKS</th></tr>
<tr><td>CHECK, ORIFICE</td><td></td><td></td><td></td></tr>
<tr><td>PILOT OPERATED CHECK VALVE, PILOT OPENED</td><td></td><td></td><td>FLOW PASSES THRU FROM RIGHT TO LEFT WHEN PRESSURE INTRODUCED THRU CONTROL CONNECTION.</td></tr>
<tr><td>MANUAL SHUT-OFF</td><td></td><td></td><td></td></tr>
<tr><td colspan="4">MULTIPLE ENVELOPE VALVE SYMBOLS CONSIST OF:
1. ENVELOPES FOR EACH OPERATING POSITION OF THE VALVE.
2. INTERNAL FLOW PATHS FOR EACH VALVE POSITION.
3. ARROWS INDICATING FLOW DIRECTION THRU THE VALVE.
4. EXTERNAL PORTS AT NORMAL OR NEUTRAL VALVE POSITION.</td></tr>
<tr><td>2-POSITION, 3-CONNECTION

NORMAL

ACTUATED</td><td colspan="2">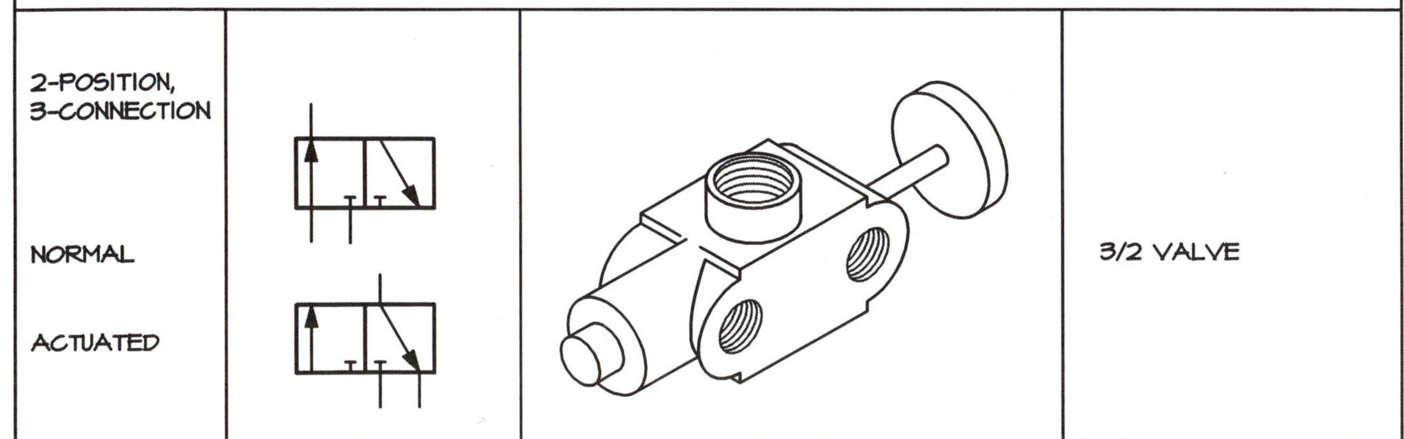</td><td>3/2 VALVE</td></tr>
</table>

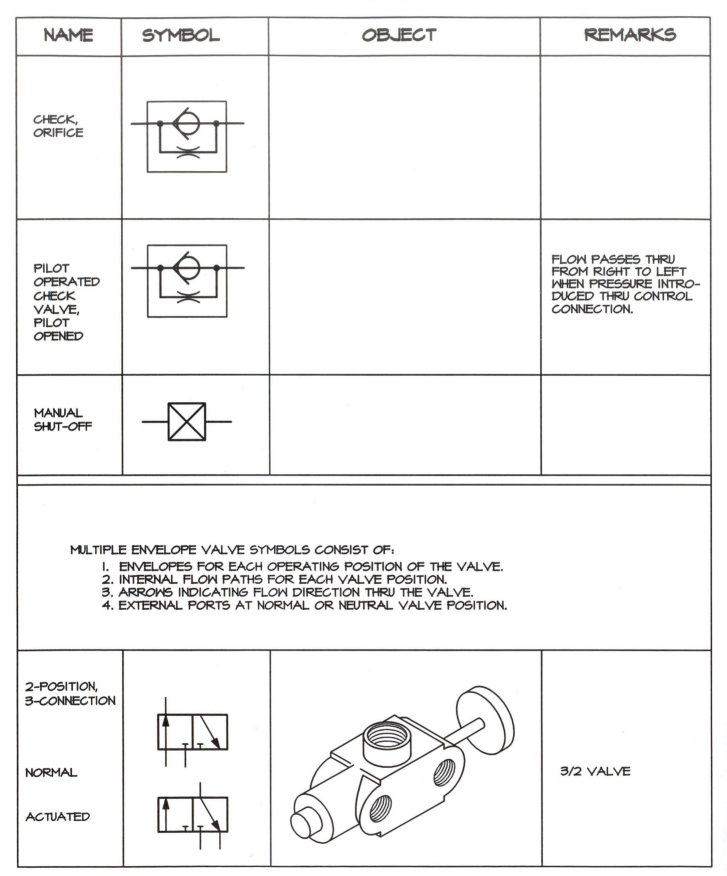

NAME	SYMBOL	OBJECT	REMARKS
2-POSITION, 4-CONNECTION NORMAL ACTUATED ALTERNATE SCHEMATIC SYMBOL	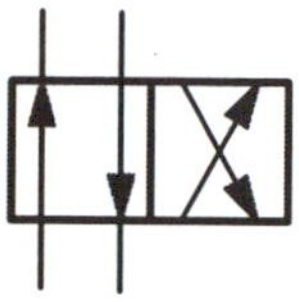 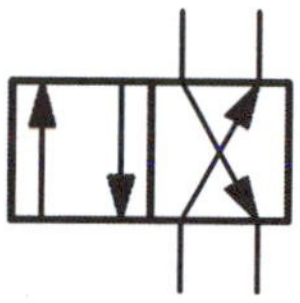 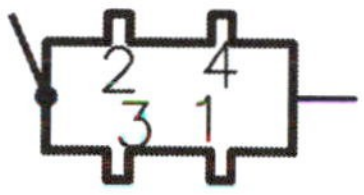 2 4 3 1	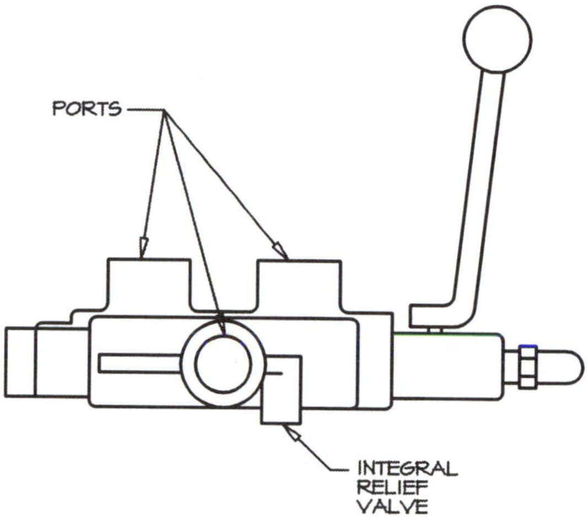 PORTS INTEGRAL RELIEF VALVE (FOURTH PORT NOT VISIBLE IN THIS VIEW.)	4/2 VALVE
3-POSITION, 4-CONNECTION NORMAL ACTUATED ACTUATED	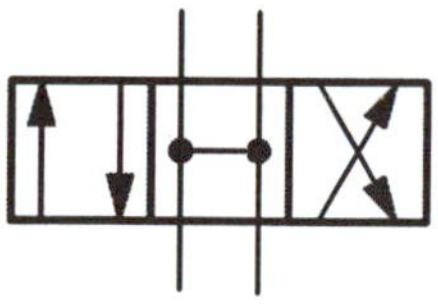 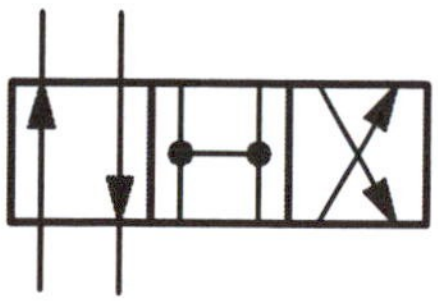 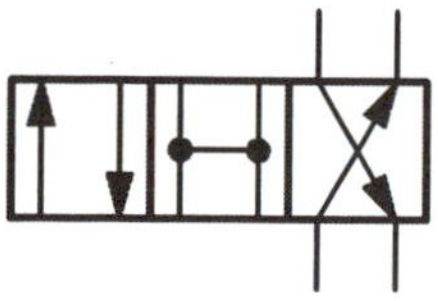	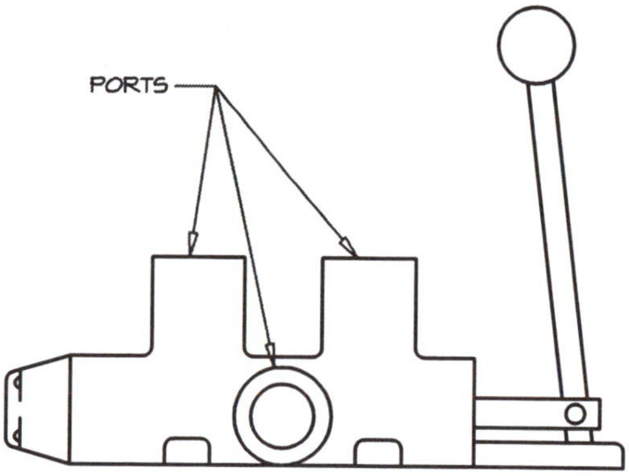 PORTS (FOURTH PORT NOT VISIBLE IN THIS VIEW.)	4/3 VALVE

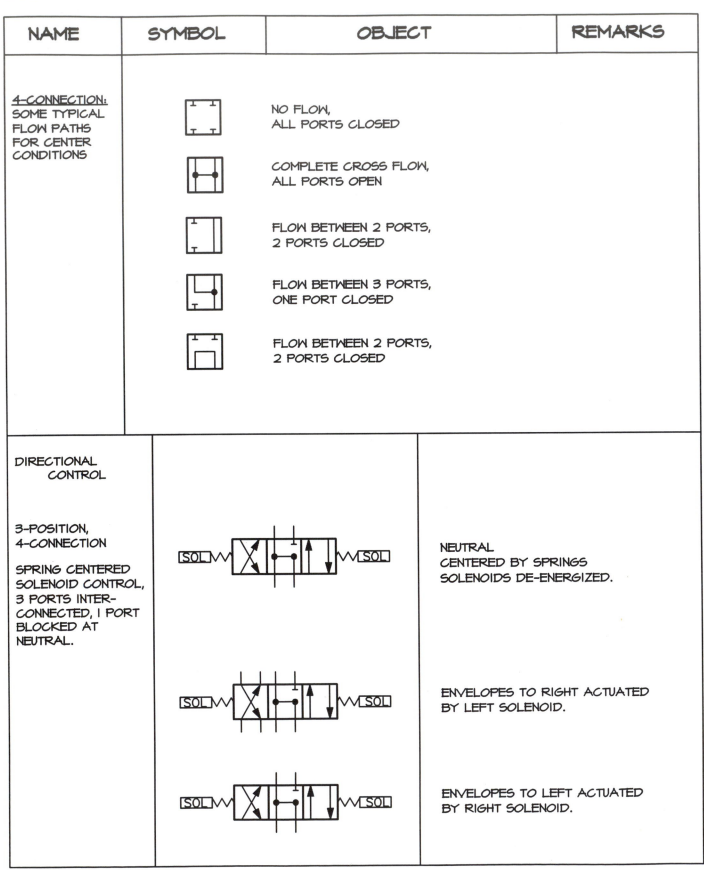

NAME	SYMBOL	OBJECT	REMARKS
4-CONNECTION: SOME TYPICAL FLOW PATHS FOR CENTER CONDITIONS		NO FLOW, ALL PORTS CLOSED	
	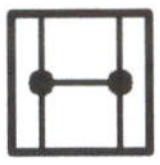	COMPLETE CROSS FLOW, ALL PORTS OPEN	
		FLOW BETWEEN 2 PORTS, 2 PORTS CLOSED	
		FLOW BETWEEN 3 PORTS, ONE PORT CLOSED	
		FLOW BETWEEN 2 PORTS, 2 PORTS CLOSED	

DIRECTIONAL CONTROL 3-POSITION, 4-CONNECTION SPRING CENTERED SOLENOID CONTROL, 3 PORTS INTER-CONNECTED, 1 PORT BLOCKED AT NEUTRAL.	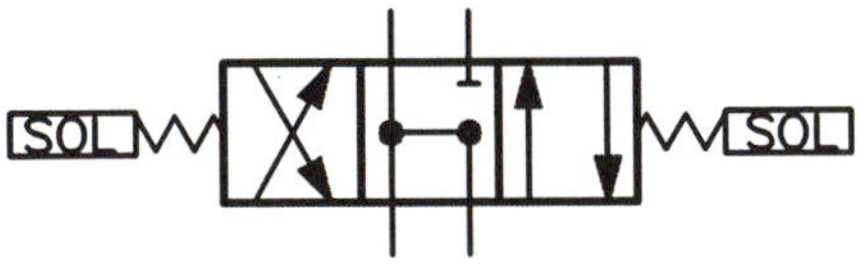 SOL SOL	NEUTRAL CENTERED BY SPRINGS SOLENOIDS DE-ENERGIZED.
	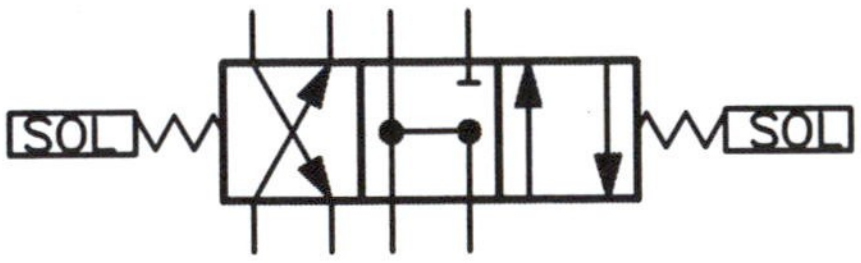 SOL SOL	ENVELOPES TO RIGHT ACTUATED BY LEFT SOLENOID.
	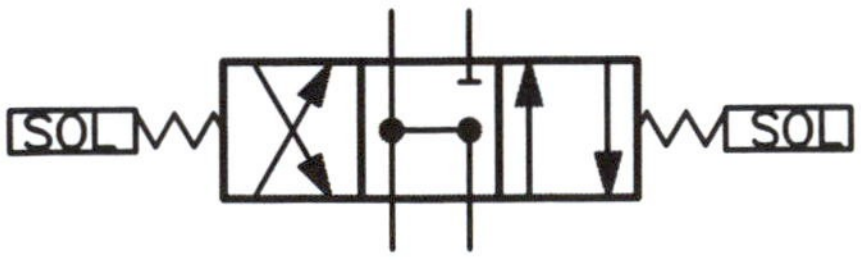 SOL SOL	ENVELOPES TO LEFT ACTUATED BY RIGHT SOLENOID.

NAME	SYMBOL	ILLUSTRATION	REMARKS
2-POSITION, 3-CONNECTION SPRING OFFSET SOLENOID CONTROL	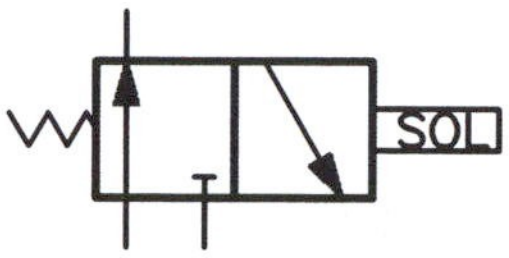 SOL		
2-POSITION, 4-CONNECTION SOLENOID CONTROL	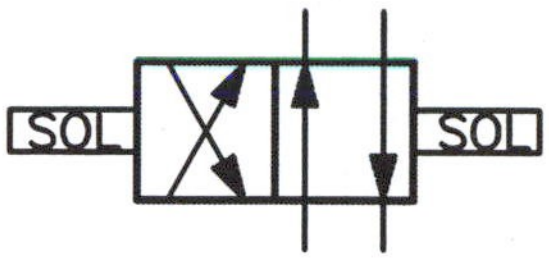 SOL SOL	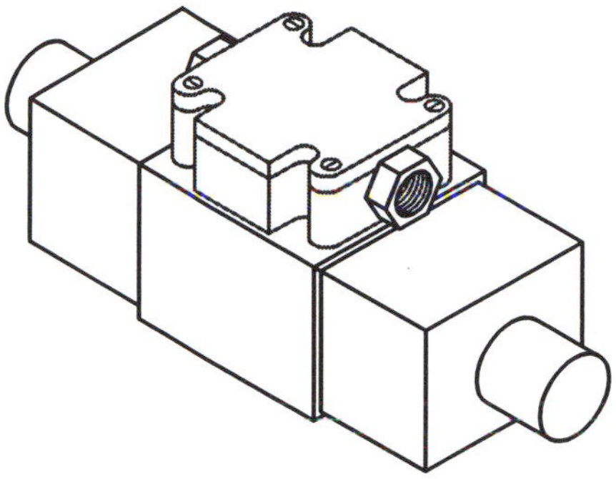	4/2 VALVE
2-POSITION, 4-CONNECTION SPRING OFFSET PILOT OPERATED	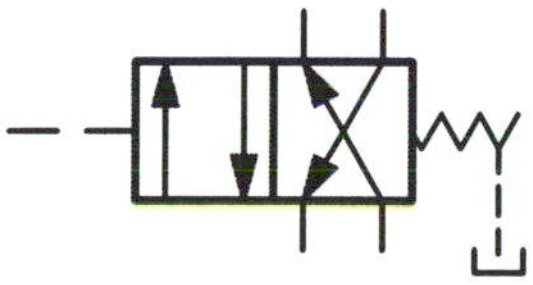		4/2 VALVE
3-POSITION, 4-CONNECTION MECHANICAL CONTROL, ONE PORT PLUGGED	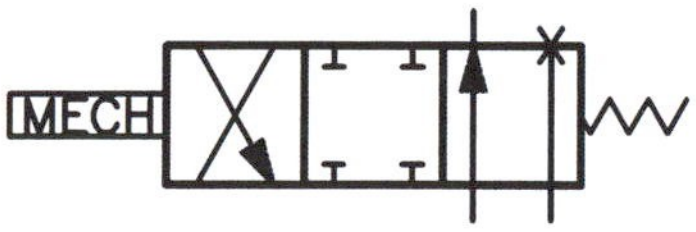 MECH		
3-POSITION, 5-CONNECTION SPRING CENTERED, SOLENOID CONTROL	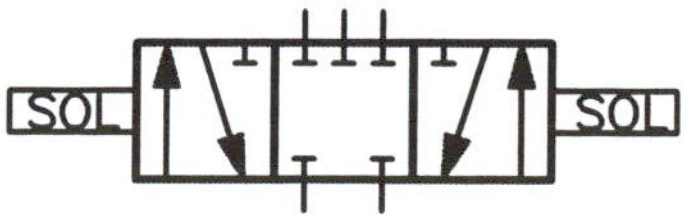 SOL SOL		

NAME	SYMBOL	ILLUSTRATION	REMARKS
2-POSITION, 8-CONNECTION, MECHANICAL CONTROL	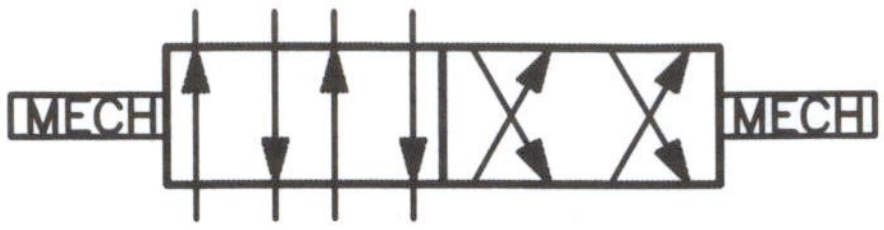 MECH MECH	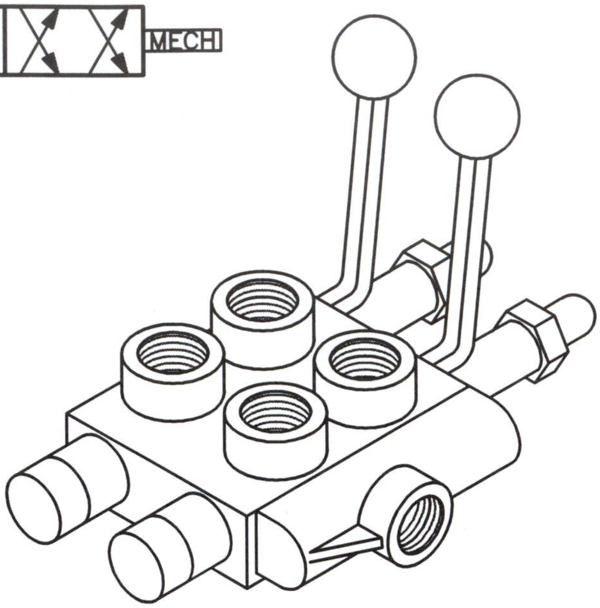	8/2 VALVE
5 WAY VALVE	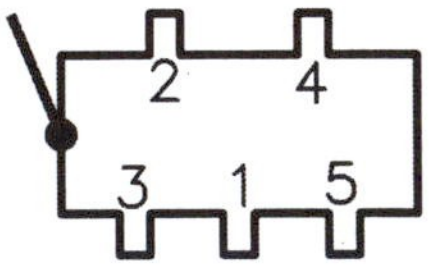 2 4 3 1 5		ALTERNATIVE SCHEMATIC REPRESENTATION.
ROTATING CONNECTION			NUMBER OF LINES THRU CIRCLE INDICATES NO. OF FLOW PATHS THRU COMPONENT.

KEYS

TYPES OF CONTROLS

KEY

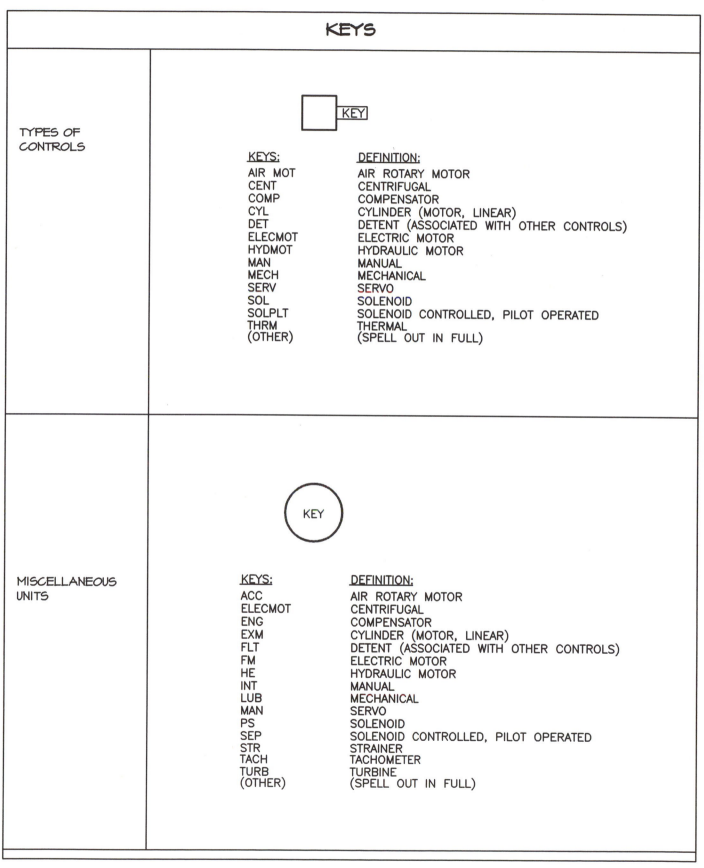

KEYS:	DEFINITION:
AIR MOT	AIR ROTARY MOTOR
CENT	CENTRIFUGAL
COMP	COMPENSATOR
CYL	CYLINDER (MOTOR, LINEAR)
DET	DETENT (ASSOCIATED WITH OTHER CONTROLS)
ELECMOT	ELECTRIC MOTOR
HYDMOT	HYDRAULIC MOTOR
MAN	MANUAL
MECH	MECHANICAL
SERV	SERVO
SOL	SOLENOID
SOLPLT	SOLENOID CONTROLLED, PILOT OPERATED
THRM	THERMAL
(OTHER)	(SPELL OUT IN FULL)

MISCELLANEOUS UNITS

KEY

KEYS:	DEFINITION:
ACC	AIR ROTARY MOTOR
ELECMOT	CENTRIFUGAL
ENG	COMPENSATOR
EXM	CYLINDER (MOTOR, LINEAR)
FLT	DETENT (ASSOCIATED WITH OTHER CONTROLS)
FM	ELECTRIC MOTOR
HE	HYDRAULIC MOTOR
INT	MANUAL
LUB	MECHANICAL
MAN	SERVO
PS	SOLENOID
SEP	SOLENOID CONTROLLED, PILOT OPERATED
STR	STRAINER
TACH	TACHOMETER
TURB	TURBINE
(OTHER)	(SPELL OUT IN FULL)

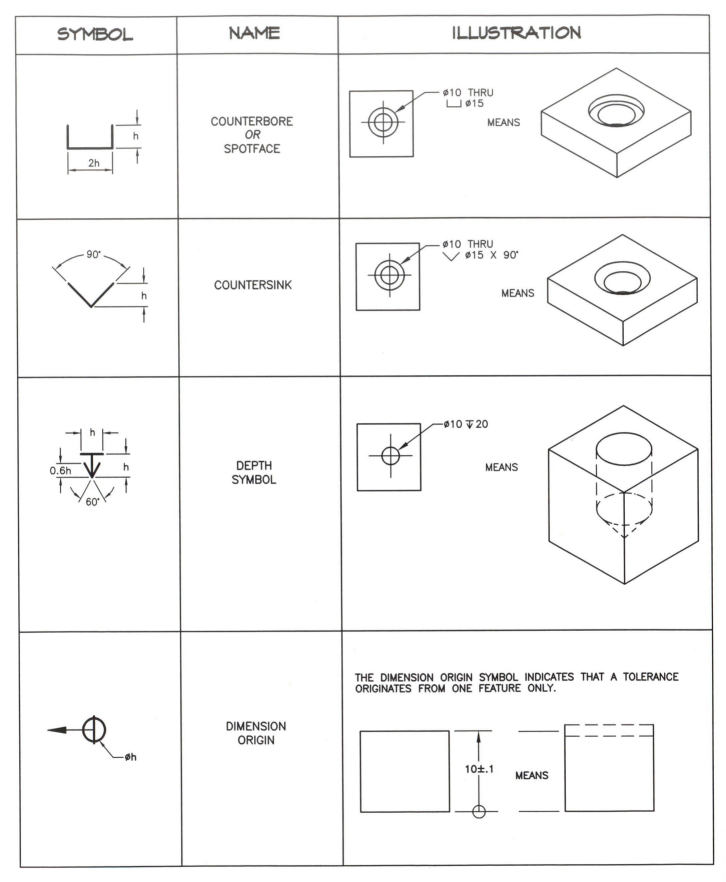

SYMBOL	NAME	ILLUSTRATION
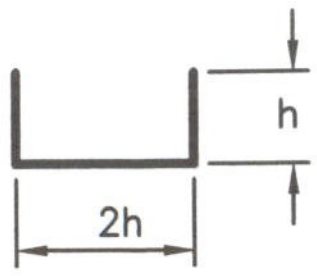 h, 2h	COUNTERBORE *OR* SPOTFACE	⌀10 THRU ⌴ ⌀15 MEANS
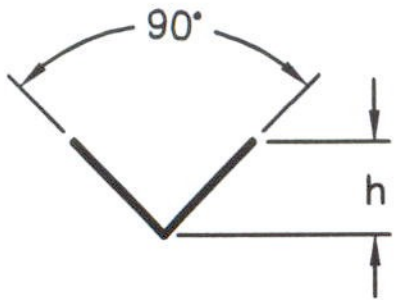 90°, h	COUNTERSINK	⌀10 THRU ⌵ ⌀15 X 90° MEANS
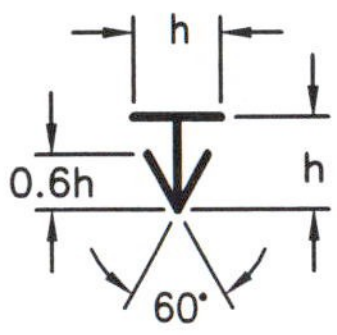 h, 0.6h, h, 60°	DEPTH SYMBOL	⌀10 ↧20 MEANS
⌀h	DIMENSION ORIGIN	THE DIMENSION ORIGIN SYMBOL INDICATES THAT A TOLERANCE ORIGINATES FROM ONE FEATURE ONLY. 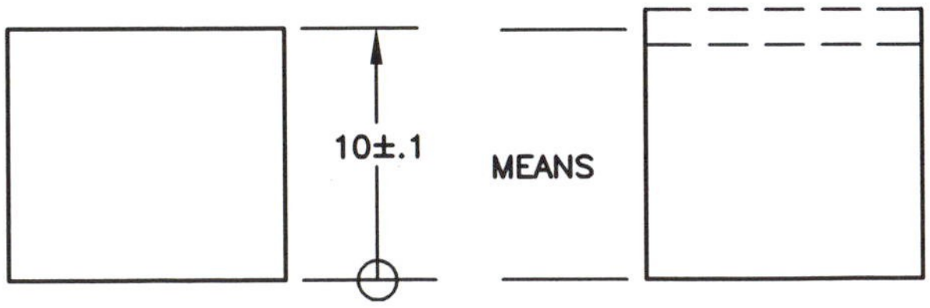 10±.1 MEANS

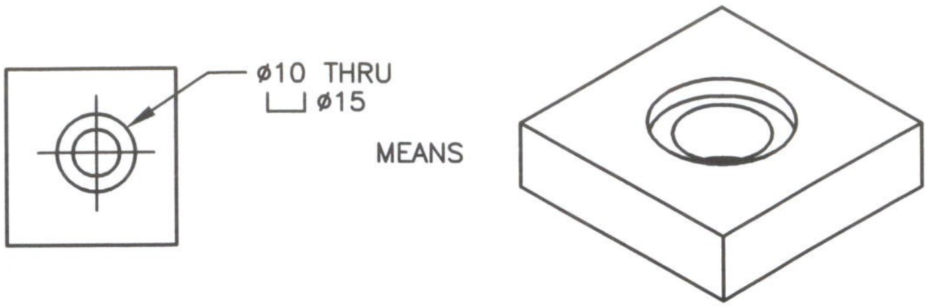

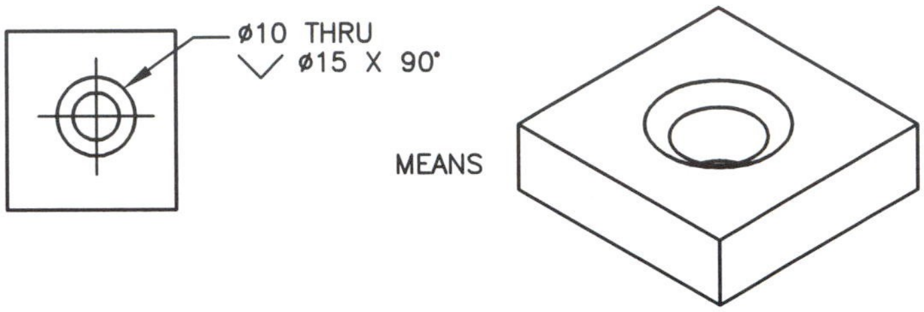

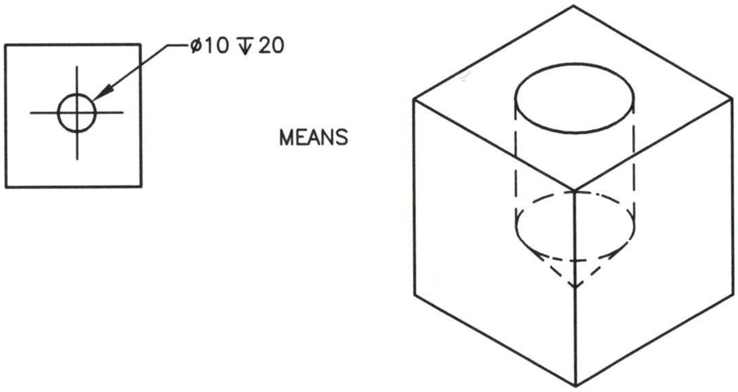

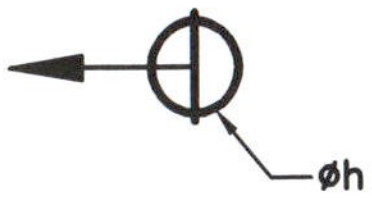

SYMBOL	NAME	ILLUSTRATION
h, h	SQUARE SHAPE	□10 MEANS
0.5h, 0.5h, 2h, 30°	TAPER CONICAL	0.3:1
0.3h, 1.5h	REFERENCE	
1.5h, 0.3h	ARC LENGTH	
h, 15°	SLOPE	0.2±0.01:1 SLOPE IS EXPRESSED AS A RATIO OF HEIGHTS.
2.5h, 30°, 0.8h, ST, 1.5h	STATISTICAL TOLERANCE	STATISTICAL TOLERANCING IS A METHOD USING MATHEMATICS TO TOLERANCE MULTIPLE PARTS. THE SYMBOL APPEARS AT THE END OF A TOLERANCE.

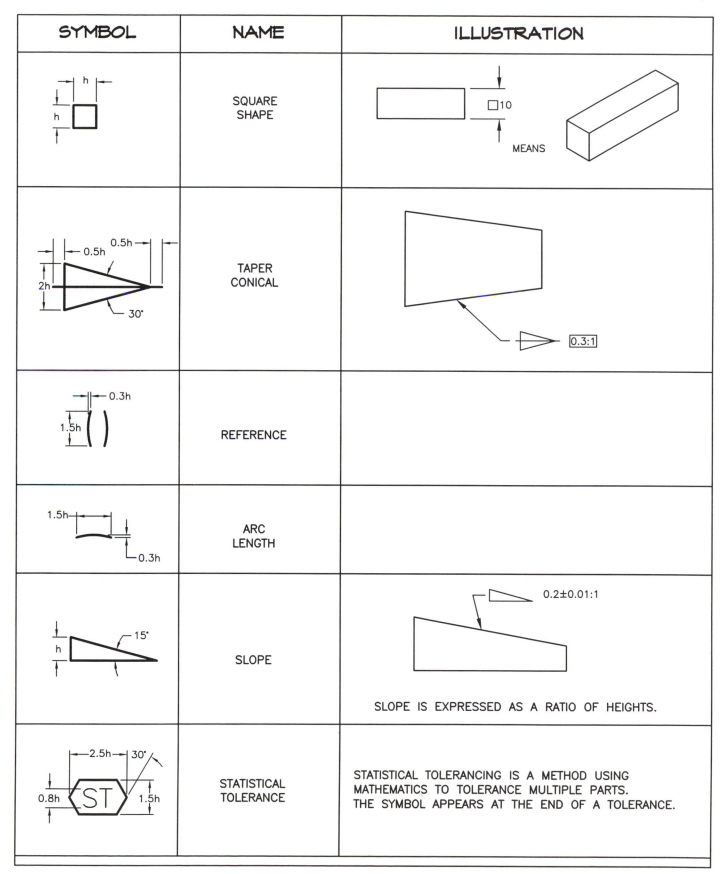

IN 1994 THE AMERICAN NATIONAL STANDARDS INSTITUTE PUBLISHED THE MOST RECENT VERSION OF WHAT IS THE BIBLE OF MECHANICAL DRAFTING & TOLERANCING: ANSI/ASME Y14.5M: "DIMENSIONING AND TOLERANCING".

THIS REVISION CONTINUES THE TREND TOWARD INTERNATIONALIZATION OF STANDARDS WHICH WILL SURELY BE THE NORM IN A WORLD PROGRESSING TOWARD GLOBALIZATION.

WHILE NUMEROUS CHANGES WERE INCORPORATED INTO THE STANDARD TO MAKE IT MORE COMPATIBLE WITH THE ISO 9000, ARGUABLY THE MOST DRAMATIC CHANGE WAS THE INTRODUCTION OF THE ISO DATUM SYMBOL. NO DOUBT THE OLD SYMBOL (SHOWN BELOW) WILL BE WITH US A LONG TIME ON "OLD" BLUELINES AND DOCUMENTS, BUT THE RAPID ADOPTION OF THE NEW SYMBOL BODES WELL FOR THE GOAL OF WORLD WIDE STANDARDIZATION.

FOR MORE DETAILS ABOUT HOW THE STANDARD HAS BEEN MODIFIED, AND A SENSE OF THE COMPLEXITY AND INTENSITY OF THE PRINCIPLES AND ISSUES INVOLVED, SEE THE FOREWARD TO ANSI Y14.5M ITSELF.

SYMBOL	NAME	REMARKS
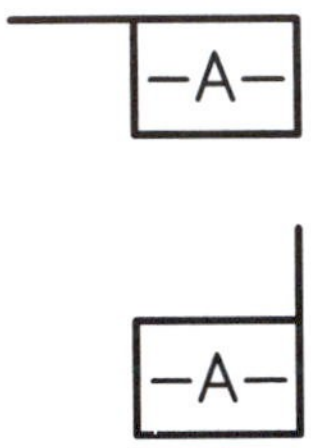 –A– –A– ETC.	OLD STYLE DATUM SYMBOL	A DATUM IS A THEORETICALLY EXACT POINT, AXIS, OR PLANE DERIVED FROM THE TRUE GEOMETRIC COUNTERPART OF A SPECIFIED DATUM FEATURE. A DATUM IS THE ORIGIN FROM WHICH THE LOCATION OR GEOMETRIC CHARACTERISTICS OF FEATURE OF A PART ARE ESTABLISHED. FOR EXAMPLE, THE SURFACE OF A BLOCK MAY BE DESIGNATED AS A DATUM. FOR THE PURPOSES OF CREATING A FEATURE ON OR IN THIS BLOCK, THE SURFACE IS ASSUMED OR IMAGINED TO BE A PERFECTLY FLAT PLANE.
[A] 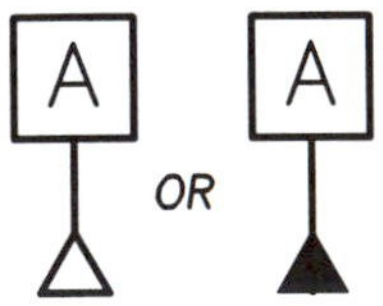 A *OR* A	DATUM SYMBOL	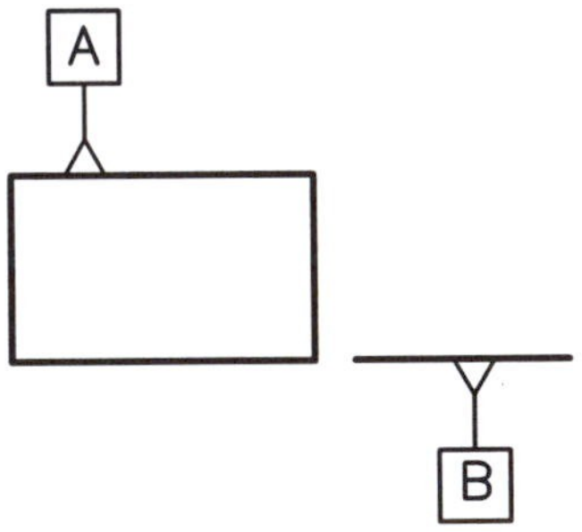 A B DATUM SYMBOL USED ON A FEATURE & ON AN EXTENSION LINE. THE TRIANGULAR BASE SHAPE MAY BE FILLED SOLID OR ONLY OUTLINED.
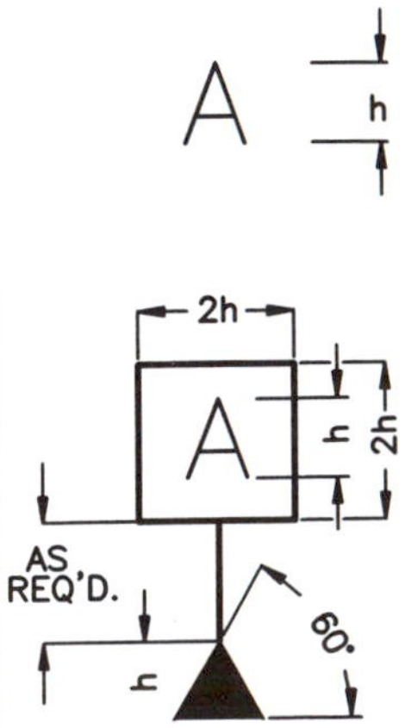 A h 2h A h 2h AS REQ'D. h 60°		THE BASIS FOR DEFINING THE PROPORTIONS OF THIS AND ALL OTHER GEOMETRIC TOLERANCING SYMBOLS.

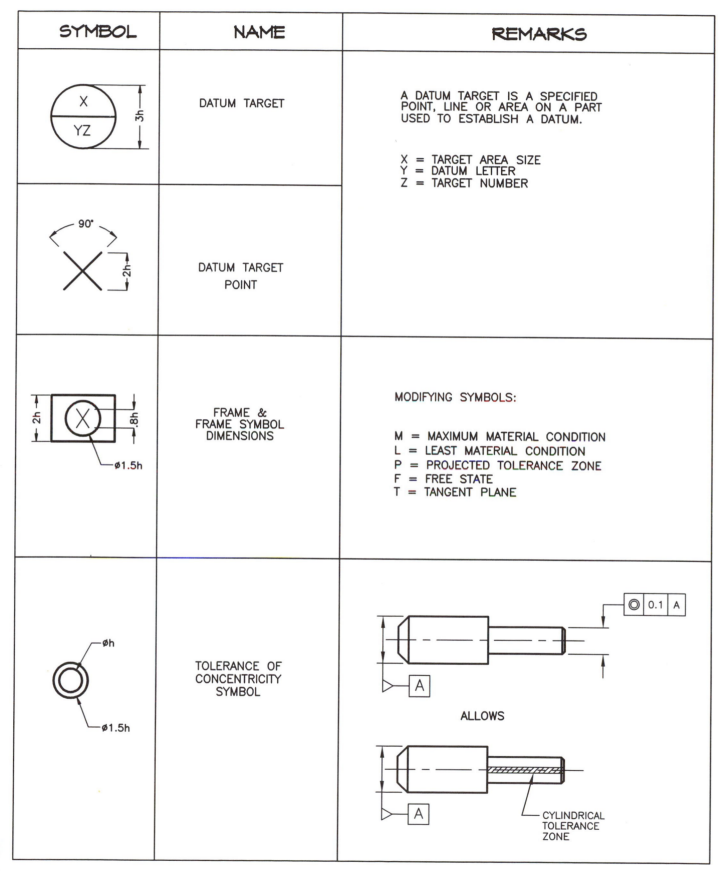

SYMBOL	NAME	REMARKS
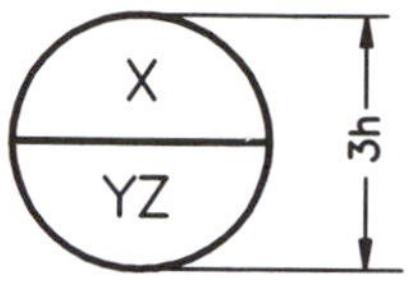 X / YZ, 3h	DATUM TARGET	A DATUM TARGET IS A SPECIFIED POINT, LINE OR AREA ON A PART USED TO ESTABLISH A DATUM. X = TARGET AREA SIZE Y = DATUM LETTER Z = TARGET NUMBER
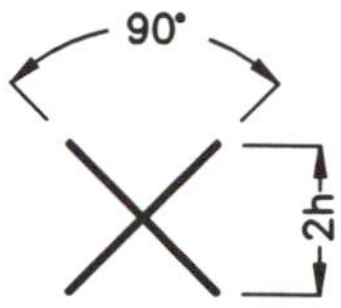 90°, 2h	DATUM TARGET POINT	
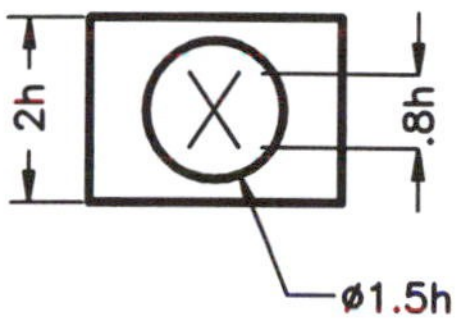 2h, X, .8h, ø1.5h	FRAME & FRAME SYMBOL DIMENSIONS	MODIFYING SYMBOLS: M = MAXIMUM MATERIAL CONDITION L = LEAST MATERIAL CONDITION P = PROJECTED TOLERANCE ZONE F = FREE STATE T = TANGENT PLANE
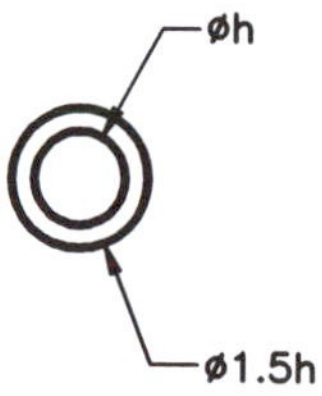 øh, ø1.5h	TOLERANCE OF CONCENTRICITY SYMBOL	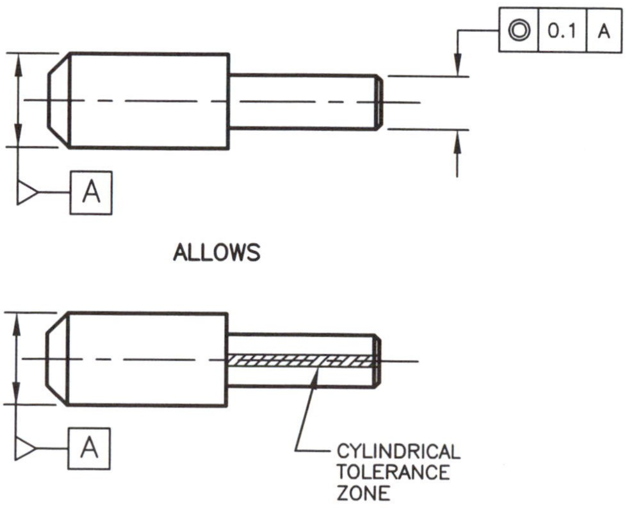 0.1 A A ALLOWS A CYLINDRICAL TOLERANCE ZONE

SYMBOL	NAME	ILLUSTRATION
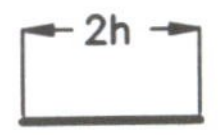 2h	TOLERANCE OF STRAIGHTNESS SYMBOL	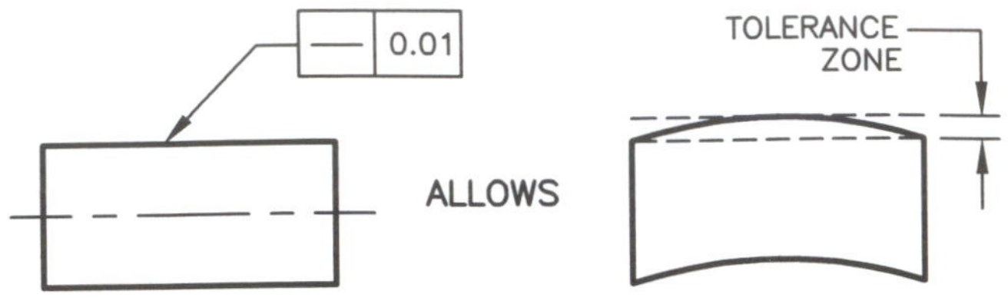 0.01 ALLOWS TOLERANCE ZONE
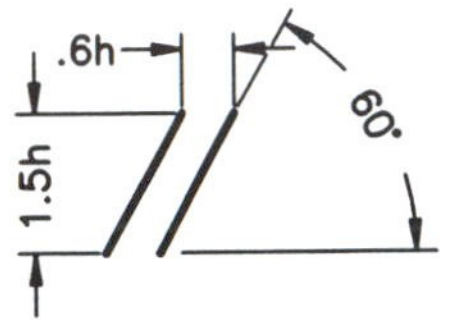 .6h 1.5h 60°	TOLERANCE OF PARALLELISM SYMBOL	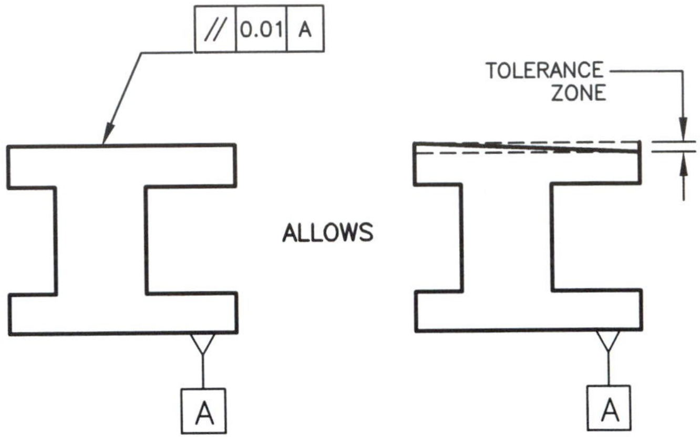 // 0.01 A ALLOWS TOLERANCE ZONE A A
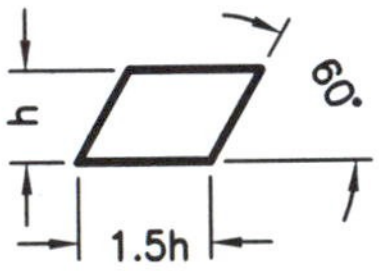 h 1.5h 60°	TOLERANCE OF FLATNESS SYMBOL	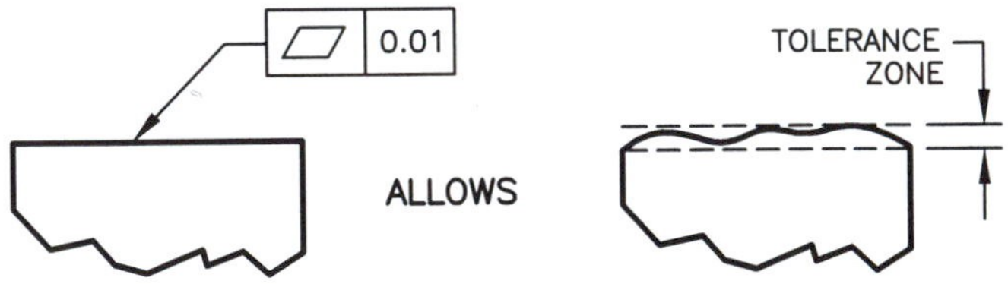 0.01 ALLOWS TOLERANCE ZONE
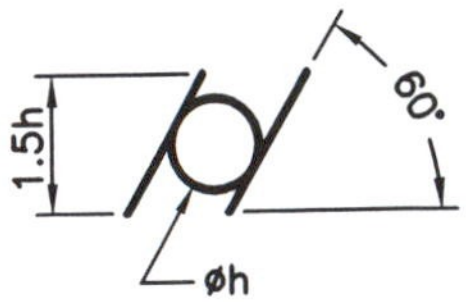 1.5h 60° øh	TOLERANCE OF CYLINDRICITY SYMBOL	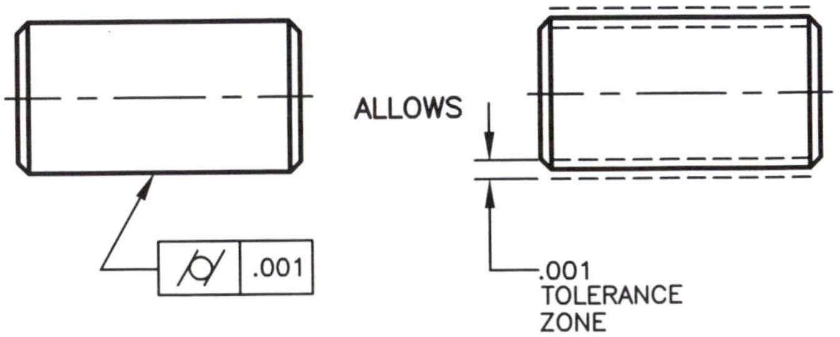 .001 ALLOWS .001 TOLERANCE ZONE

SYMBOL	NAME	ILLUSTRATION
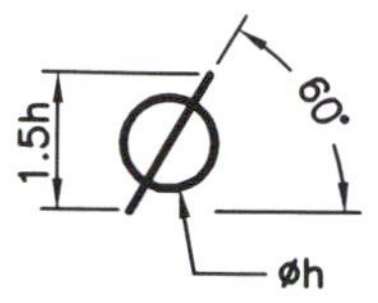 1.5h, 60°, øh	TOLERANCE OF DIAMETER SYMBOL	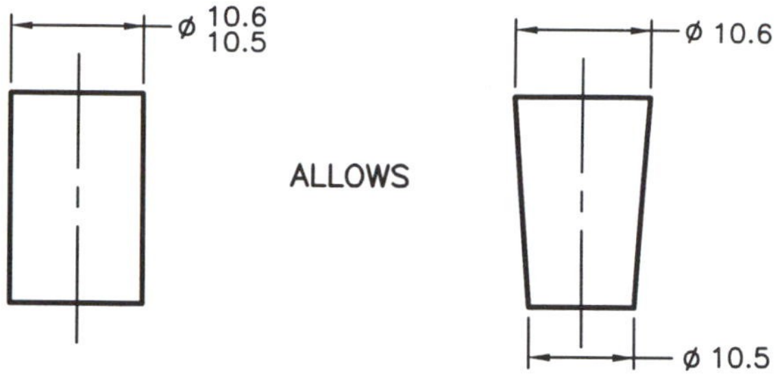 ø 10.6 10.5 ALLOWS ø 10.6 ø 10.5
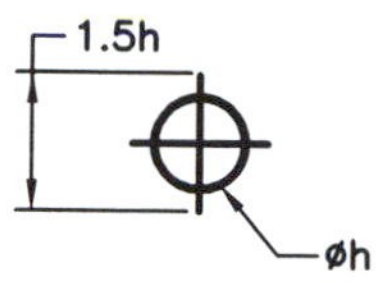 1.5h, øh	TOLERANCE OF POSITION SYMBOL	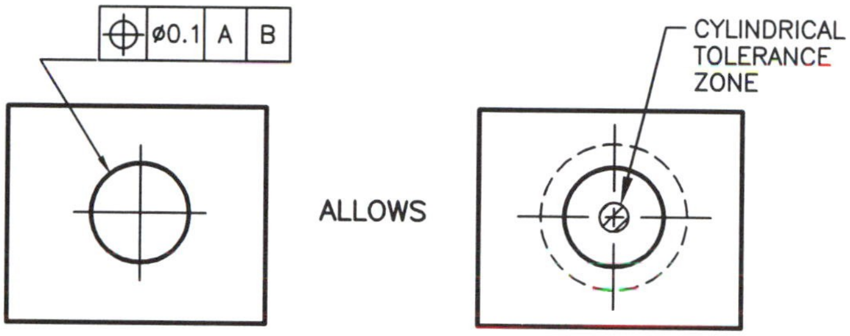 ⌖ ø0.1 | A | B ALLOWS CYLINDRICAL TOLERANCE ZONE
CR	CONTROLLED RADIUS SYMBOL	NORMALLY TOLERANCED RADIUS: R1±0.1 ALLOWS ZONE OF TOLERANCE MAX. RADIUS 1.1 MIN. RADIUS 0.9
		CONTROLLED RADIUS: CR1±0.1 ALLOWS ZONE OF TOLERANCE MAX. RADIUS 1.1 MIN. RADIUS 0.9 IN A CONTROLLED RADIUS, THE CONTOUR WITHIN THE TOLERANCE ZONE MUST BE A "FAIR" CURVE.

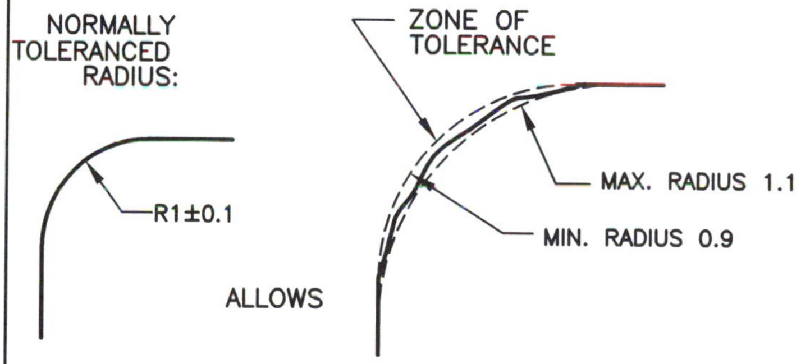

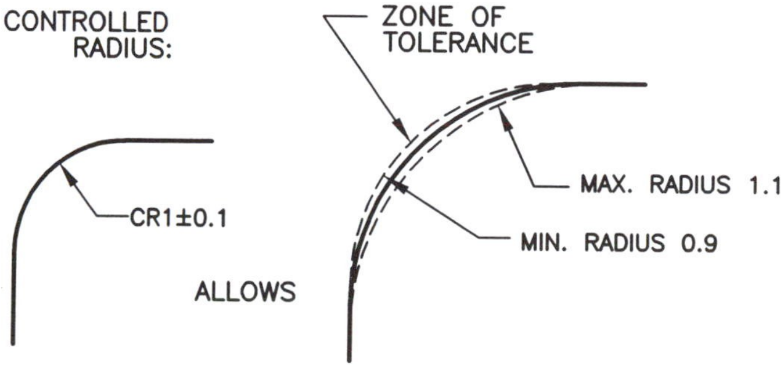

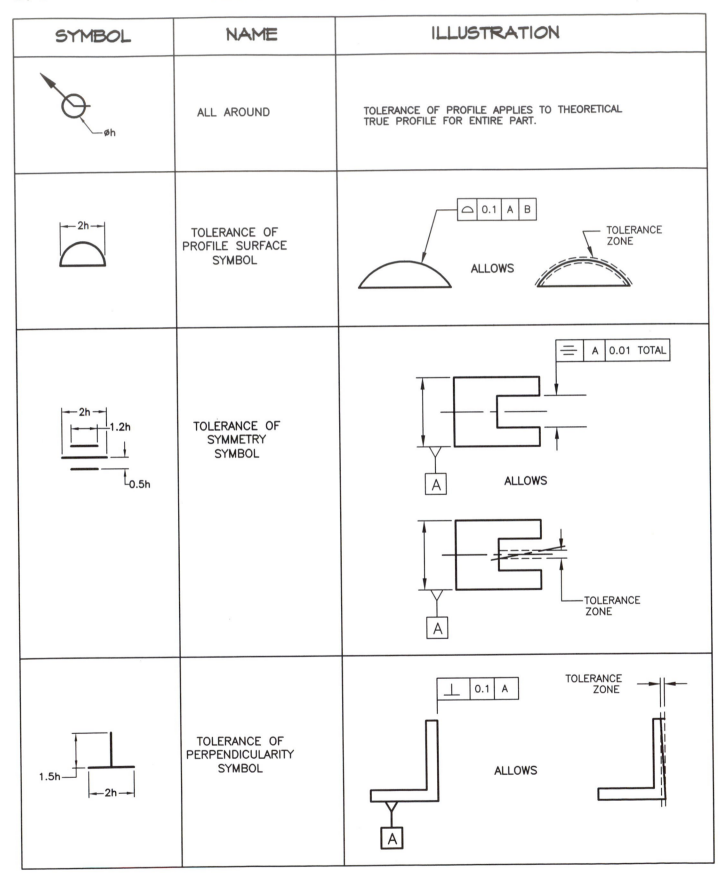

SYMBOL	NAME	ILLUSTRATION
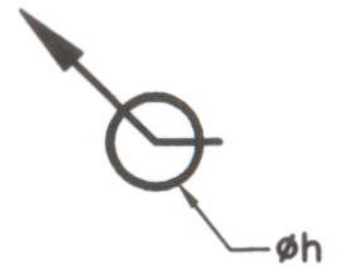 øh	ALL AROUND	TOLERANCE OF PROFILE APPLIES TO THEORETICAL TRUE PROFILE FOR ENTIRE PART.
 2h	TOLERANCE OF PROFILE SURFACE SYMBOL	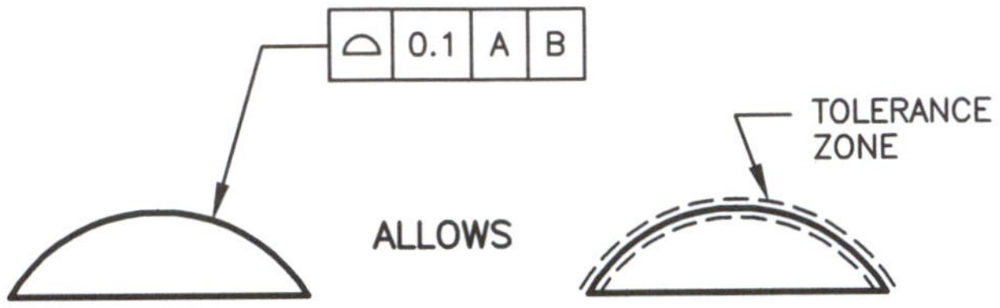 0.1 A B ALLOWS TOLERANCE ZONE
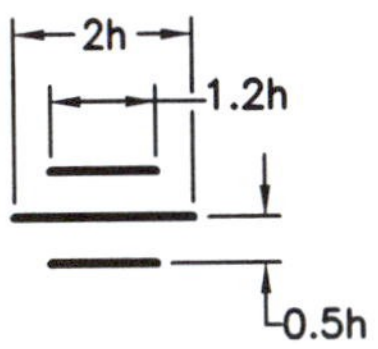 2h 1.2h 0.5h	TOLERANCE OF SYMMETRY SYMBOL	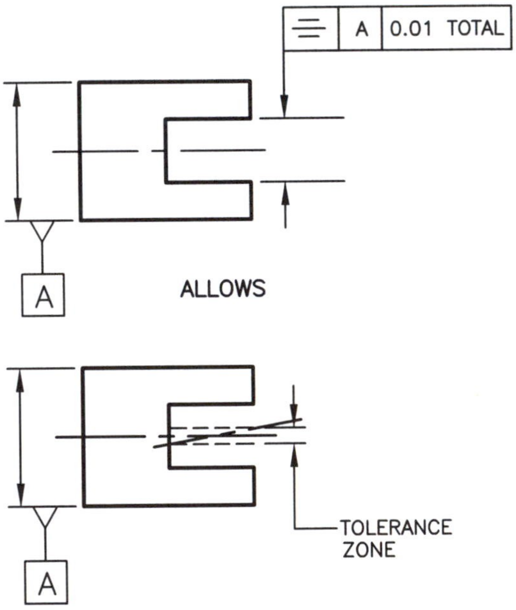 A 0.01 TOTAL A ALLOWS A TOLERANCE ZONE
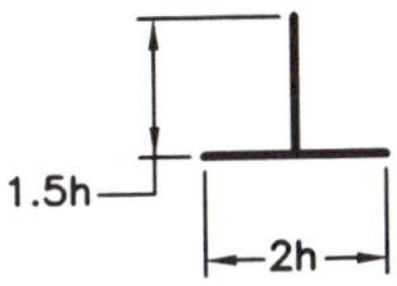 1.5h 2h	TOLERANCE OF PERPENDICULARITY SYMBOL	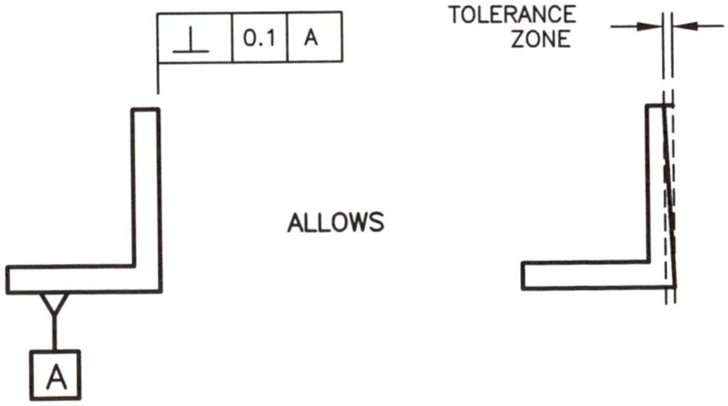 0.1 A A ALLOWS TOLERANCE ZONE

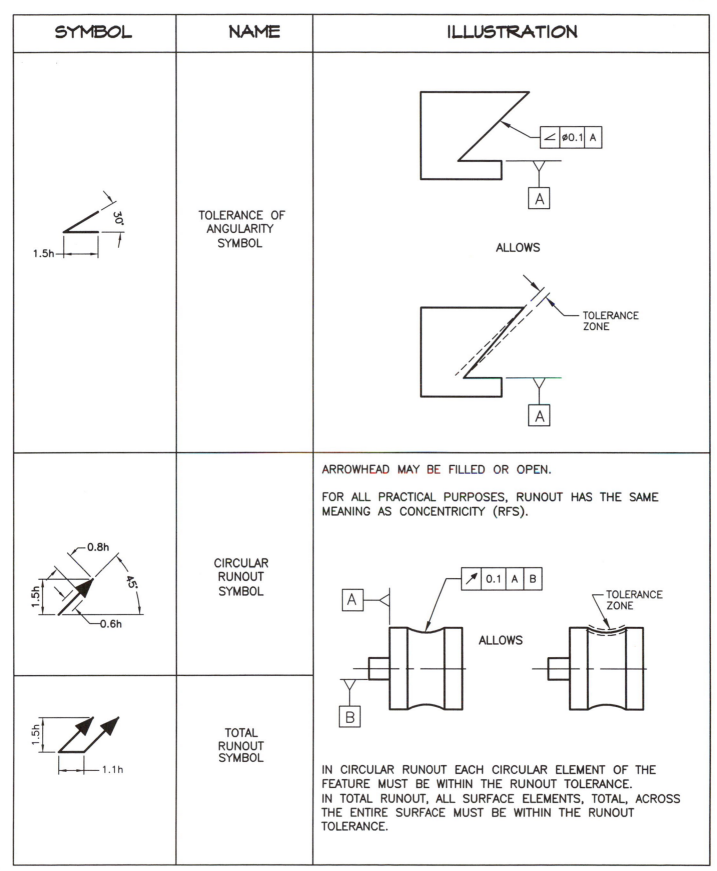

SYMBOL	NAME	ILLUSTRATION
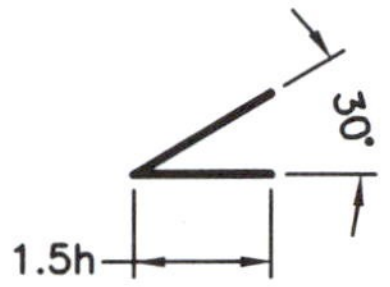 30° 1.5h	TOLERANCE OF ANGULARITY SYMBOL	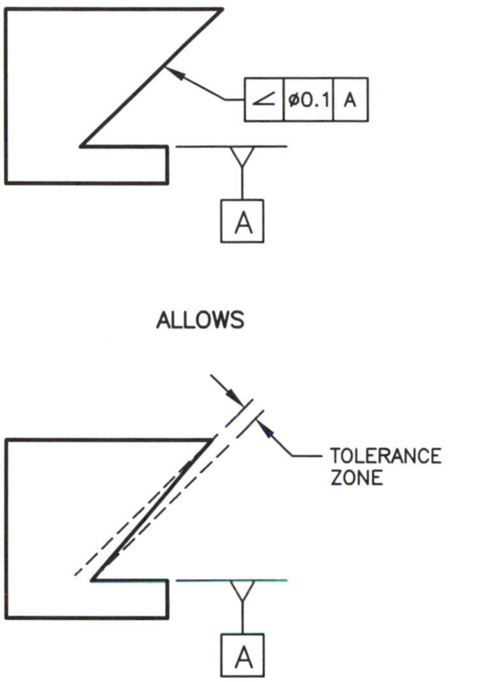 ∠ ø0.1 A; A; ALLOWS; TOLERANCE ZONE; A
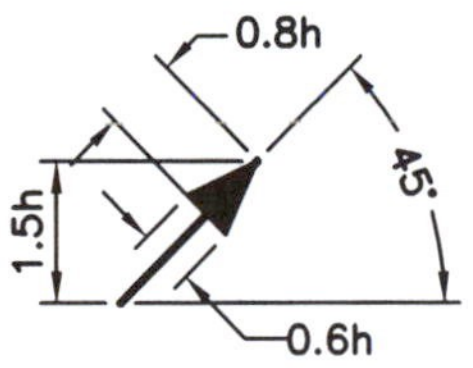 0.8h 45° 1.5h 0.6h	CIRCULAR RUNOUT SYMBOL	ARROWHEAD MAY BE FILLED OR OPEN. FOR ALL PRACTICAL PURPOSES, RUNOUT HAS THE SAME MEANING AS CONCENTRICITY (RFS). 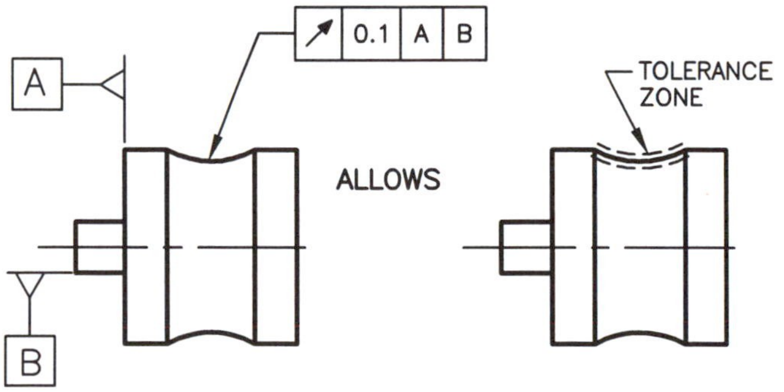 ↗ 0.1 A B; A; ALLOWS; TOLERANCE ZONE; B
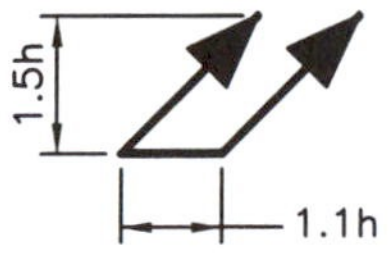 1.5h 1.1h	TOTAL RUNOUT SYMBOL	IN CIRCULAR RUNOUT EACH CIRCULAR ELEMENT OF THE FEATURE MUST BE WITHIN THE RUNOUT TOLERANCE. IN TOTAL RUNOUT, ALL SURFACE ELEMENTS, TOTAL, ACROSS THE ENTIRE SURFACE MUST BE WITHIN THE RUNOUT TOLERANCE.

SYMBOL	NAME	ILLUSTRATION
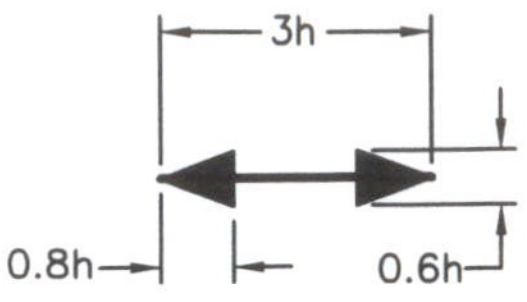 3h; 0.8h; 0.6h	BETWEEN SYMBOL	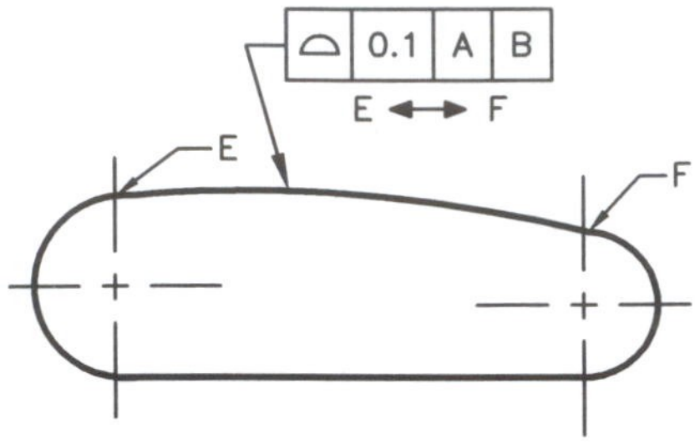 0.1 A B; E ↔ F; E; F
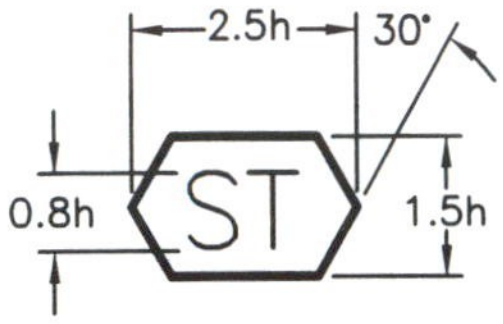 2.5h; 30°; 0.8h; ST; 1.5h	STATISTICAL TOLERANCE	STATISTICAL TOLERANCING IS A METHOD USING MATHEMATICS TO TOLERANCE MULTIPLE PARTS. THE SYMBOL APPEARS AT THE END OF A TOLERANCE.
Ⓣ	TANGENT SYMBOL	TOLERANCE APPLIES TO THE CONTACTING (TANGENTIAL) ELEMENT.
CR	CONTROLLED RADIUS SYMBOL	NORMALLY TOLERANCED RADIUS: R1±0.1 ALLOWS; ZONE OF TOLERANCE; MAX .RADIUS 1.1; MIN. RADIUS 0.9
		CONTROLLED RADIUS: CR1±0.1 ALLOWS; ZONE OF TOLERANCE; MAX. RADIUS 1.1; MIN. RADIUS 0.9 IN A CONTROLLED RADIUS, THE CONTOUR WITHIN THE TOLERANCE ZONE MUST BE A FAIR CURVE.

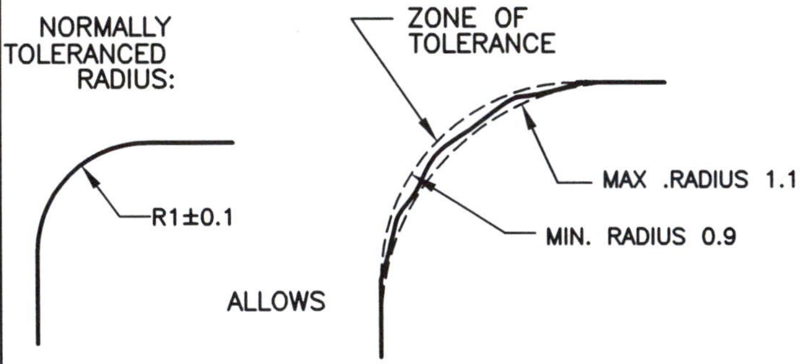

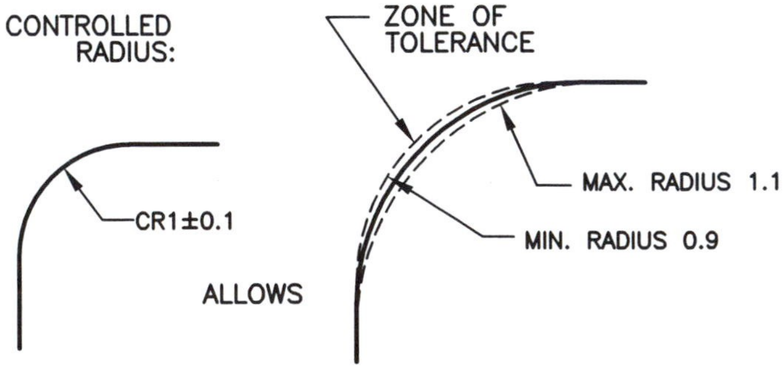

NAME	SYMBOL	OBJECT	REMARKS
COMPRESSOR	[A]		
COMPRESSOR: ENCLOSED CRANKCASE, ROTARY, BELTED	[A]		
COMPRESSOR: OPEN CRANK-CASE, RECIP-ROCATING, BELTED	[A]		
COMPRESSOR: OPEN CRANK-CASE, RECIP-ROCATING, DIRECT DRIVE	[A]		
CONDENSER: AIR COOLED, FINNED, STATIC	[A]		
CONDENSER: AIR COOLED, FINNED, FORCED AIR.	[A]		

NAME	SYMBOL	OBJECT	REMARKS
CONDENSER: WATER COOLED, CONCENTRIC TUBE IN A TUBE.	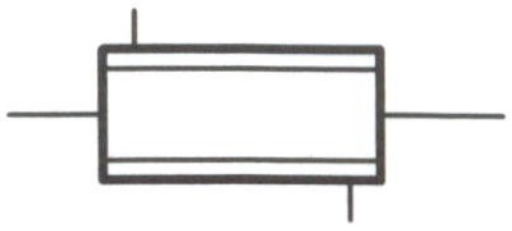		
CONDENSER: WATER COOLED, CONCENTRIC TUBE IN A TUBE.	[A] 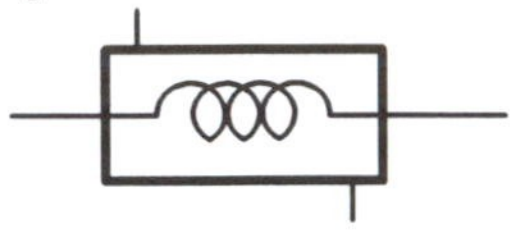		
CONDENSER: WATER COOLED, SHELL & TUBE.	[A] 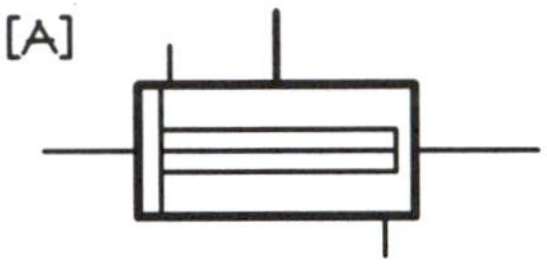		
CONDENSING UNIT, AIR COOLED	[A] 		
CONDENSING UNIT, WATER COOLED	[A] 		
MOTOR-COMPRESSOR: ENCLOSED CRANKCASE, RECIPROCATING, DIRECT CONNECTING	[A] 		

NAME	SYMBOL	OBJECT	REMARKS
MOTOR-COMPRESSOR: SEALED CRANKCASE, RECIPROCATING	[A]		
MOTOR-COMPRESSOR: SEALED CRANKCASE, ROTARY	[A]		

NAME	SYMBOL	OBJECT	REMARKS
MOTORIZED AGITATING UNIT	 [I]	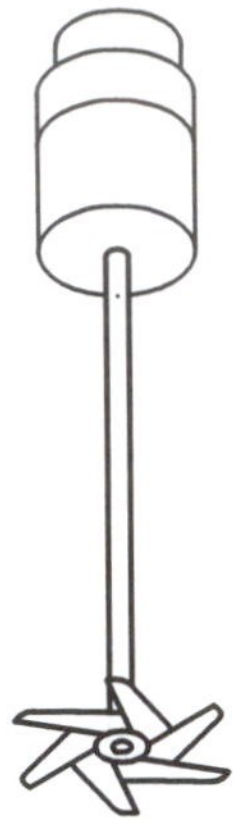	
AGITATOR	 [I]		PROPELLOR, BLADE OR PADDLE MOUNTED ON A ROD.
DRUM MIXER	 [I]		
HELICAL MIXER	[I]		

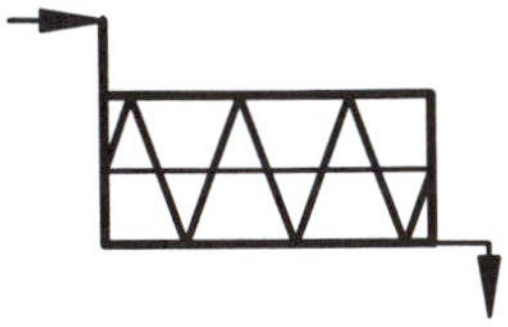

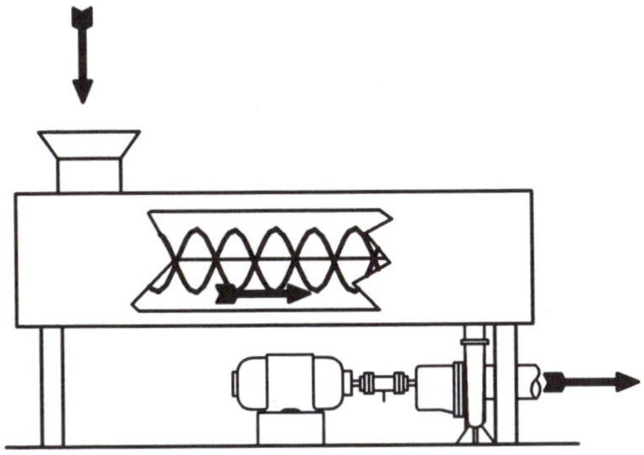

NAME	SYMBOL	OBJECT/REMARKS
IN-LINE MIXER	[I] 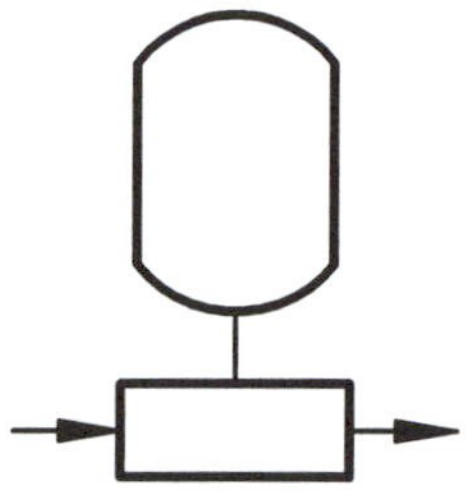	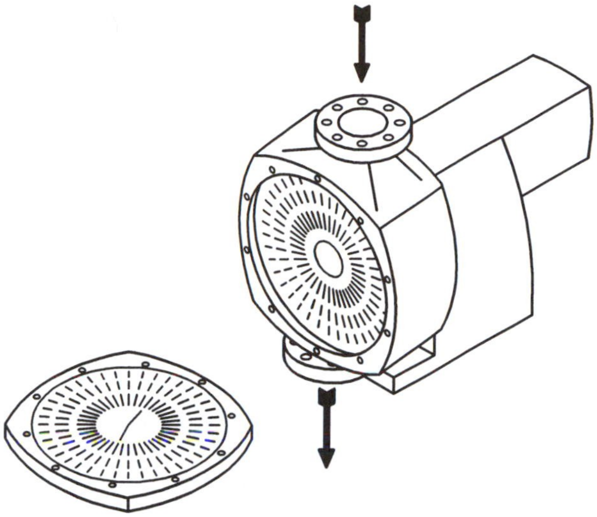 USED FOR IN-LINE MIXING, EMULSIFICATION, DISPERSION, HOMOGENIZATION, MILLING AND PROCESS REACTIONS. (SHOWN WITH COVER REMOVED.)
RIBBON BLENDER	[I] 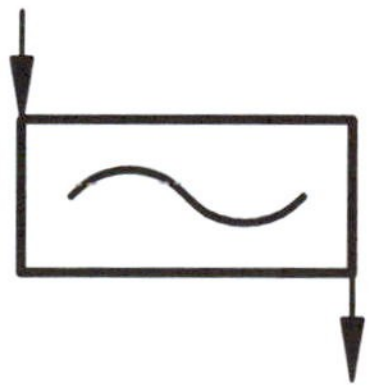	ILLUSTRATION SHOWS CONTINUOUS RIBBON. 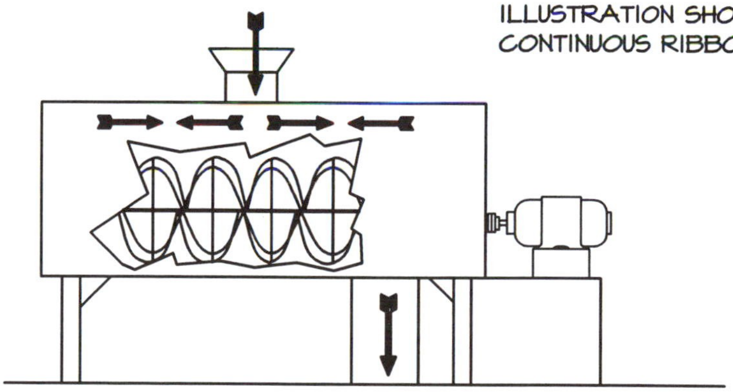
ROLL BLENDER	[I] 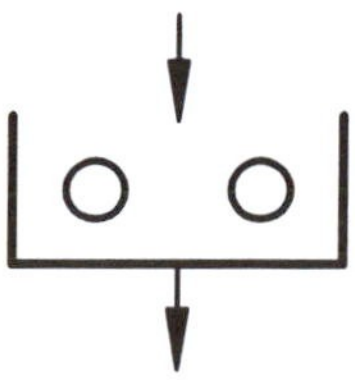	

NAME	SYMBOL	OBJECT	REMARKS
STATIC MIXER OR MOTIONLESS MIXER	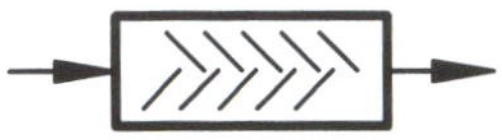	 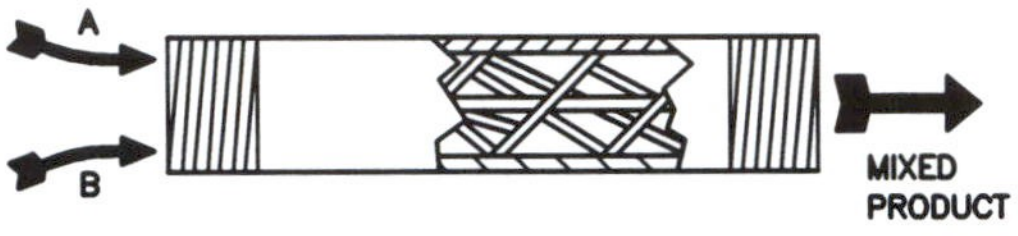 A B MIXED PRODUCT	MOVEMENT OF FLUID OVER BLADES MIXES IT. ILLUSTRATIONS SHOW SMALL DIAMETER MIXER AND LARGE DIAMETER FLANGED MIXER.
MIXING TEE	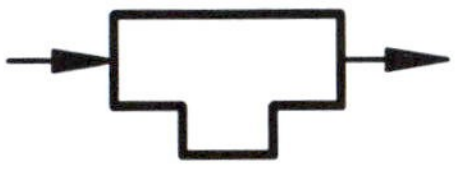		
MIXER	[A] 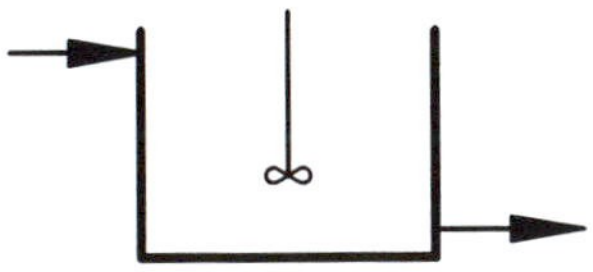		
KETTLE	[I] 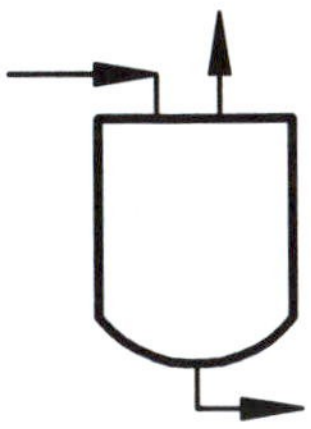		

NAME	SYMBOL	OBJECT	REMARKS
AGITATED KETTLE OR VESSEL	[I] 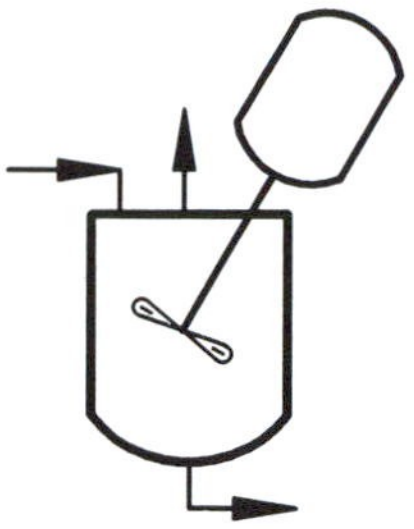	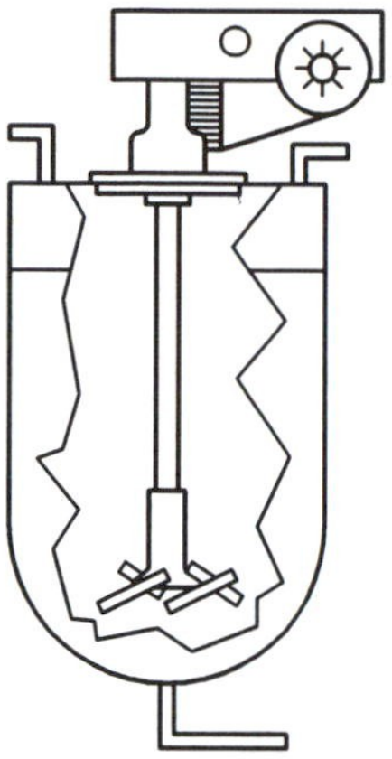	
JACKETED KETTLE	[I] 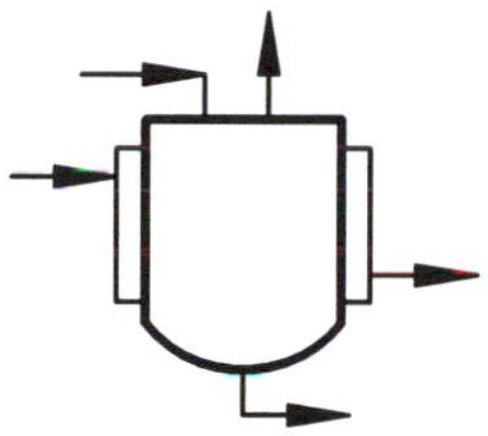 [A] 		JACKET PROVIDES EITHER HEATING OR COOLING FOR CONTENTS.
CHEMICAL REACTOR	[I] 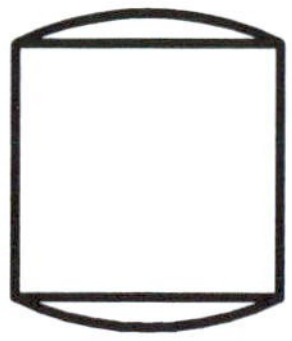		
VESSEL	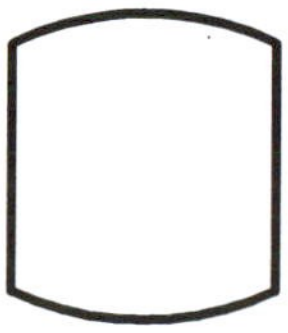		VESSEL OR SEPARATOR, ALSO PRESSURIZED VERTICAL OR HORI-ZONTAL VESSEL. ENDS MAY BE SLOPED, DISHED, OR CONICAL.

NAME	SYMBOL	OBJECT	REMARKS
PARTIALLY JACKETED VESSEL	[A]	OUTLET INLET	ILLUSTRATION SHOWS PARTIALLY JACKETED VERTICAL VESSEL WITH CONICAL BOTTOM.

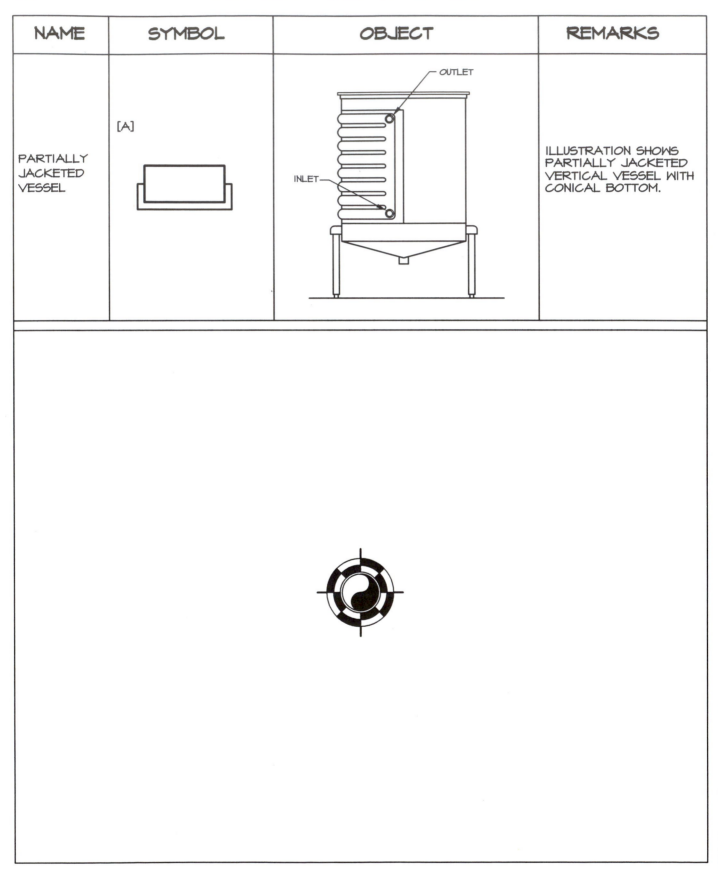

NAME	SYMBOL	OBJECT	REMARKS
ROTARY DRUM DRYER OR KILN	[I] [A]		
ROTARY SHELF DRYER	[I]		
BATCH TRAY DRYER	[A]		
SPRAY DRYER	[I] [A]		SEPARATES LIQUIDS FROM MIXTURES OF SOLIDS & LIQUIDS BY EVAPORATION.

NAME	SYMBOL	OBJECT	REMARKS
COLUMNS OR TOWERS PACKED	[A] 		
PLATE	[A] 		
SECTIONED	[A] 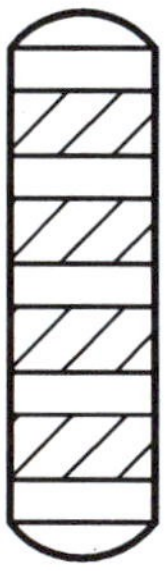		
DISK & DONUT	[A] 		

NAME	SYMBOL	OBJECT
DESSICANT DRYER	[A] 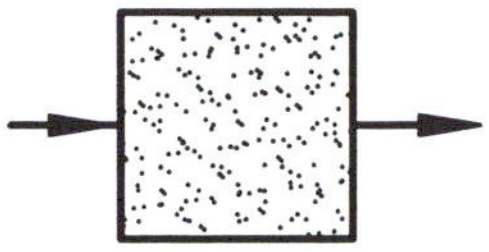	
CONTINUOUS TUNNEL DRYER	[A] 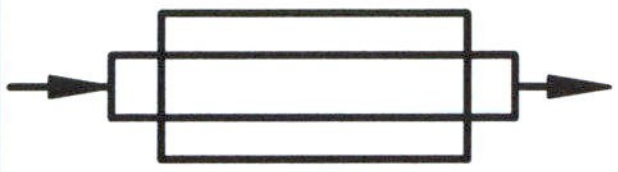	
FORCED CIRCULATION EVAPORATOR	[I] 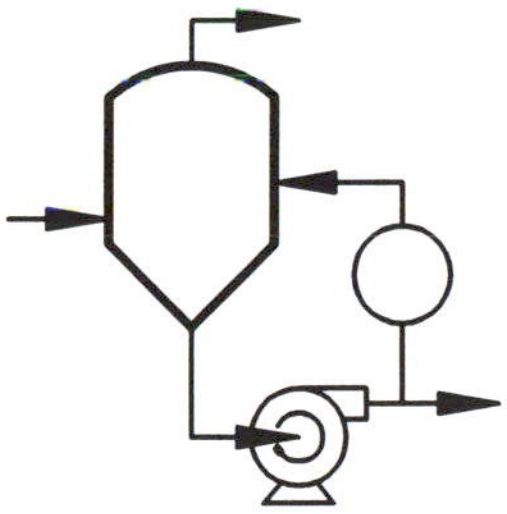	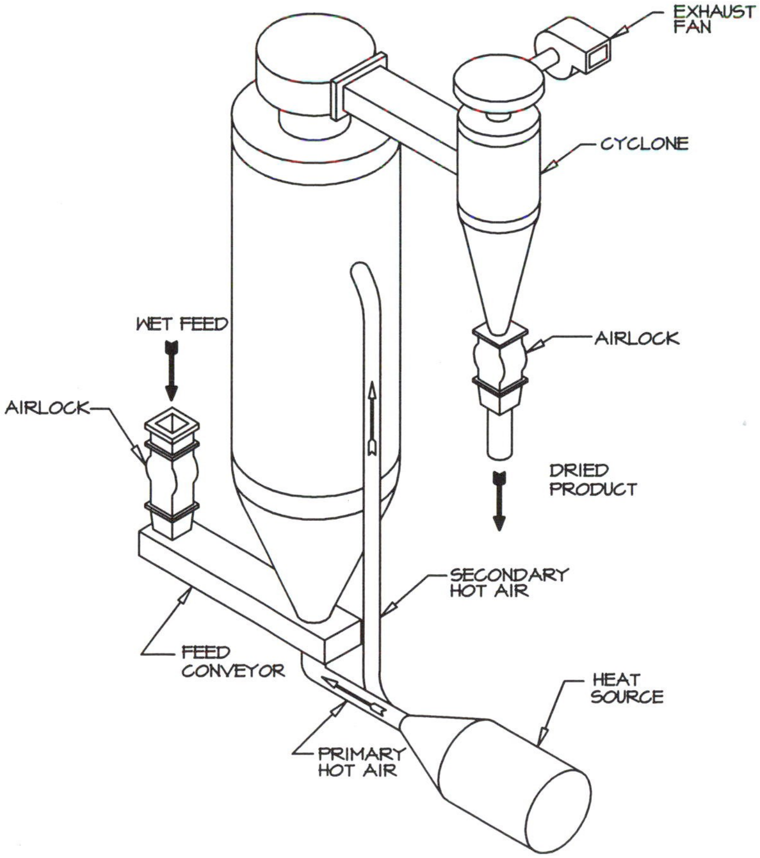

NAME	SYMBOL	OBJECT	REMARKS
NATURAL CONVECTION EVAPORATOR	[I] 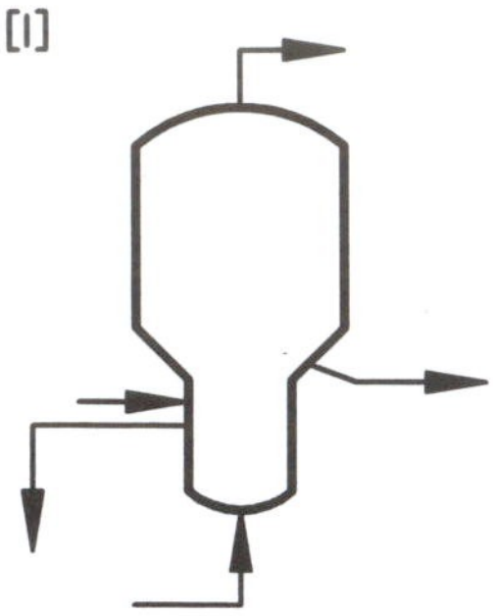		
CENTRIFUGAL LIQUID/LIQUID EXTRACTOR	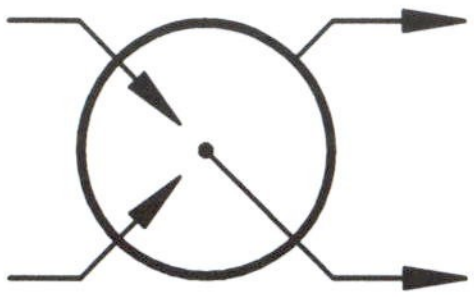		
CYCLONE SEPARATOR	[I] 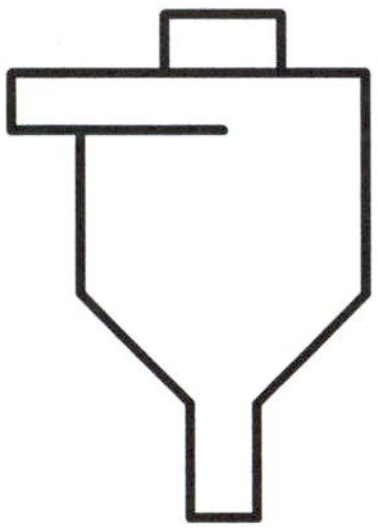 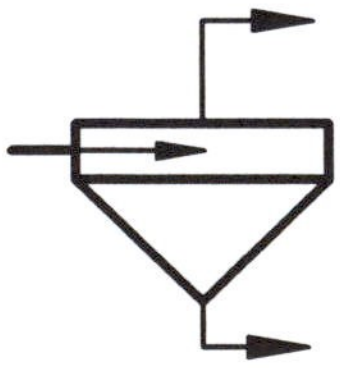 [A] 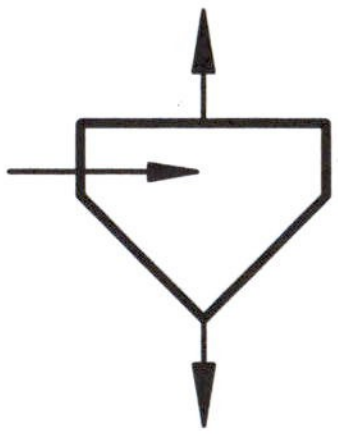		SEPARATES SOLIDS, LIQUIDS & GASES.
ROTARY SEPARATOR	[I] 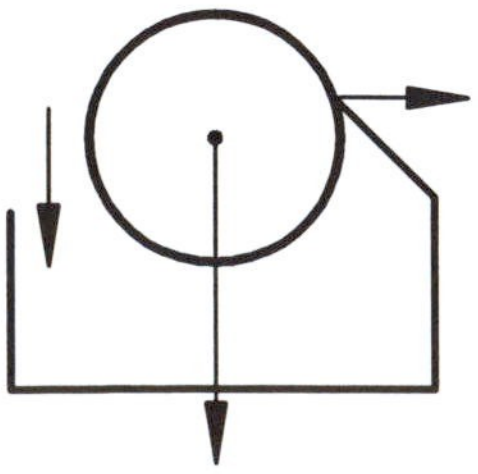		SEPARATES SOLIDS FROM LIQUIDS BY EVAPORATION.

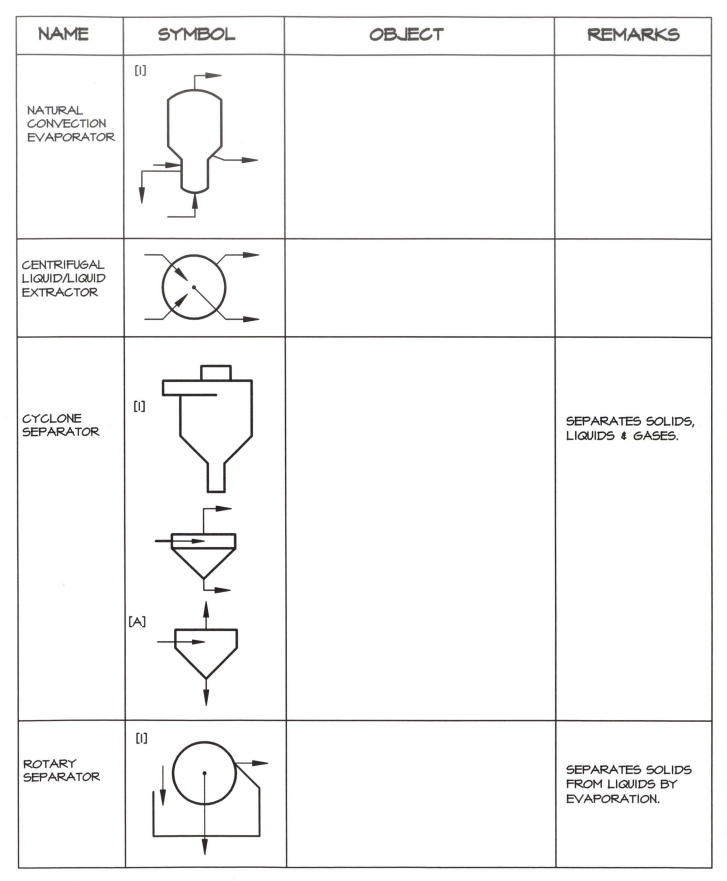

NAME	SYMBOL	OBJECT	REMARKS
CENTRIFUGAL LIQUID/LIQUID EXTRACTOR	[I] 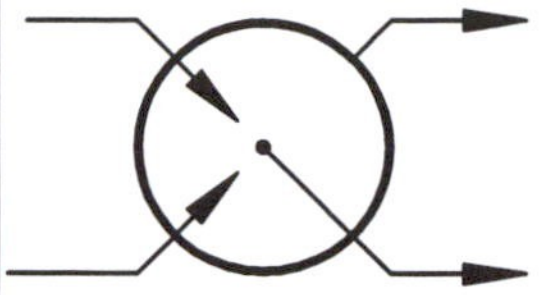		
PACKED COLUMN EXTRACTOR	[I] 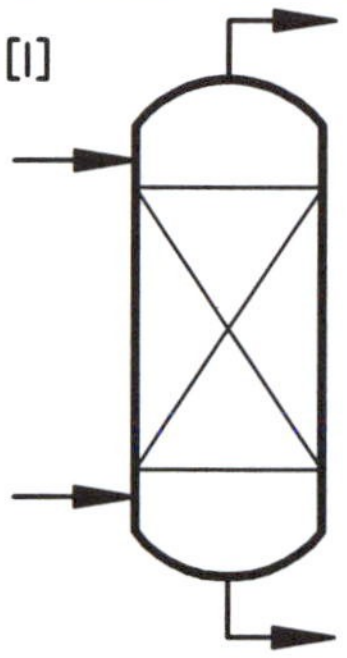		
BAG FILTER	[I] 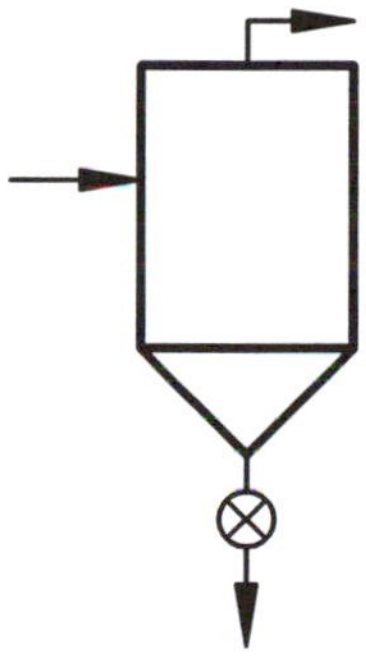	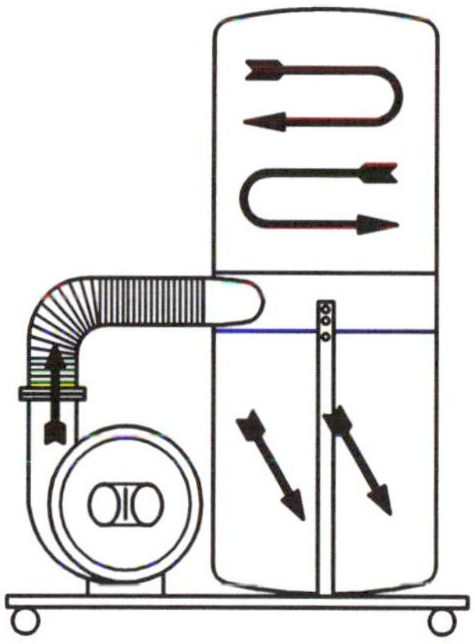	ILLUSTRATION SHOWS MOBILE DUST COLLECTION TYPE BAG FILTER.
DRUM SETTLER	[A] 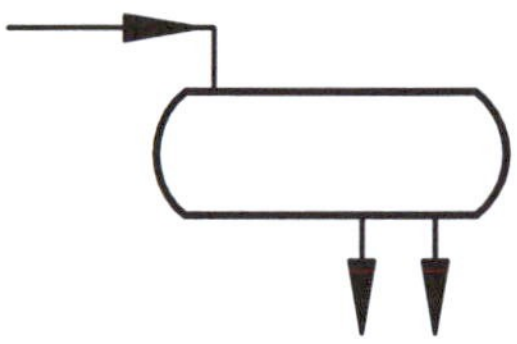		
OPEN SETTLING TANK	[A] 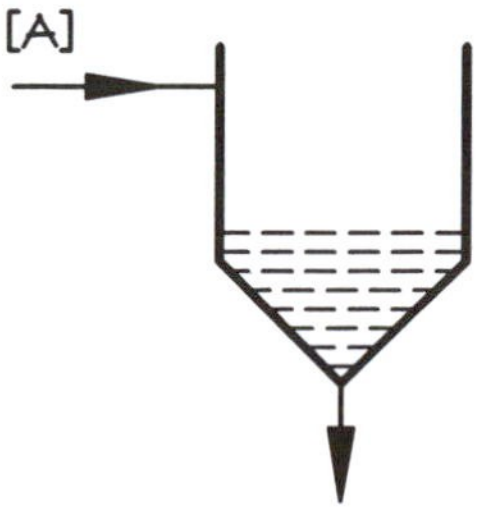		

NAME	SYMBOL	OBJECT
PLATE AND FRAME FILTER PRESS	[I] 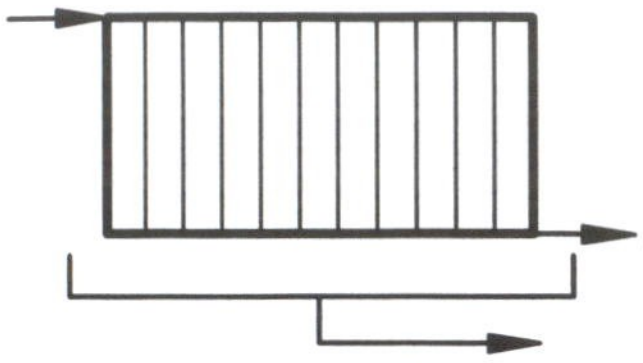 [A] 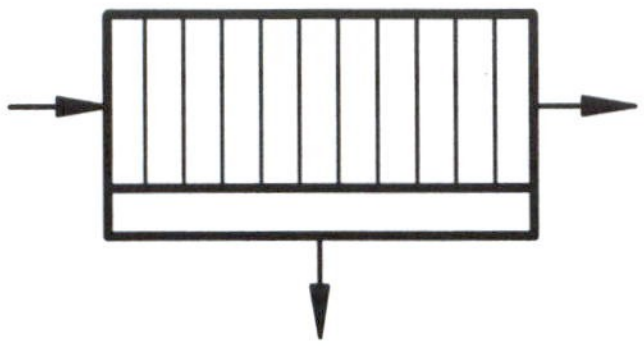	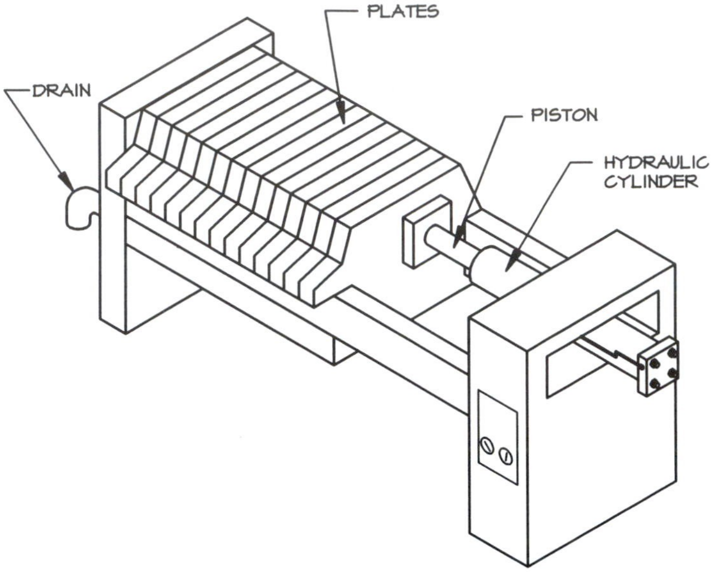 DRAIN PLATES PISTON HYDRAULIC CYLINDER
HYDROSCREEN	[I] 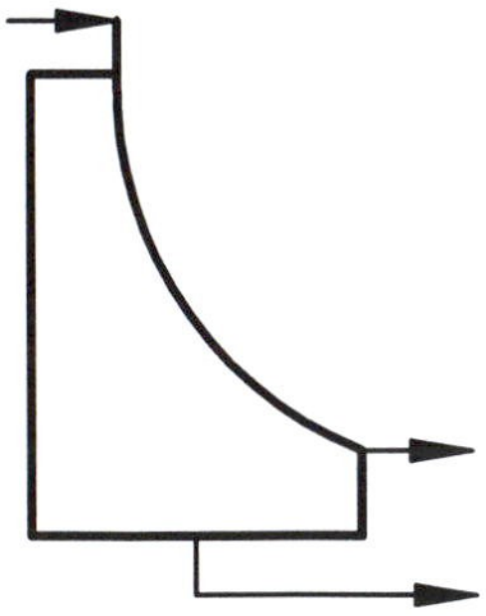	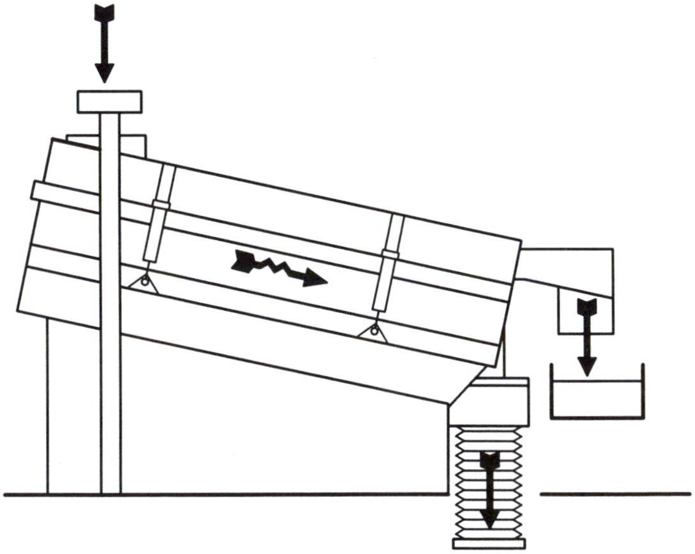

NAME	SYMBOL	OBJECT	REMARKS
VIBRATING SCREEN SEPARATOR	[I] 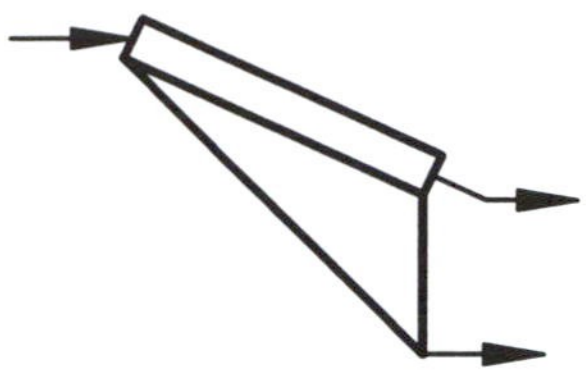	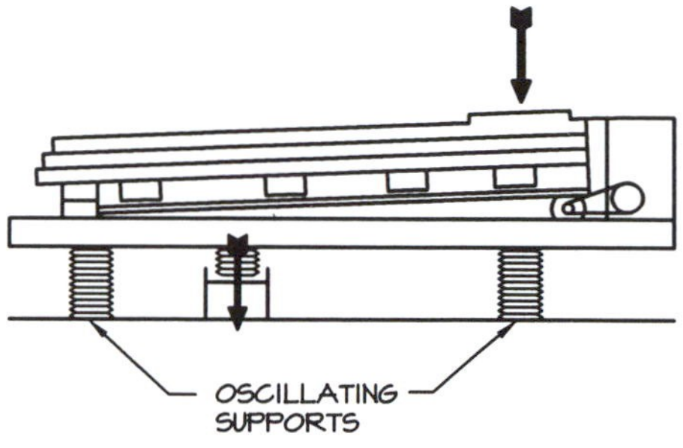 OSCILLATING SUPPORTS	
PRECIPITATOR	[I] 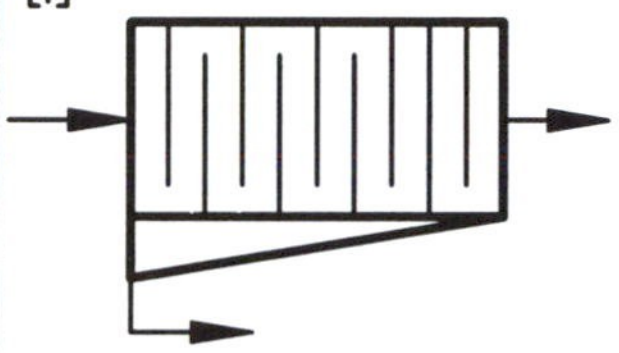		
ELECTRO-STATIC PRECIPITATOR	[I] 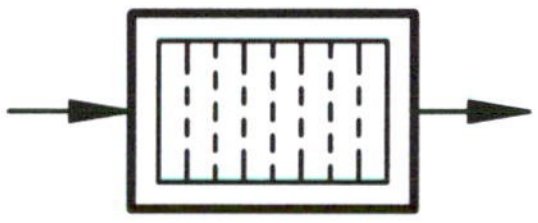 [A] 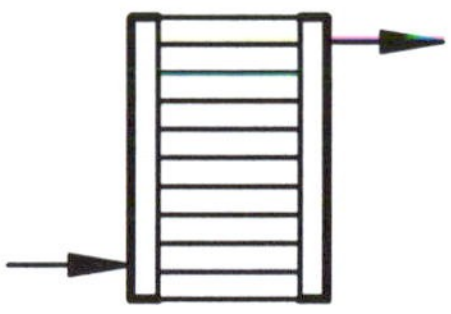	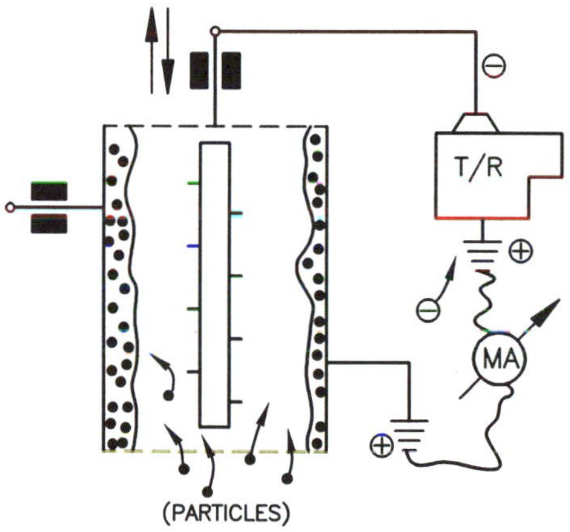 T/R MA (PARTICLES)	SEPARATES PARTICLES FROM GAS BY USE OF ELECTRIFIED GRID.
SETTLER THICKENER	[I]		
LIQUID FILTER			

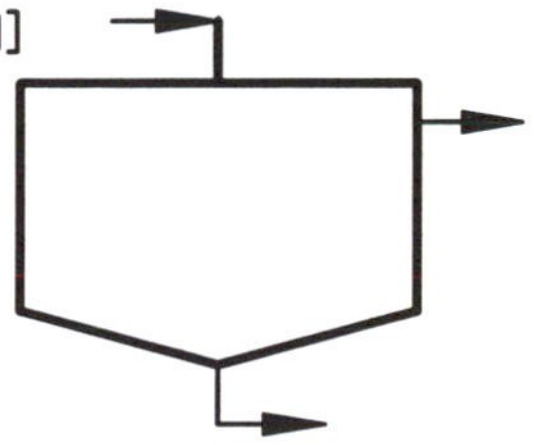

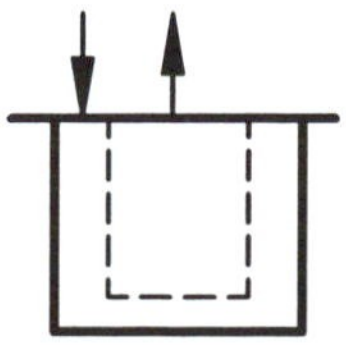

NAME	SYMBOL	OBJECT	REMARKS
SCRUBBER	[1] 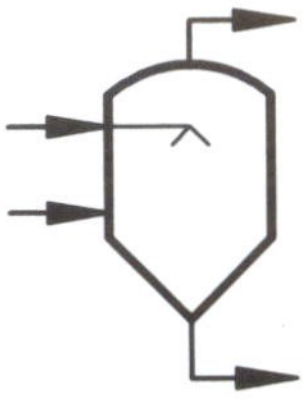 [1] 	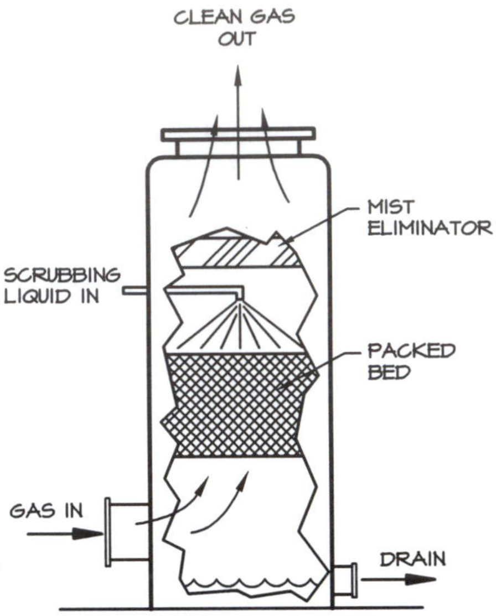	SCRUBBERS ARE INDUSTRIAL CLEANING DEVICES.
VACUUM FILTER	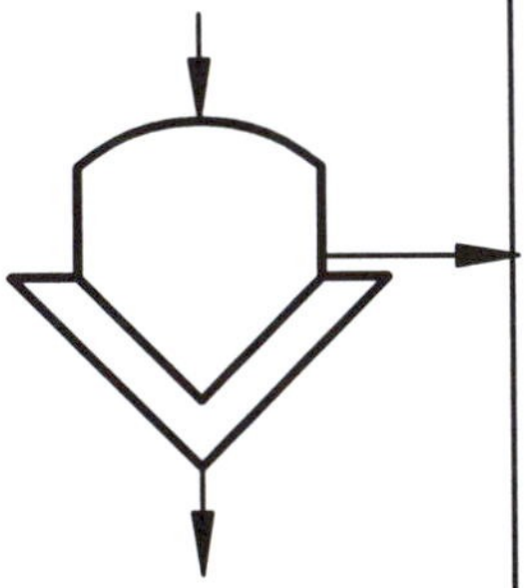		FILTRATION IS ENHANCED BY VACUUM.
ROTARY FILTER	[1] 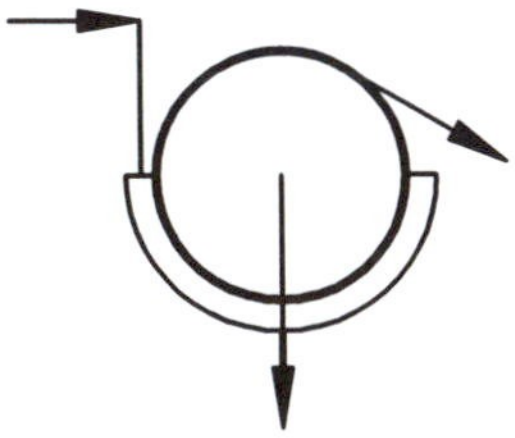	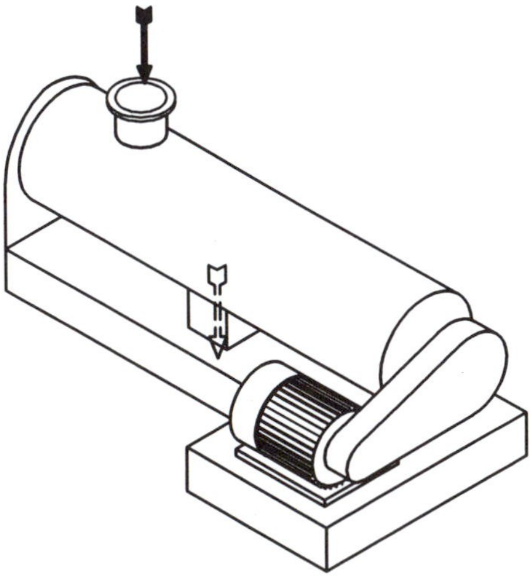	

NAME	SYMBOL	OBJECT	REMARKS
ROTARY VACUUM FILTER	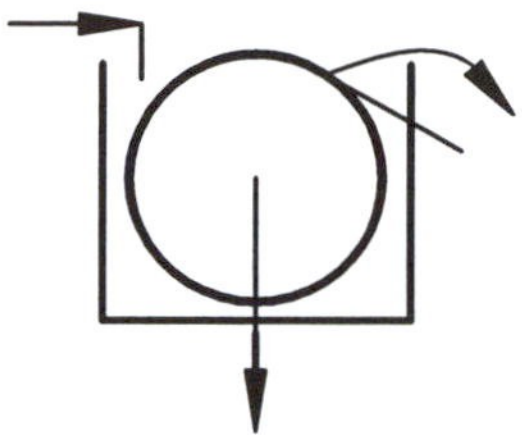		
ROTARY FILM DRYER OR FLAKER	[A] 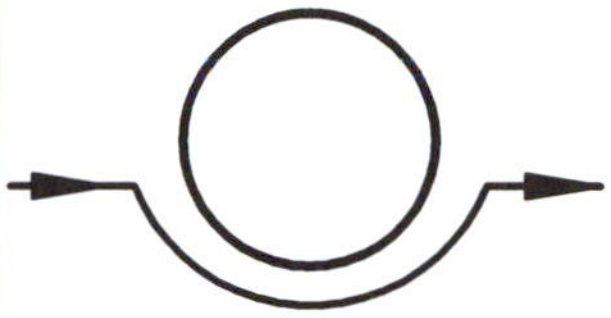	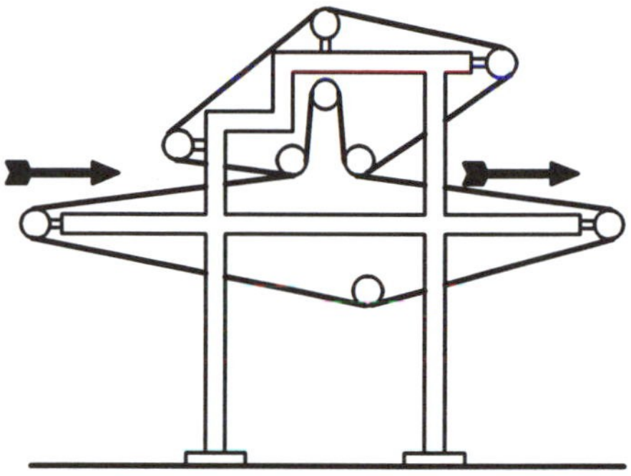	
SCREENER	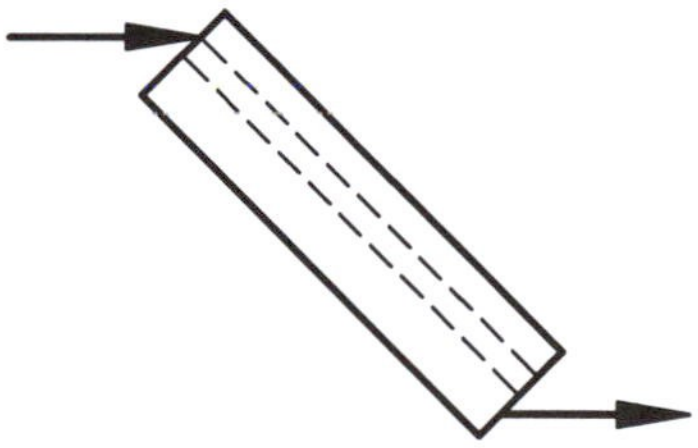		
SETTLER	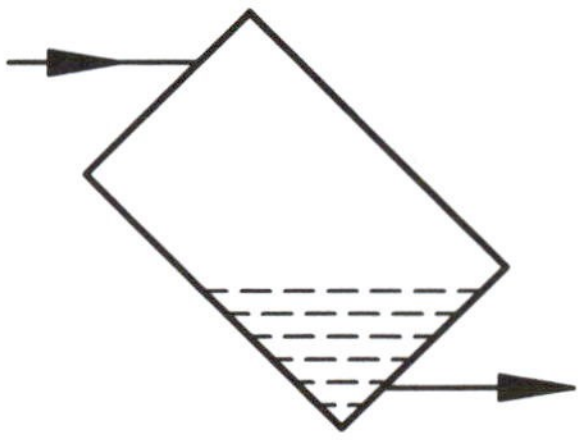		

NAME	SYMBOL	OBJECT	REMARKS
ADSORPTION COLUMN	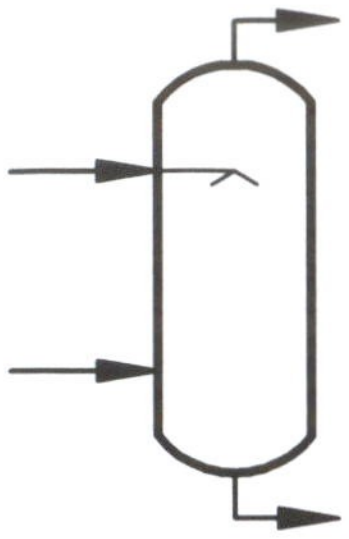		
PACKED ADSORPTION COLUMN	[I] 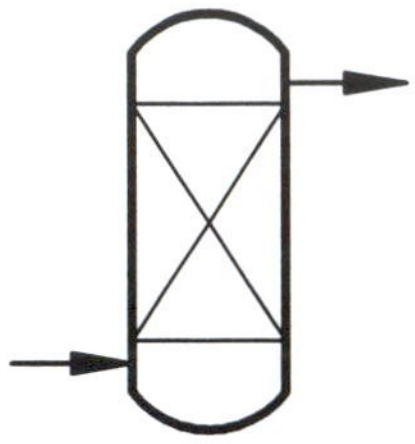		
SOLID BOWL CENTRIFUGE	[I] 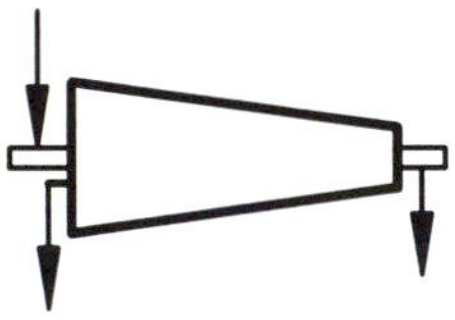		
VERTICAL CENTRIFUGE	[I] 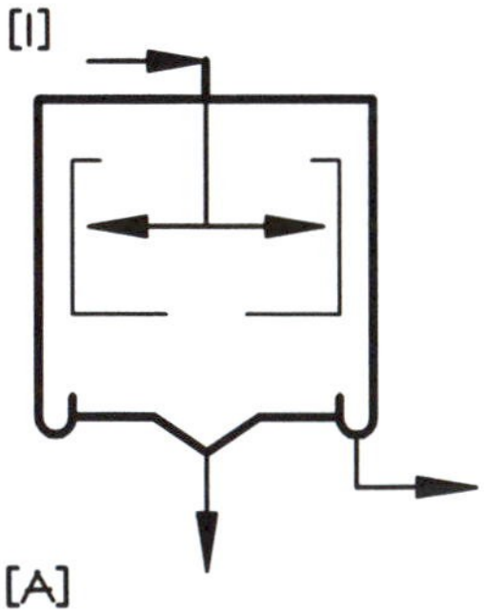 [A] 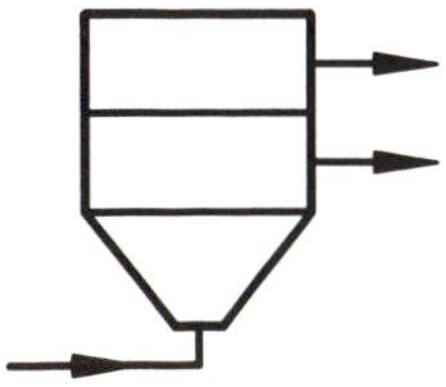	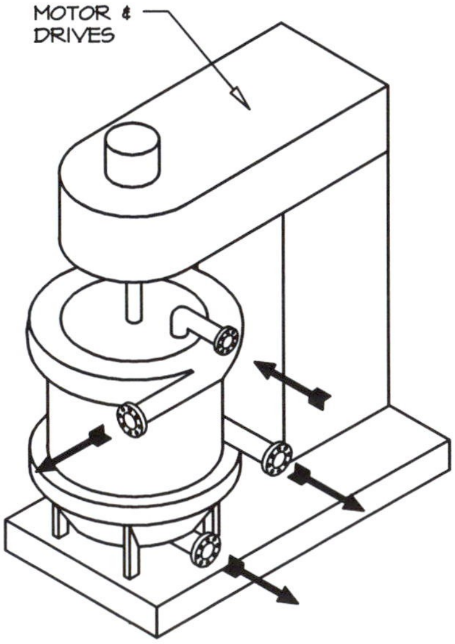 MOTOR & DRIVES	

NAME	SYMBOL	OBJECT	REMARKS
RACK OR SPIRAL CLASSIFIER	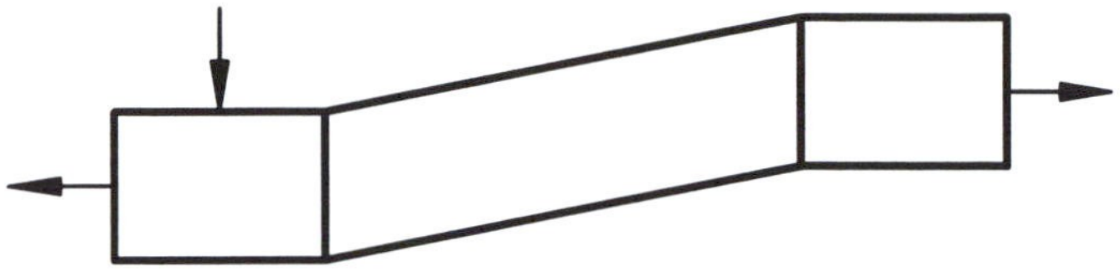 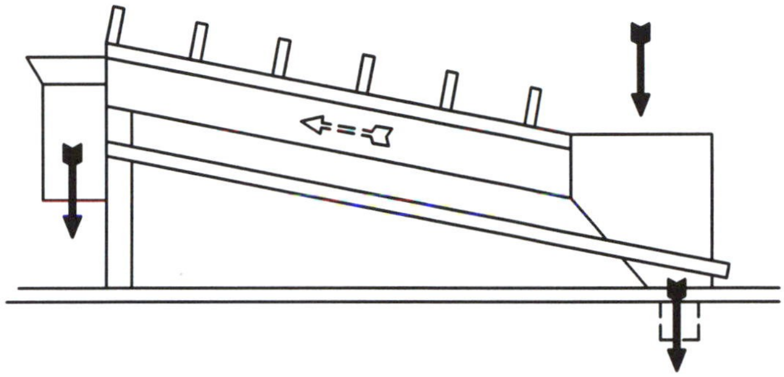		
AGITATED BATCH CRYSTALLIZER	[I] 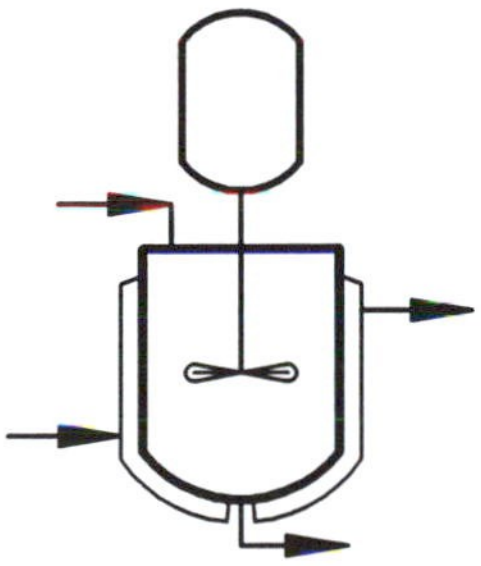		
BATCH DRYER	[I] 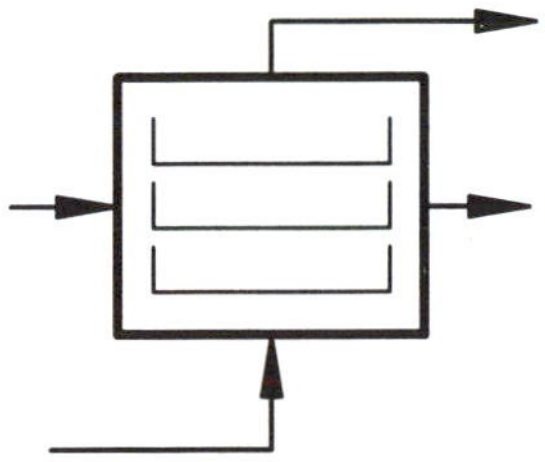		
BELT DRYER	[I] 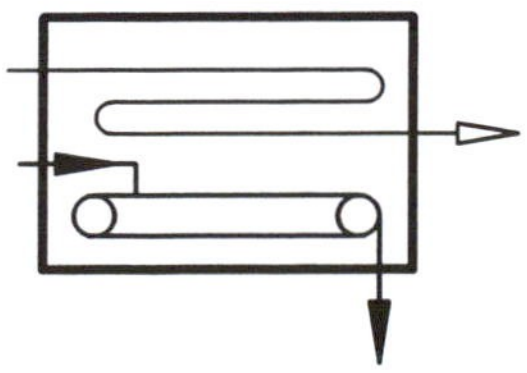		

NAME	SYMBOL	OBJECT	REMARKS
DRUM DRYER	[1]		

NAME	SYMBOL	OBJECT	REMARKS
ROLL CRUSHER	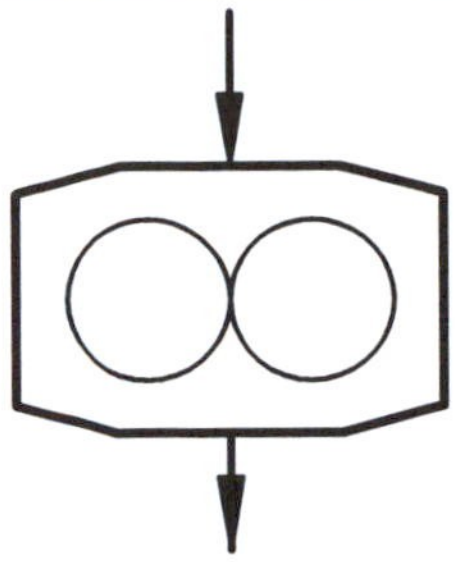 [A] 		
JAW CRUSHER	[I] 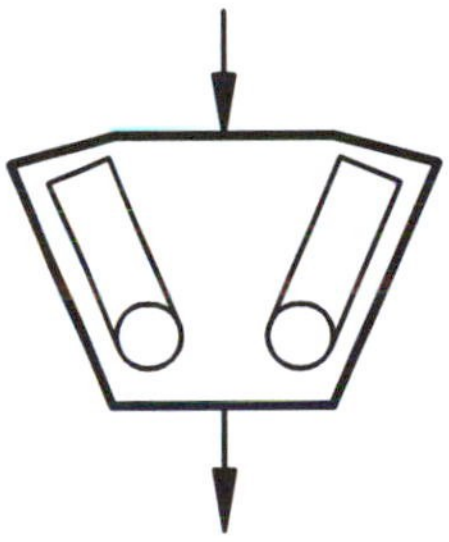	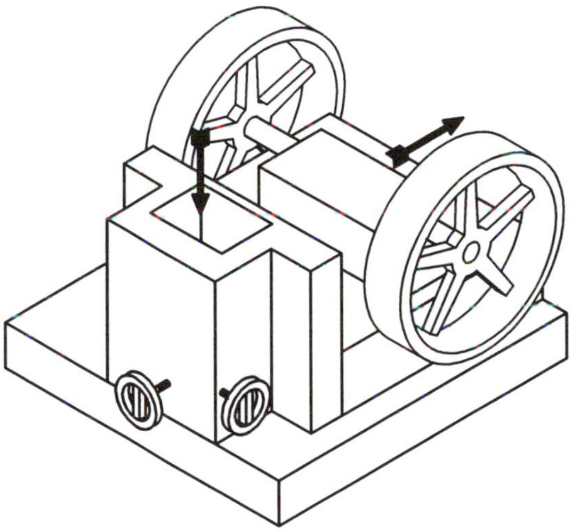	
GYRATORY CRUSHER	[I] 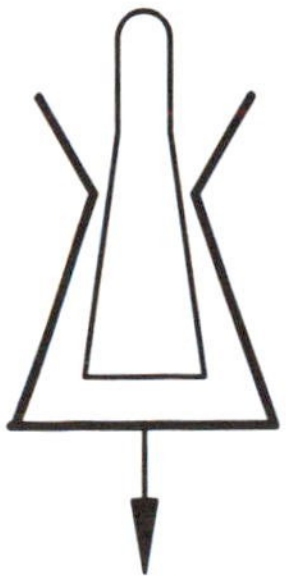		

NAME	SYMBOL	OBJECT	REMARKS
ROD, BALL AUTOGENOUS, OR SEMI-AUTOGENOUS MILL	[I] 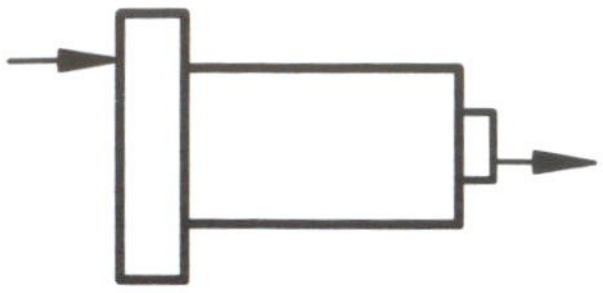 [I] 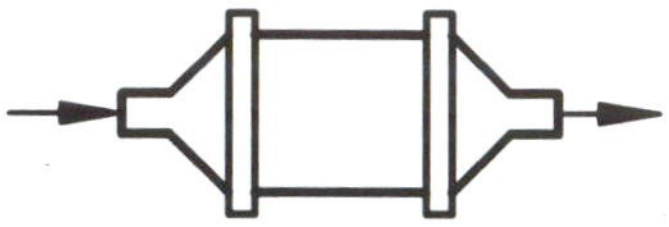		
BALL MILL	[A] 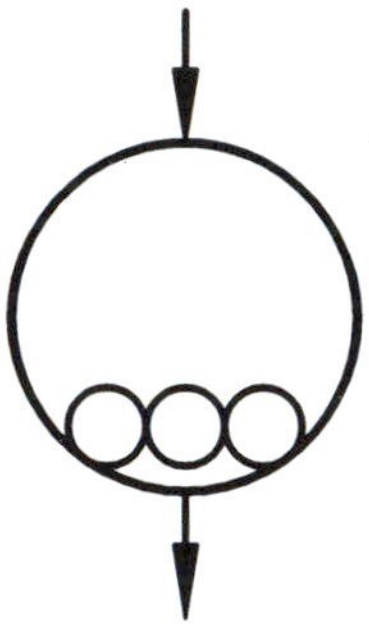	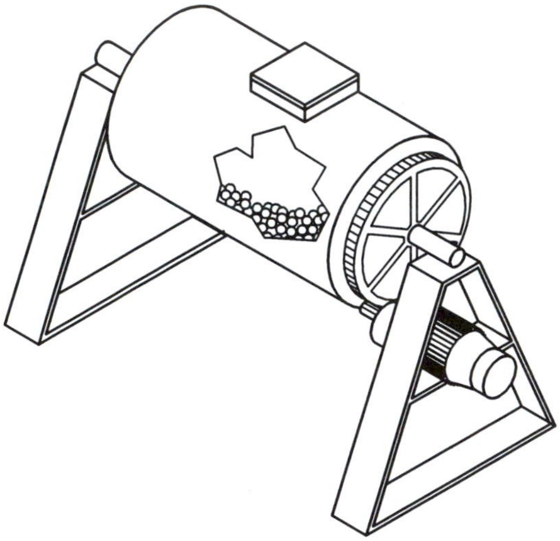	
AUTOGENOUS GRINDER	[I] 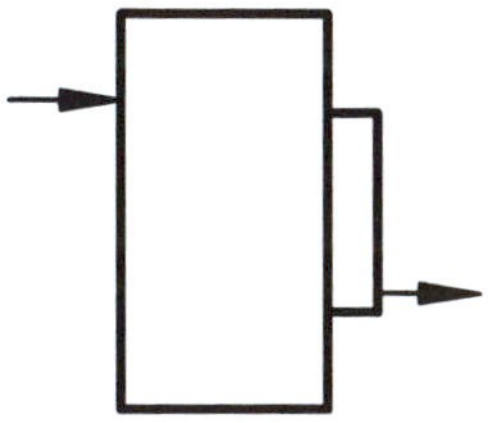		

NAME	SYMBOL	OBJECT	REMARKS
ROLL STAND			USED IN PAPER, PLASTIC, METAL, GLASS INDUSTRIES, ETC.
GRINDER			

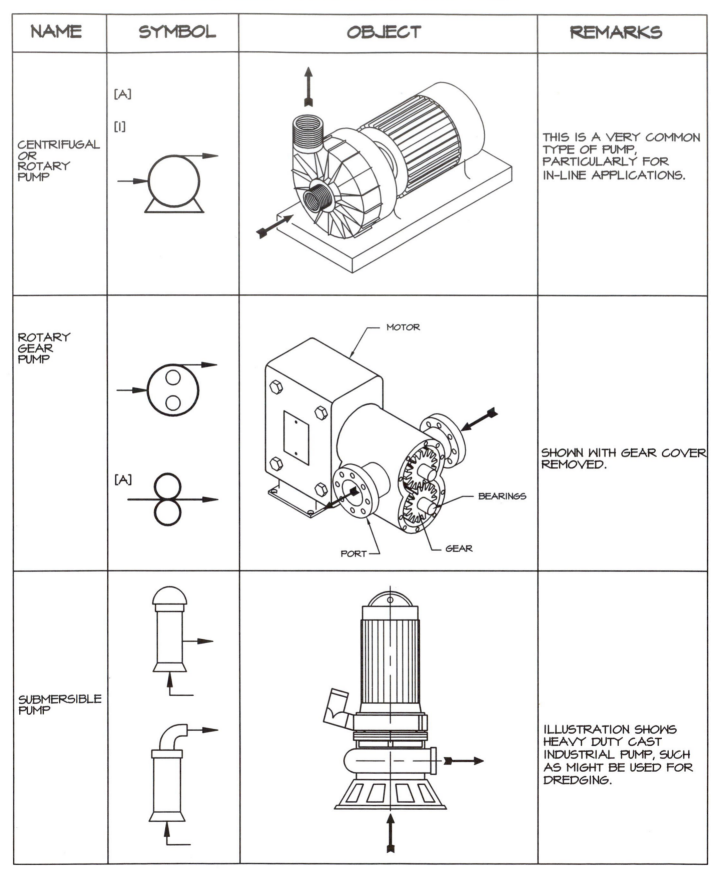

NAME	SYMBOL	OBJECT	REMARKS
CENTRIFUGAL OR ROTARY PUMP	[A] [I]		THIS IS A VERY COMMON TYPE OF PUMP, PARTICULARLY FOR IN-LINE APPLICATIONS.
ROTARY GEAR PUMP	[A]	MOTOR BEARINGS PORT GEAR	SHOWN WITH GEAR COVER REMOVED.
SUBMERSIBLE PUMP			ILLUSTRATION SHOWS HEAVY DUTY CAST INDUSTRIAL PUMP, SUCH AS MIGHT BE USED FOR DREDGING.

NAME	SYMBOL	OBJECT	REMARKS
RECIPRO-CATING PUMP OR COMPRESSOR	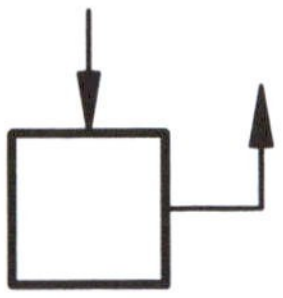 [I] [A] 		TRANSFERS SLURRIES OR LIQUIDS BY PISTON, DIAPHRAGM ACTION, ETC.
BLOWER	[I] 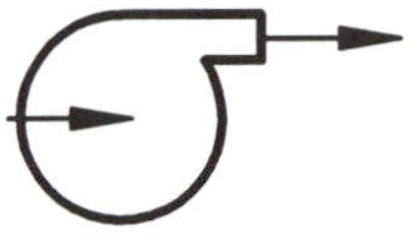 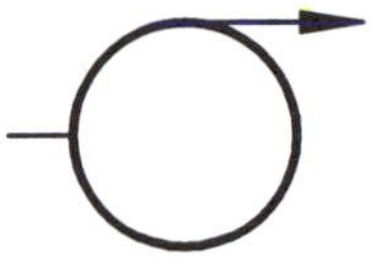 [A] 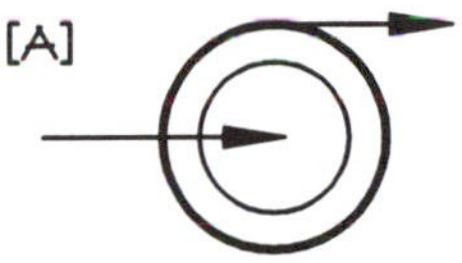	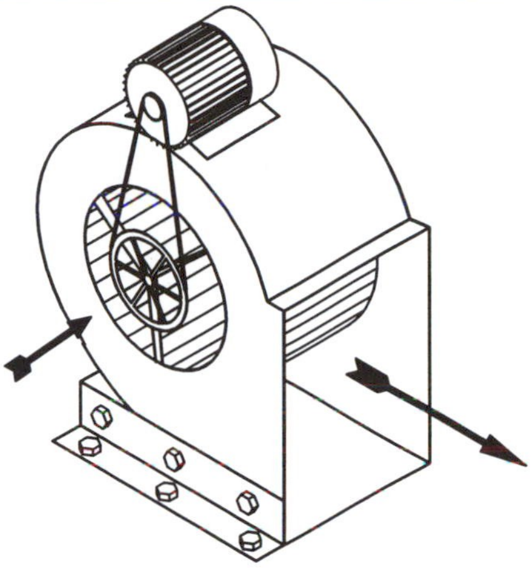	ILLUSTRATION SHOWS BOTTOM HORIZONTAL TYPE BELT-DRIVE BLOWER, USED WITH DUCTWORK.
MOTOR	[I] 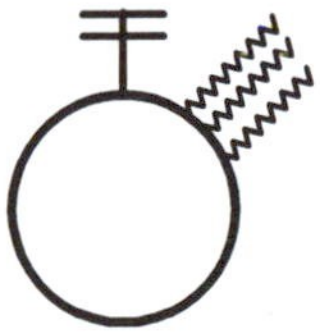	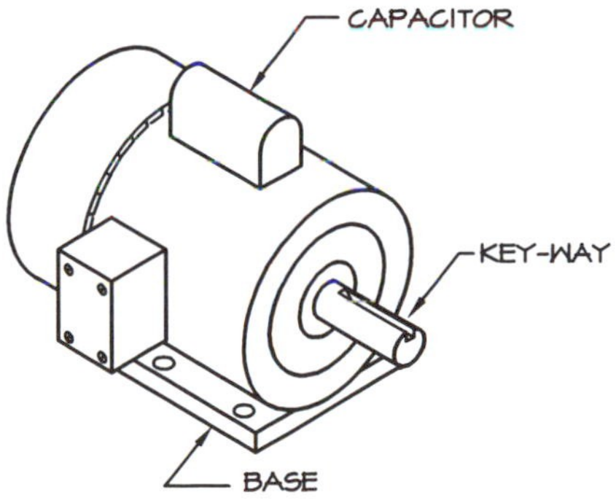 CAPACITOR KEY-WAY BASE	AC OR DC
ENGINES	[A] 		SINGLE DRIVE
	[A] 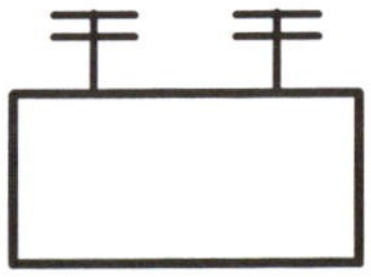		DUAL DRIVE

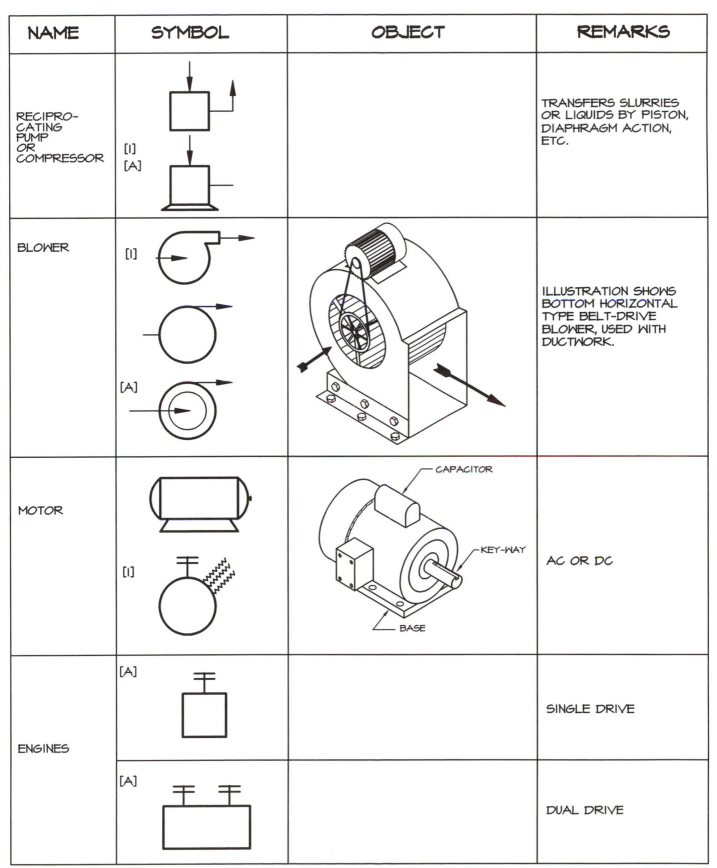

NAME	SYMBOL	OBJECT	REMARKS
CENTRIFUGAL COMPRESSOR	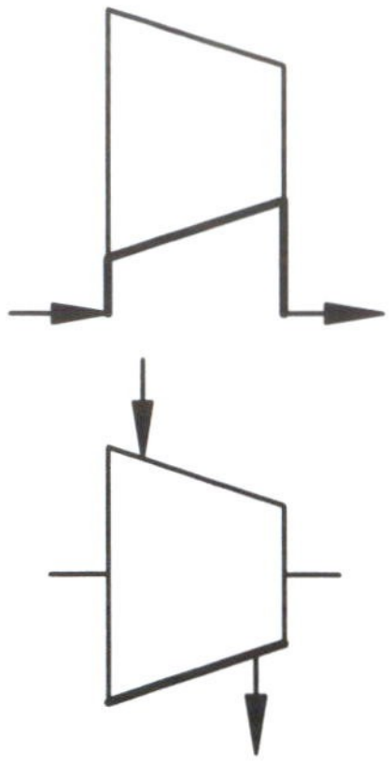 [I]		FOR GAS MOVEMENT UNDER HIGH PRESSURE.
TURBINE	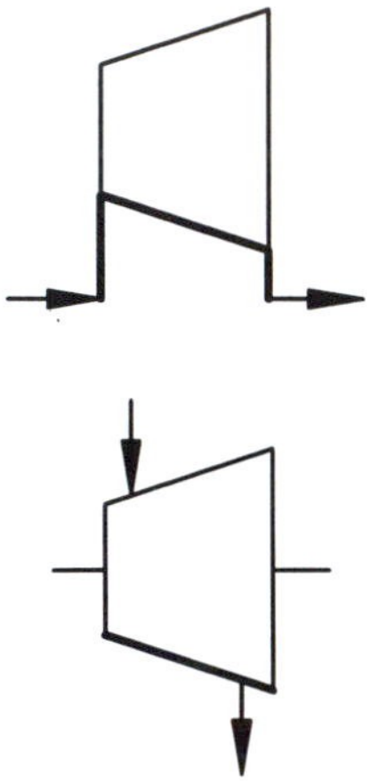 [I] [A] 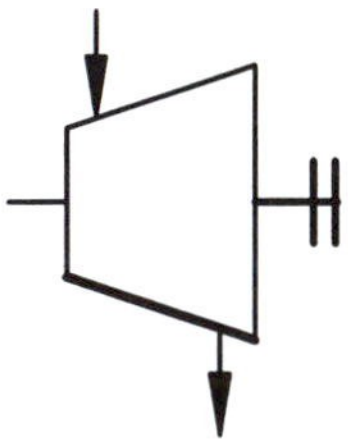		A MOTOR DRIVEN BY THE FORCE OF EXPANDING GAS OR THE MOVEMENT OF LIQUID.
STEAM TURBINE	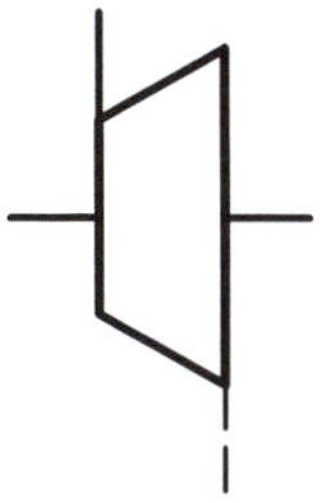		

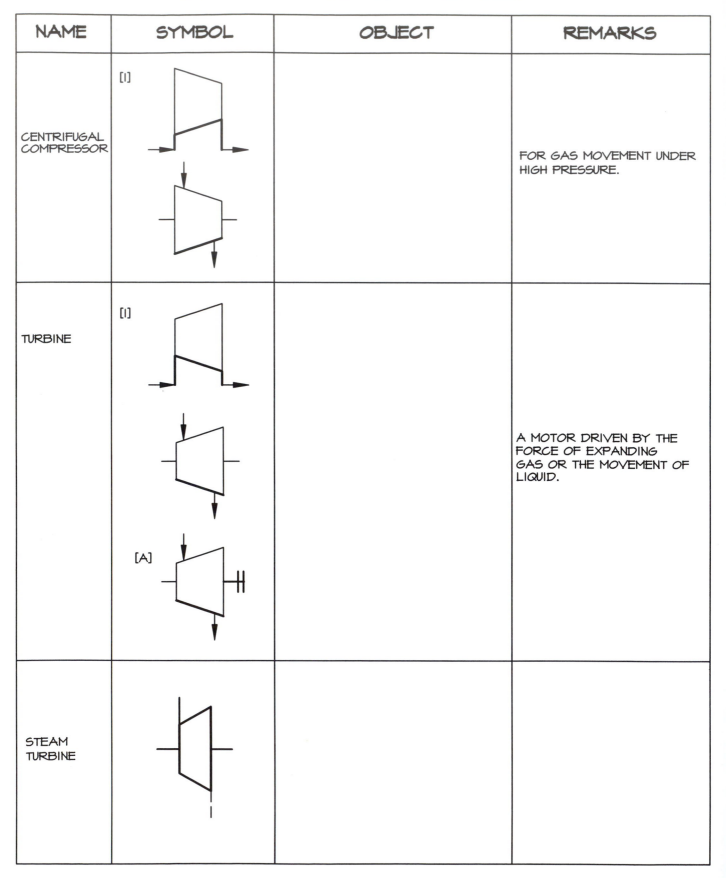

NAME	SYMBOL	OBJECT	REMARKS
CONDENSING TURBINE	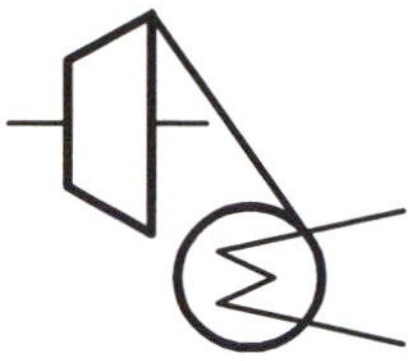		
RECIPROCATING COMPRESSOR	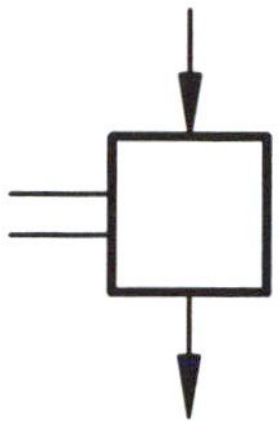 [A] 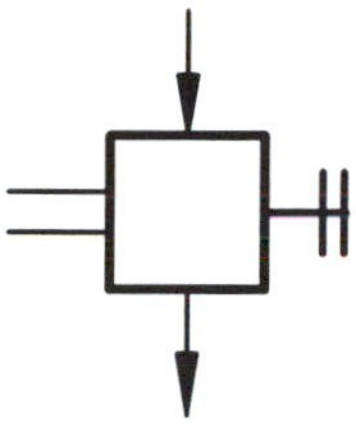		
ROTATING COMPRESSOR	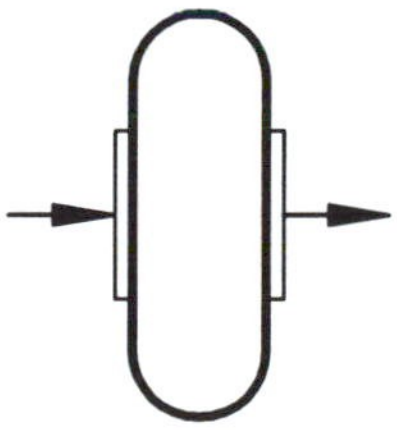		
EJECTOR	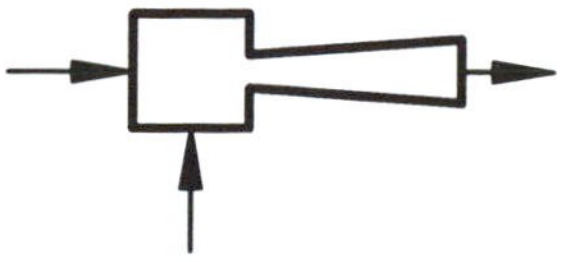	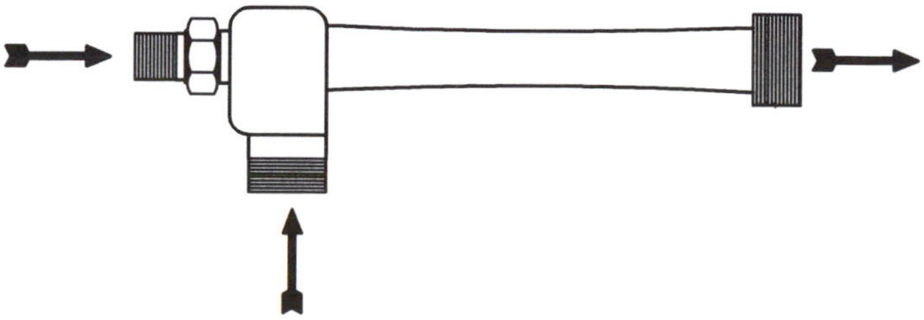	

NAME	SYMBOL	OBJECT	REMARKS
BELT CONVEYOR OR SHAKER	[I] [A] 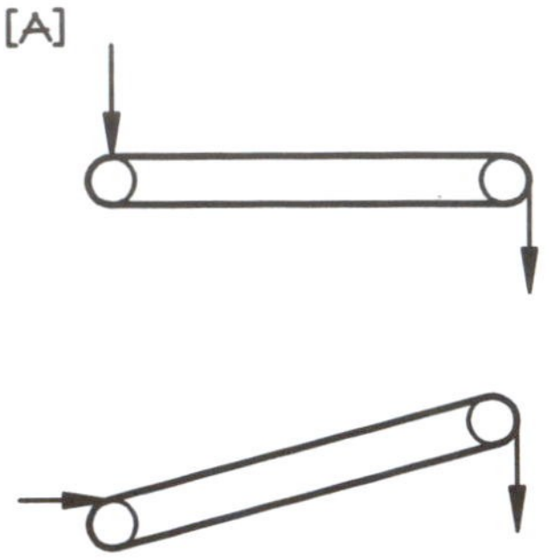	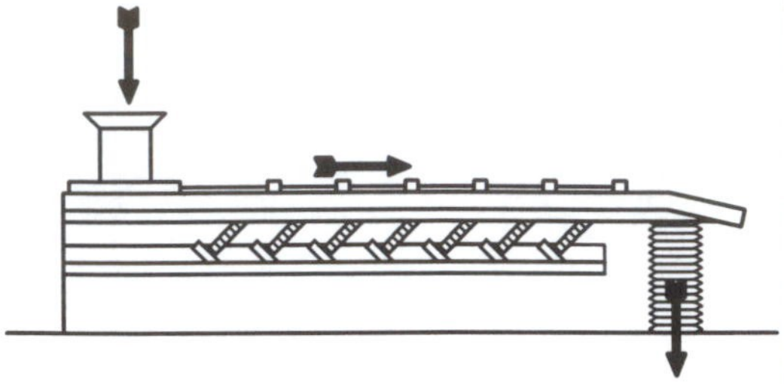	CAN MOVE MATERIAL OR ITEMS FROM ONE PLACE TO ANOTHER, BOTH HORIZONTALLY AND VERTICALLY. UPPER ILLUSTRATION SHOWS VIBRATING CONVEYOR.
SCREW CONVEYOR OR SCREW PUMP	[I] 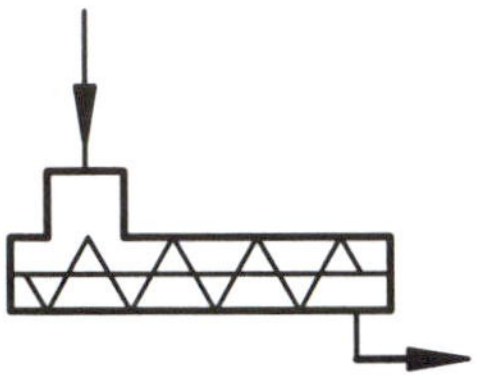 [A] 	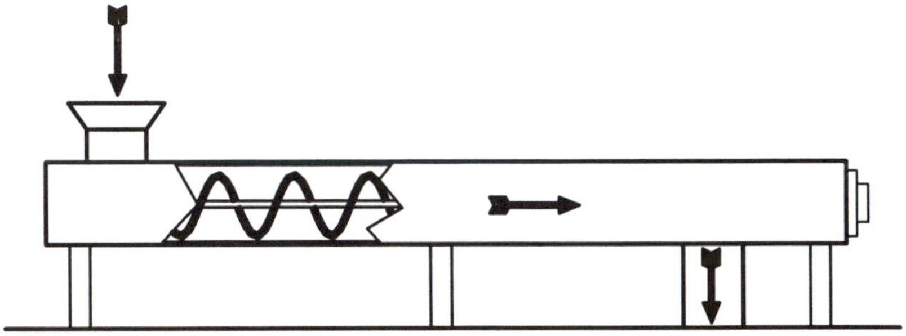	
ROTARY FEEDER	[I] [A] 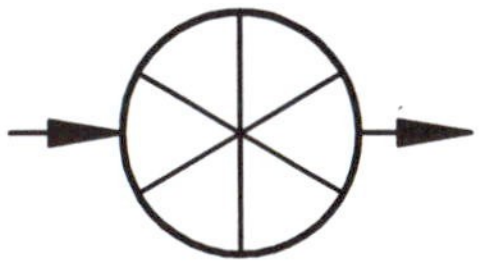		MOVES DRY POWDER.
AIR LIFT	[A] 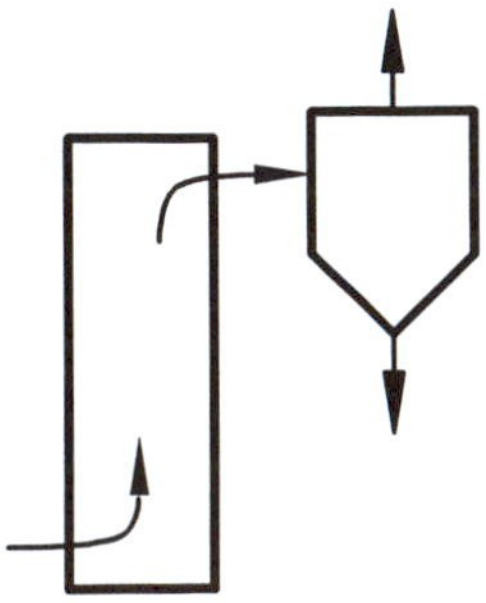		

NAME	SYMBOL	OBJECT	REMARKS
BUCKET ELEVATOR	[I] 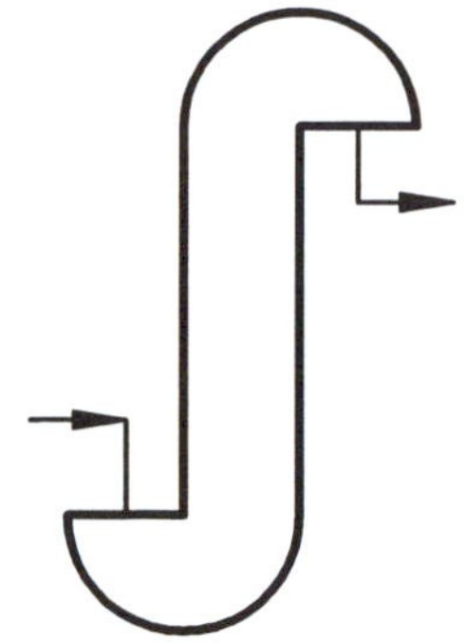 [A] 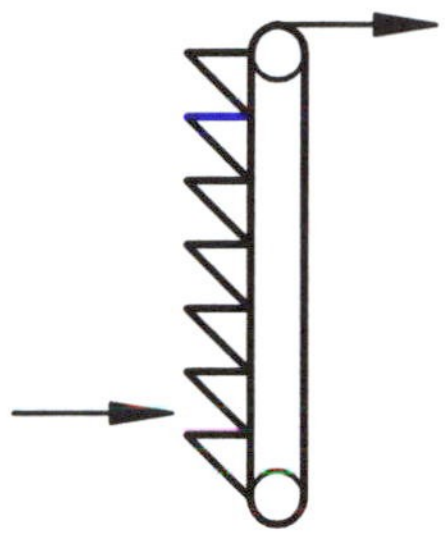	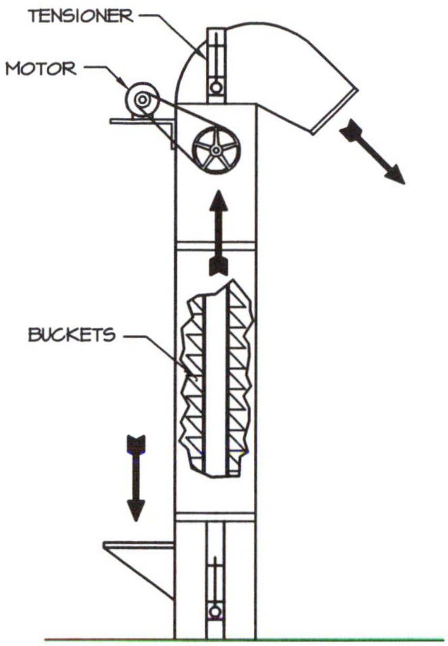 TENSIONER MOTOR BUCKETS	CAN BE AS TALL AS 30 FEET.
ROLLER CONVEYOR	[A] 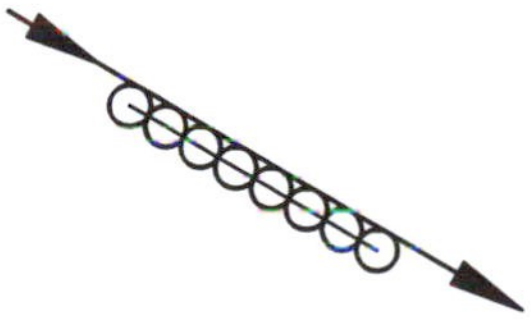		
FEEDER & HOPPER	[A] 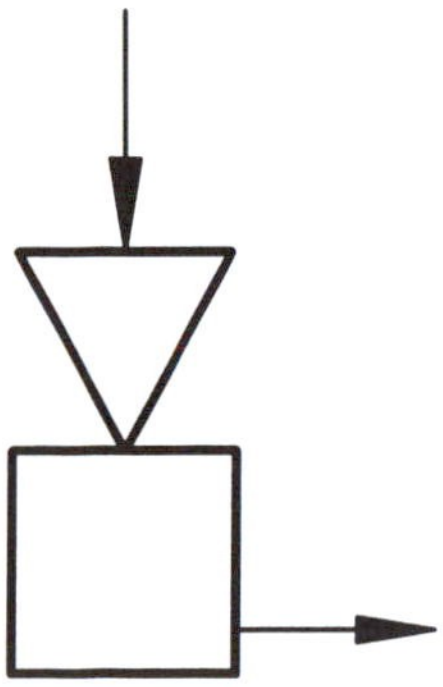	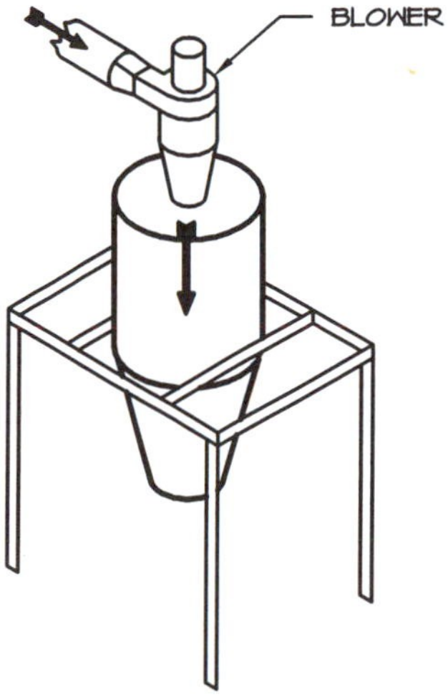 BLOWER	

PROCESS - HEAT

NAME	SYMBOL	OBJECT	REMARKS
COOLING TOWER	[I] 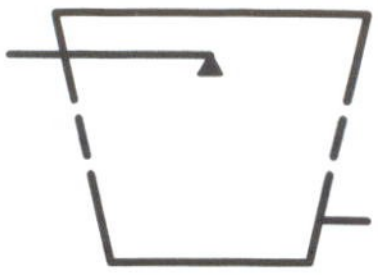 [A] 		COOLS WATER BY FORCED EVAPORATION.
EVAPORATOR	[I] 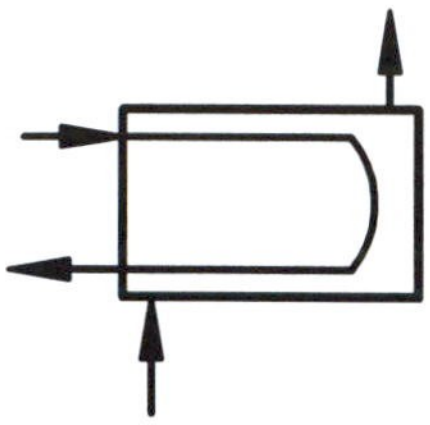		EXCHANGES HEAT BETWEEN FLUID AND REFRIGREANT.
FINNED EXCHANGER	[I] 		EXCHANGES HEAT BETWEEN FLUID AND AIR.
BOILER (STEAM GENERATOR)			
FLUE GAS REHEATER			

NAME	SYMBOL	OBJECT	REMARKS
LIVE STEAM SUPERHEATER	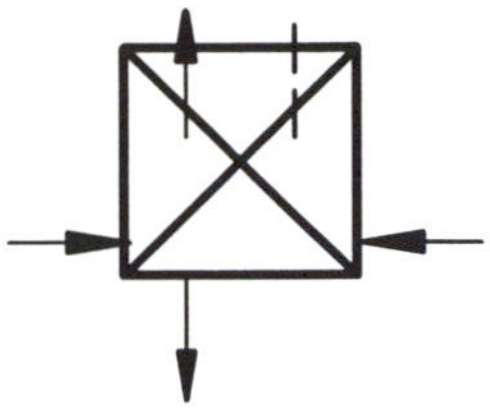		
FEED HEATER WITH AIR OUTLET	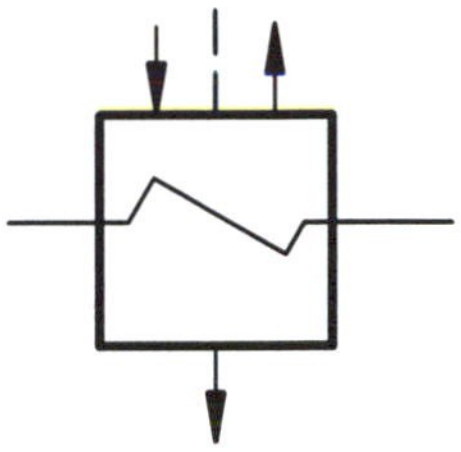		
CONDENSER, SURFACE	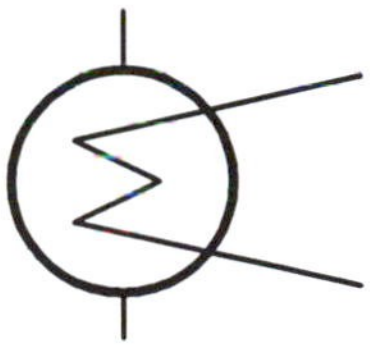		
HEAT EXCHANGER	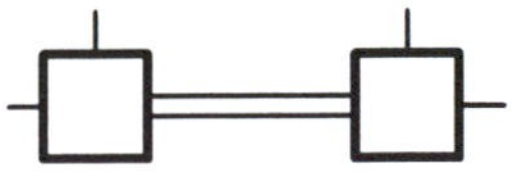		
EVAPORATIVE CONDENSER			

NAME	SYMBOL	OBJECT	REMARKS
HEATER	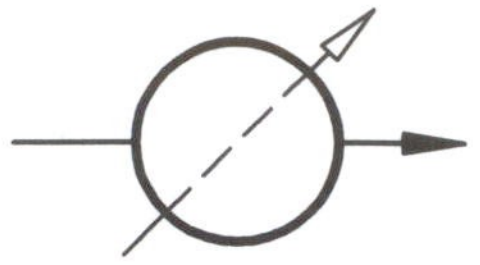		
COOLER			
PROCESS/ PROCESS EXCHANGERS	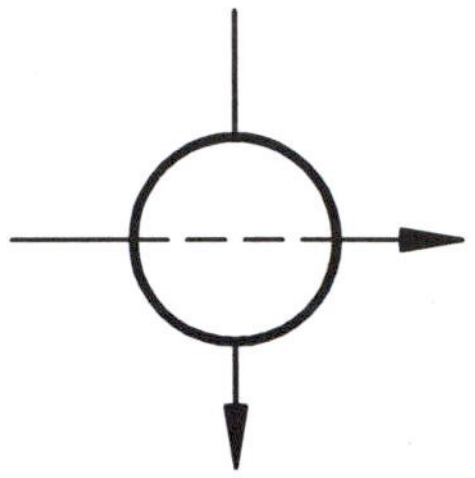		
	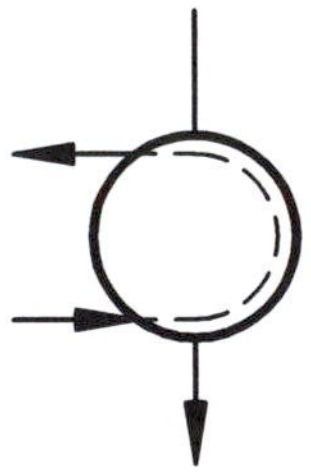		
KETTLE REBOILER	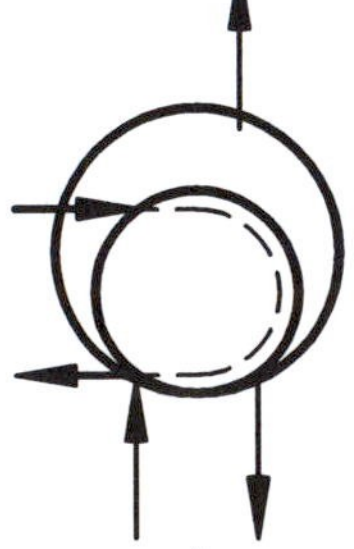 [A] 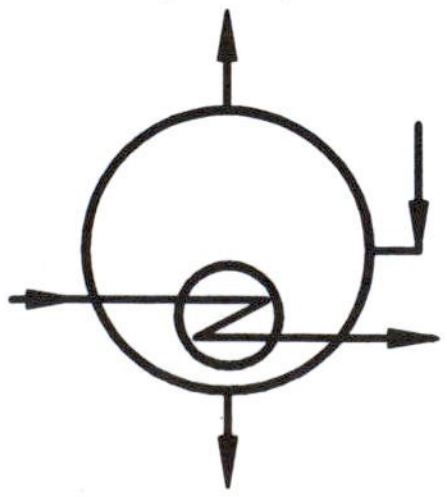		

NAME	SYMBOL	OBJECT	REMARKS
FIN FAN	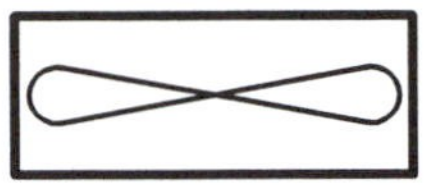		
HEAT EXCHANGER	[A] 		SHELL & TUBE TYPE EXCHANGER
	[A] 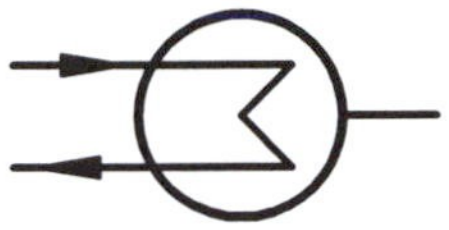		THERMO SIPHON REBOILER
BOX COOLER (SINGLE COIL)	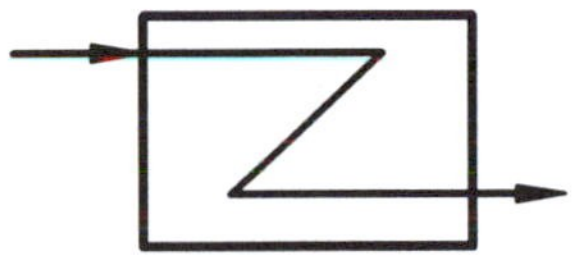		
SUPERHEATER OR REHEATER	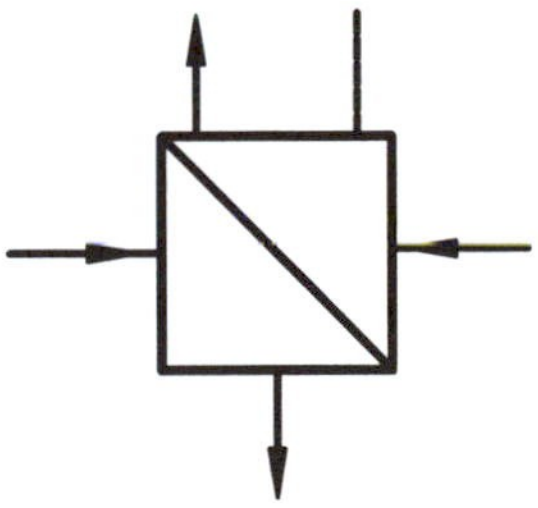		
BAROMETRIC CONDENSER	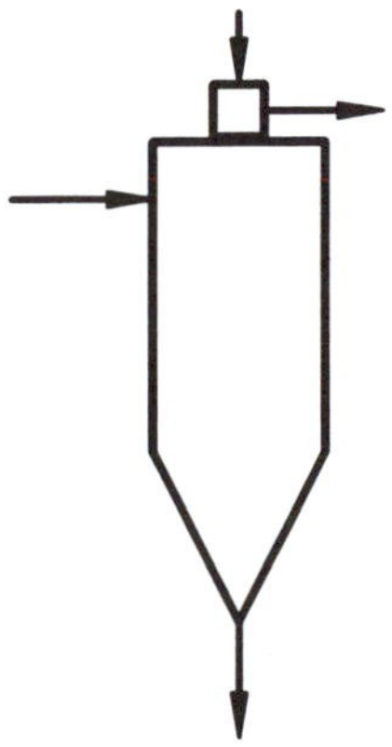		

NAME	SYMBOL	OBJECT	REMARKS
FURNACES & BOILERS	[I] 		A-FRAME TYPE 
	[A]		
	[A] 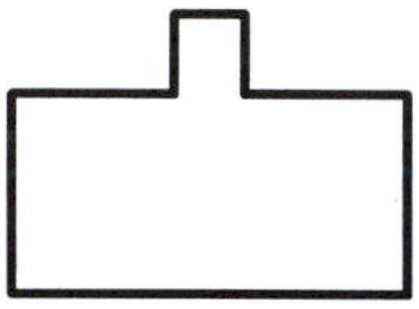		BOX TYPE
	[A] 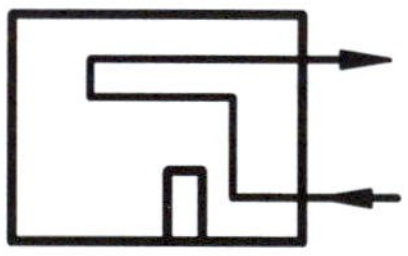		IF DUAL COIL SHOW BOTH.
	[A] 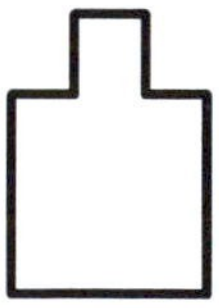		
	[A]		BOILER FIRE OR WASTE HEAT.

NAME	SYMBOL	OBJECT	REMARKS
ROTARY KILN			

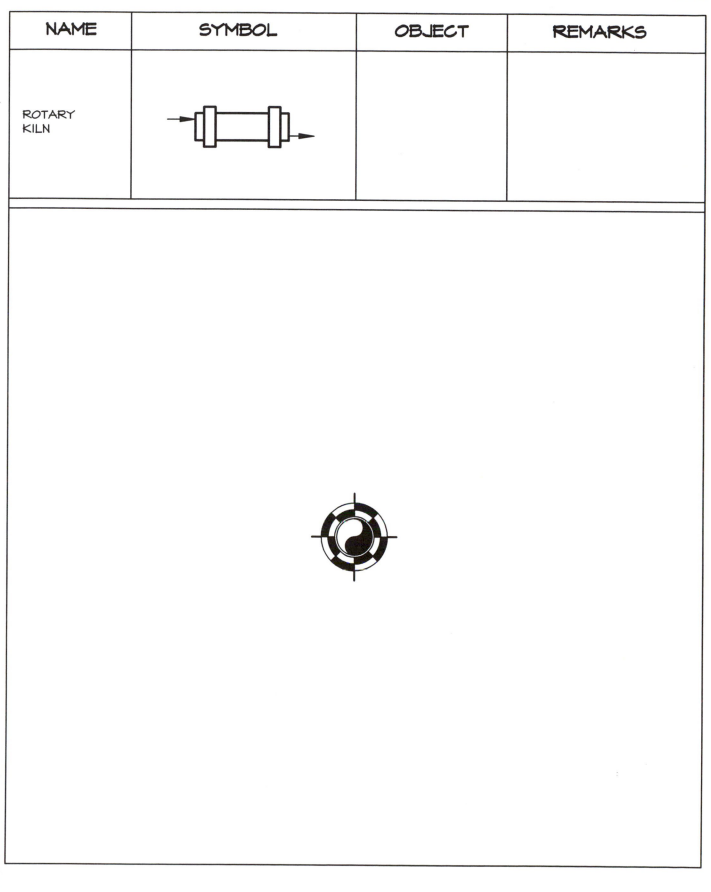

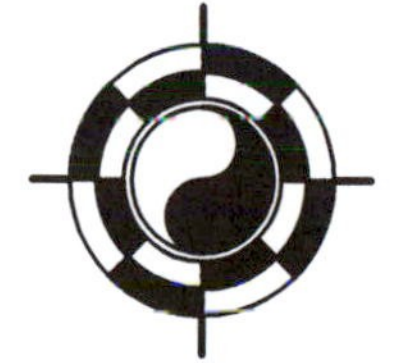

NAME	SYMBOL	OBJECT	REMARKS
ACID EGG			
BLOWING EGG	[A] 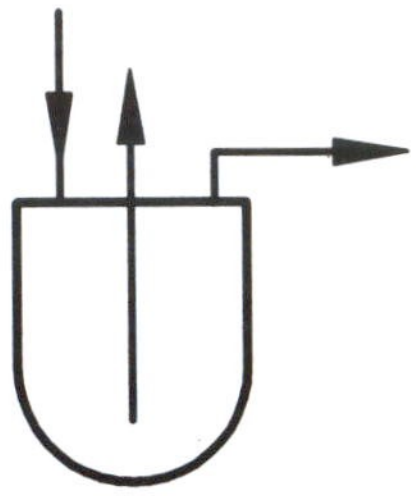		
EXTRACTOR	[A] 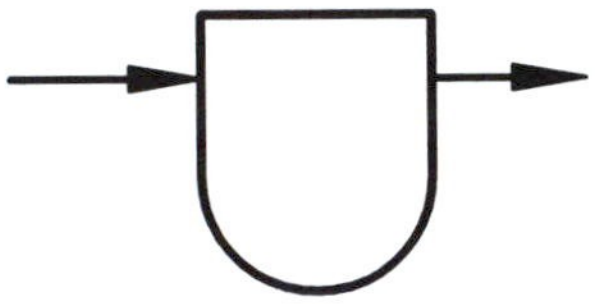		
THICKENER	[A] 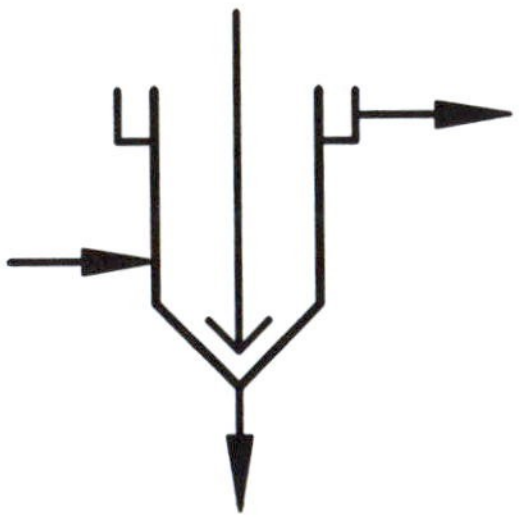		
CATALYTIC REACTOR	[A] 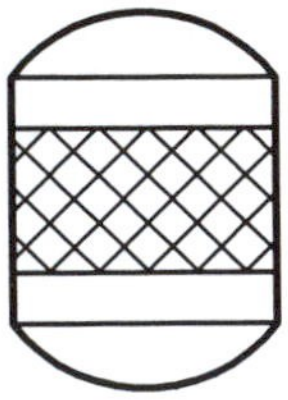		

NAME	SYMBOL	OBJECT	REMARKS
NUCLEAR REACTOR	[A] 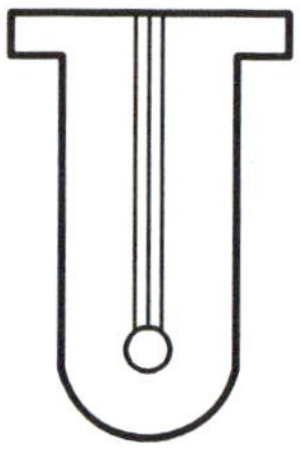		
AUTOCLAVE	[A] 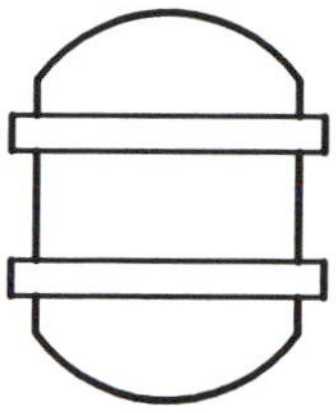		
ROTARY CALCINER	[I] 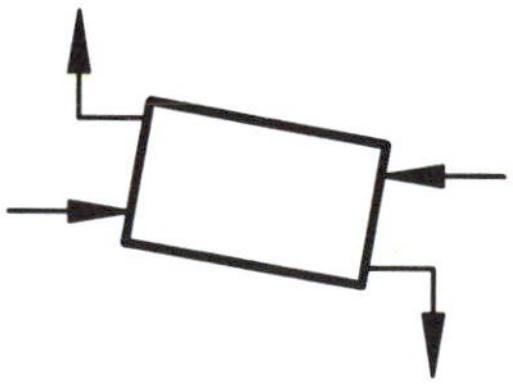	A CALCINER IS A MACHINE WHICH USES INDIRECT HEAT FOR HEATING AND PROCESSING MATERIALS. LARGE INDUSTRIAL MODEL SHOWN - APPROX. 40 FT. IN LENGTH. 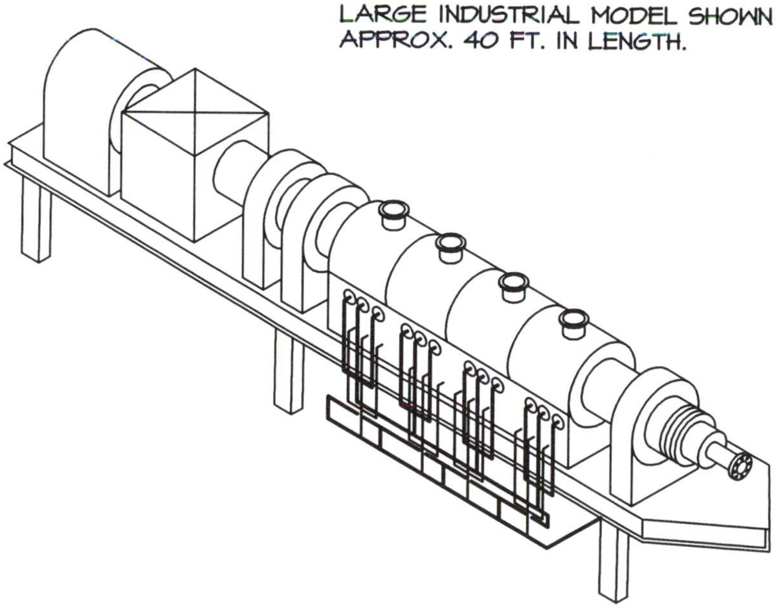	

NAME	SYMBOL	OBJECT	REMARKS
CONE ROOF TANK	[A] 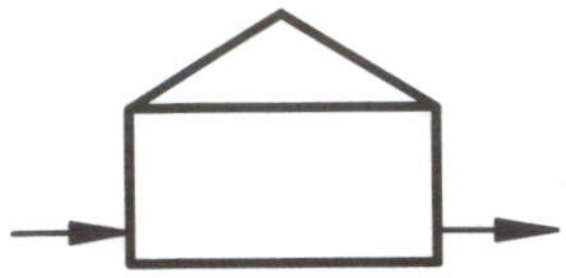		FOR STORAGE OF MATERIALS AT ATMOSPHERIC PRESSURE.
BIN	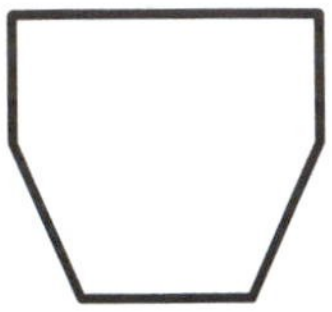		MATERIAL IS DISCHARGED FROM THE BOTTOM.
BULK STORAGE (NON-PRESSURE)	[I] [A] [A] 	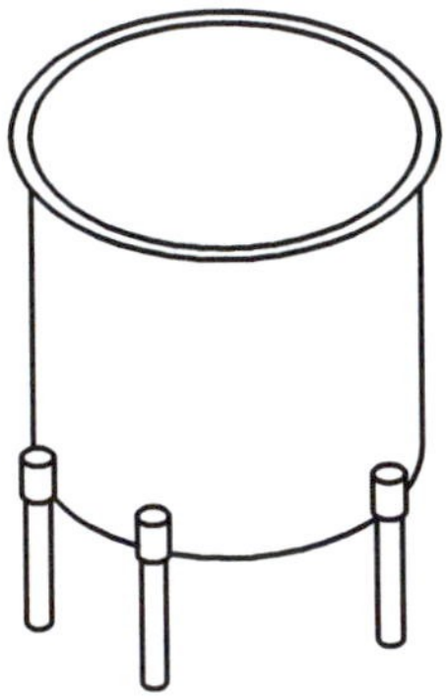	BOTTOM MAY BE SLOPED, DISHED OR CONICAL, FOR EXAMPLE.
FLOATING ROOF TANK	[I] [A] 		ROOF MOVES UP AND DOWN AS STORED VOLUME INCREASES & DECREASES.
WEIGH HOPPER	[I] 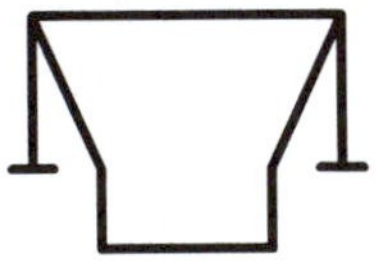		HOPPER HAS INTEGRAL SCALE FOR WEIGHING MATERIALS CONTAINED THEREIN.

NAME	SYMBOL	OBJECT	REMARKS
HORIZONTAL VESSEL	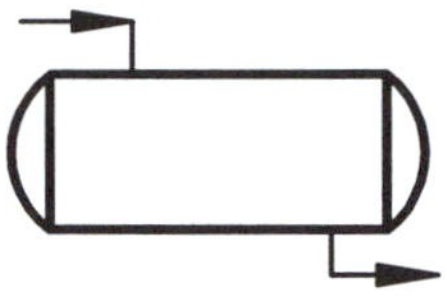	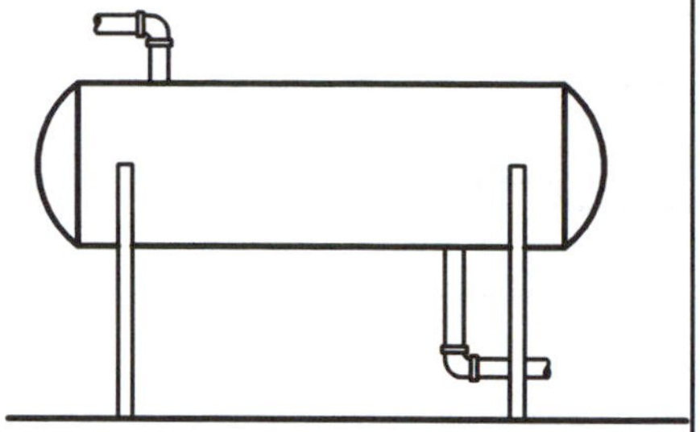	VESSEL SHOWN WITH DISHED HEADS.
HOPPER OR CONE BOTTOM BIN	 [A]	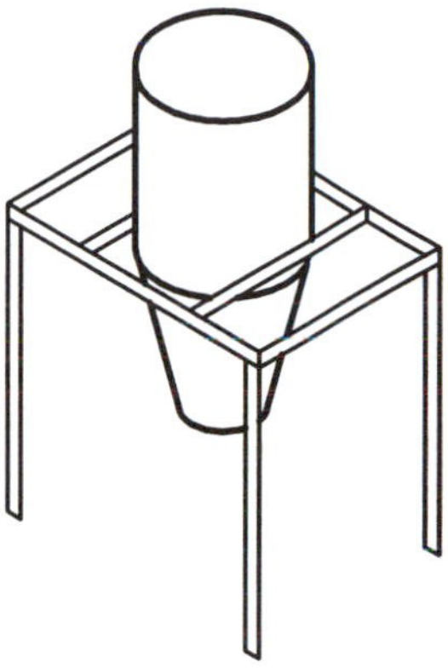	
SPHERICAL OR SPHEROID PRESSURE STORAGE	[I] 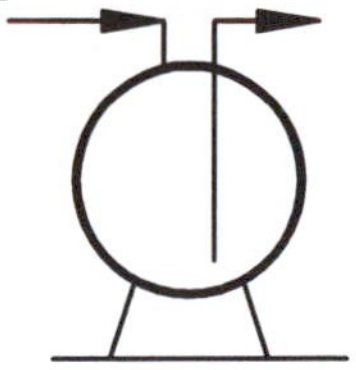 [A] 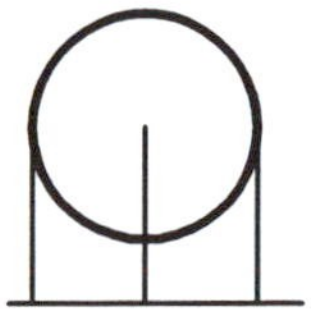		FOR GASES & LIQUIDS UNDER PRESSURE.
GAS HOLDER	[A] [I] 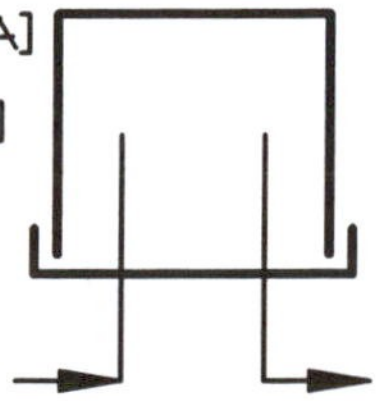		ROOF MOVES UP AND DOWN AS STORED VOLUME INCREASES & DECREASES.

NAME	SYMBOL	OBJECT	REMARKS
OPEN STORAGE PILE			

NAME	SYMBOL		REMARKS
LIQUID FLOW			
WEIGHT FLOW			
GAS FLOW			
PRESSURE			
TEMPERATURE			

NOTE: WELD SYMBOLS DENOTE THE TYPE OF WELD. WELDING SYMBOLS CONSIST OF AN ARROW, A REFERENCE LINE, A WELD SYMBOL AND ADDITIONAL ELEMENTS AS REQUIRED TO CONVEY SPECIFIC WELDING INFORMATION. THE ARROW SHOULD POINT CLEARLY TO A SPECIFIC LINE ON THE DRAWING WHICH REPRESENTS THE WELDED JOINT.

ELEMENTS OF A WELDING SYMBOL:

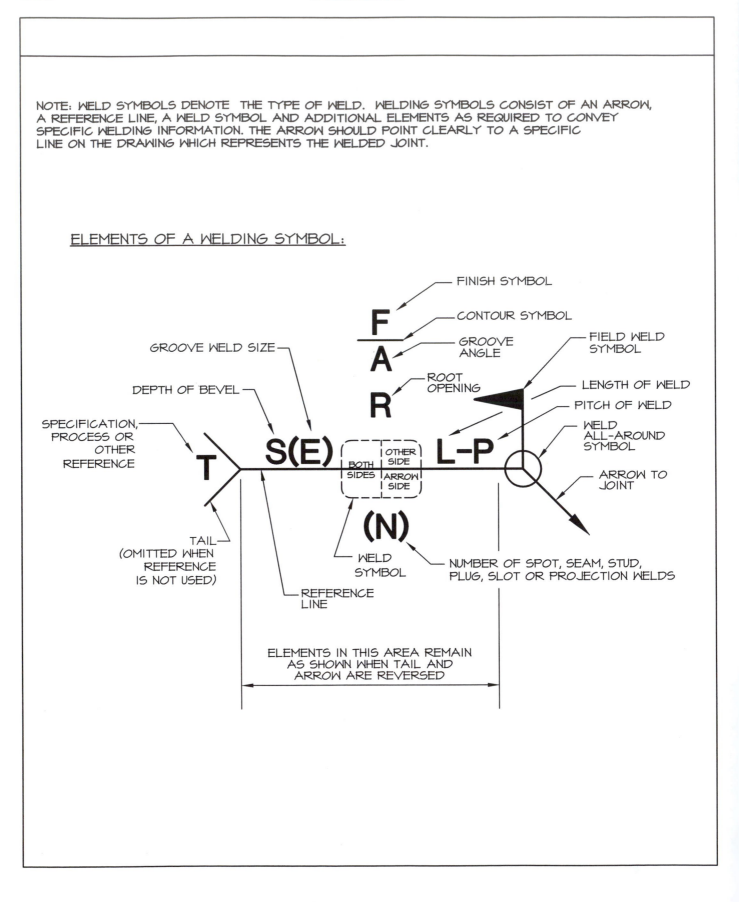

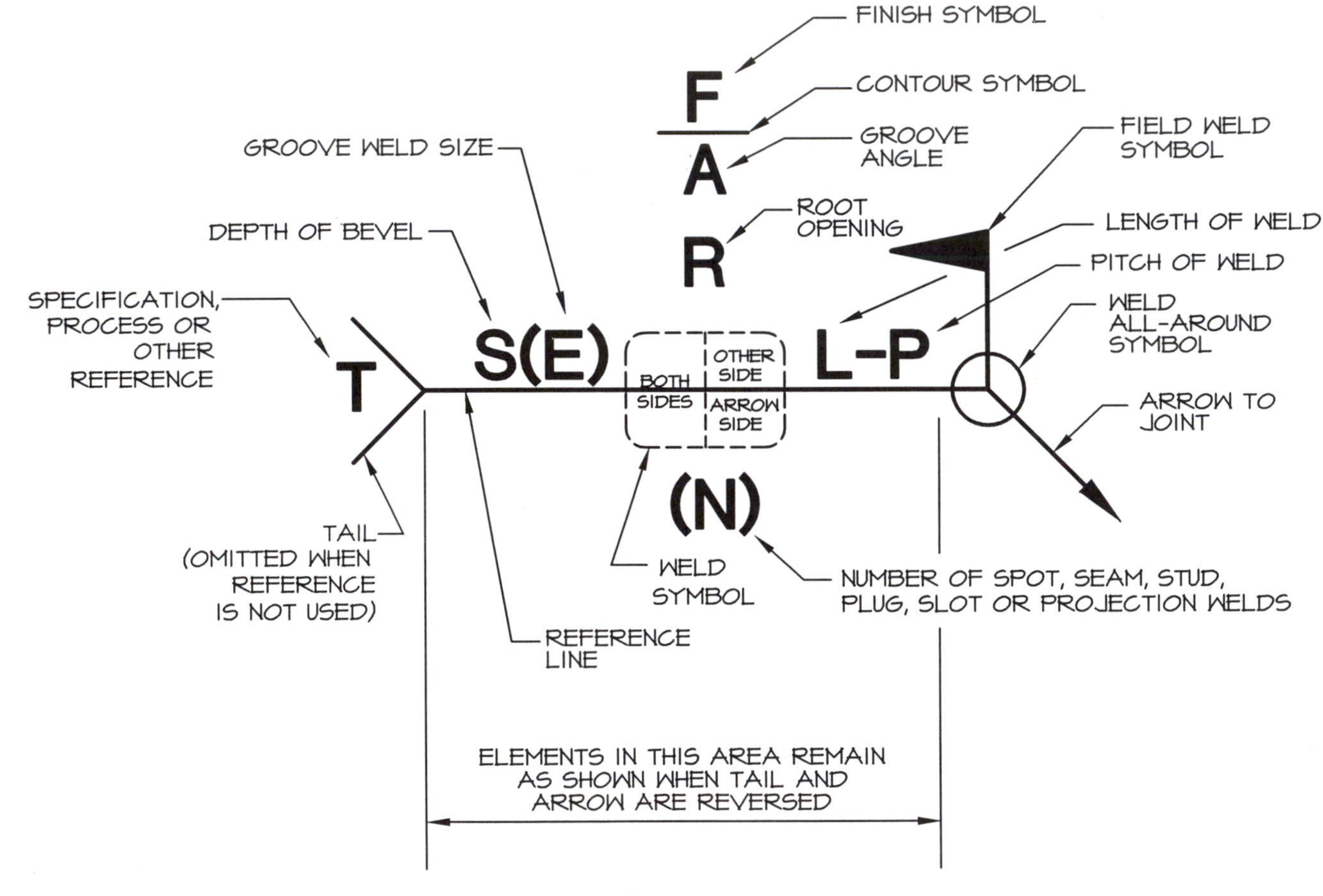

WELD	SYMBOL	WELD X-SECTION	SYMBOL DIMS.
GROOVE WELDS GROOVE WELDS ARE A TYPE OF WELD MADE IN THE GROOVE OF THE PIECES TO BE WELDED. WELD SYMBOLS, BELOW, ARE SHOWN ON A SEGMENT OF WELDING SYMBOL REFERENCE LINE.			
SQUARE WELD	[A] 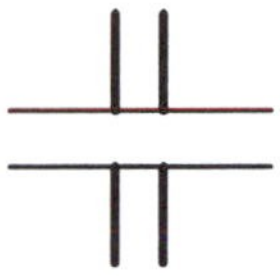	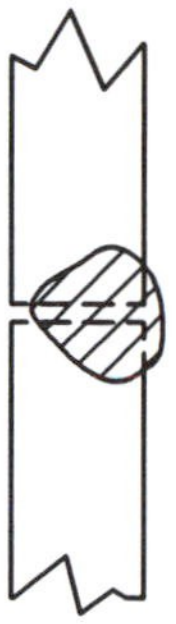	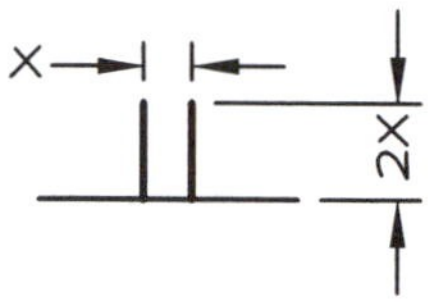 X 2X
SCARF WELD	[A] 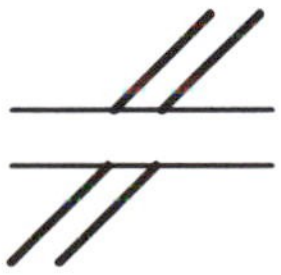	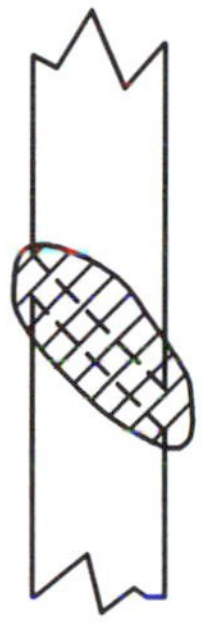	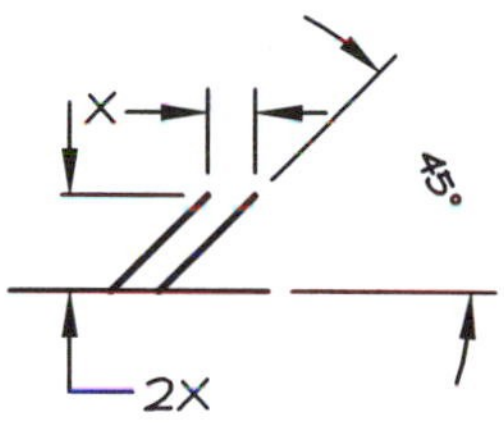 X 45° 2X
V WELD	[A] 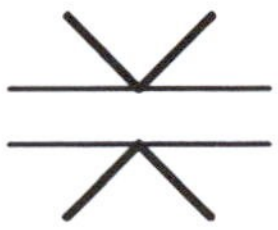	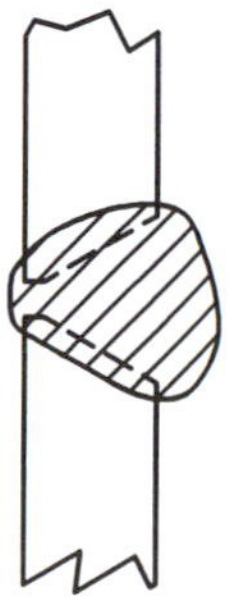	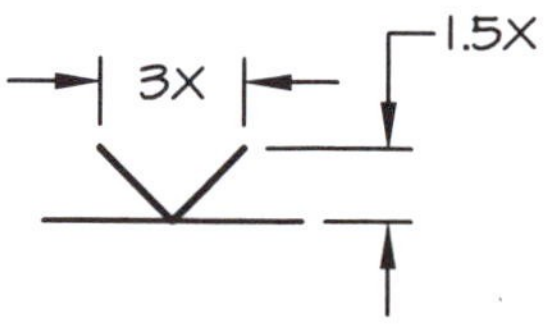 3X 1.5X

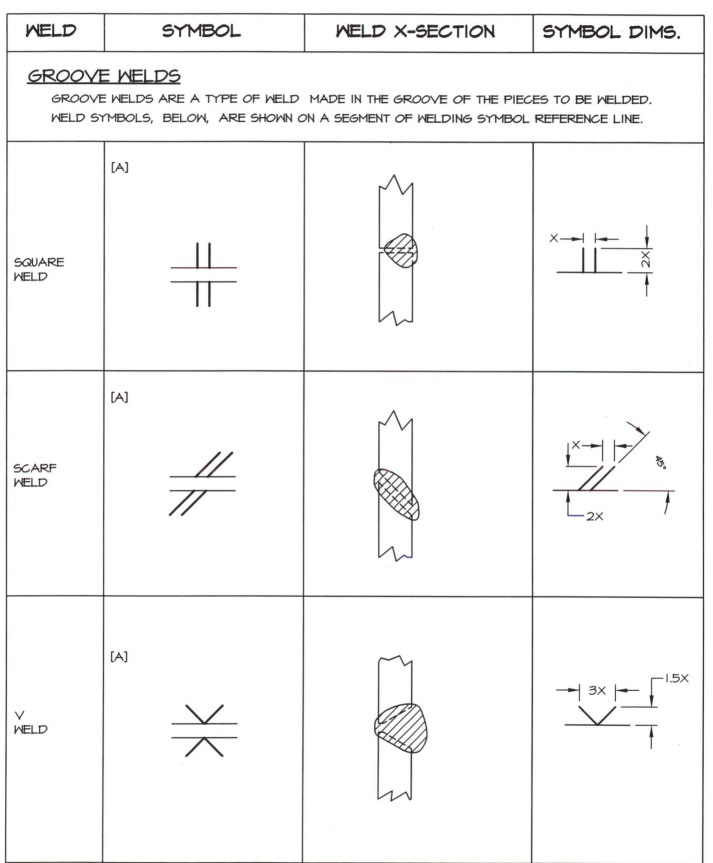

WELD	SYMBOL	WELD X-SECTION	SYMBOL DIMS.
BEVEL WELD	[A] 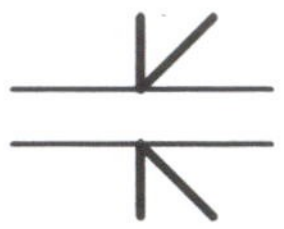		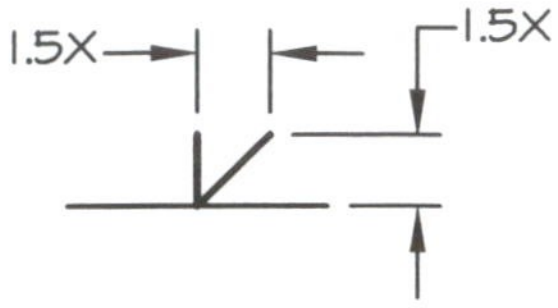 1.5X 1.5X
U WELD	[A] 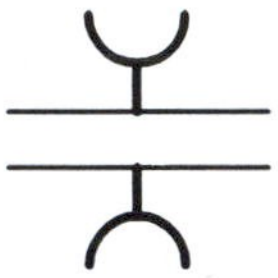	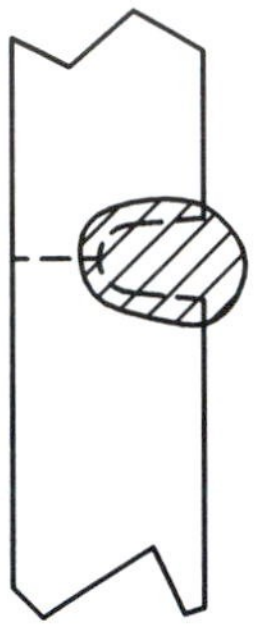	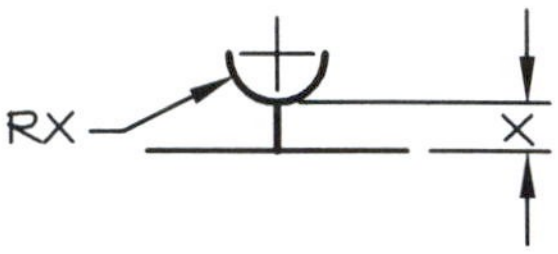 RX X
J WELD	[A] 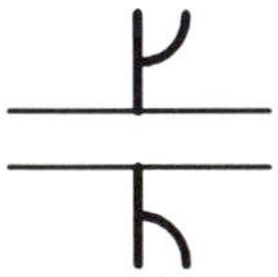	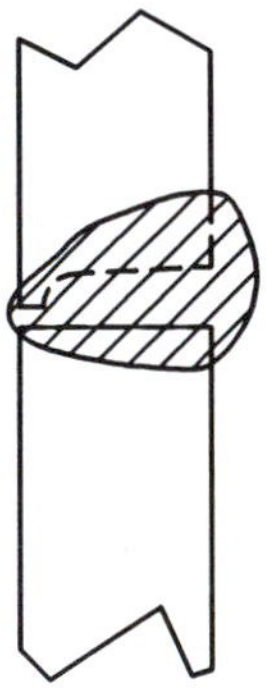	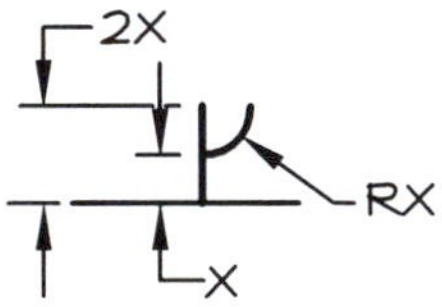 2X RX X
FLARE-V WELD	[A] 	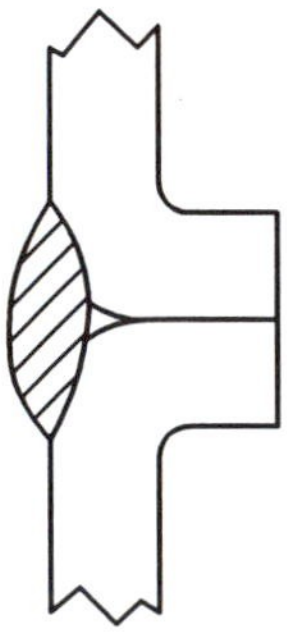	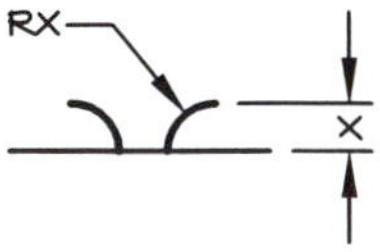 RX X

WELD	SYMBOL	WELD X-SECTION	SYMBOL DIMS.
SURFACE WELDS SURFACE WELDS ARE A WELD TYPE IN WHICH BEADS ARE DEPOSITED ON A SURFACE. WELD SYMBOLS, BELOW, ARE SHOWN ON A SEGMENT OF WELDING SYMBOL REFERENCE LINE.			
FLARE-BEVEL WELD	[A] 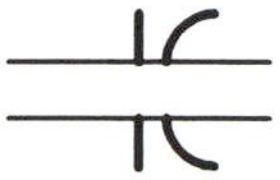	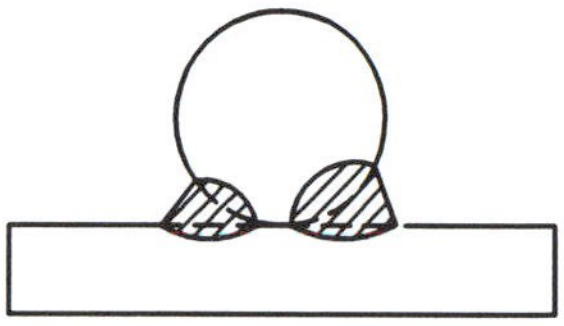	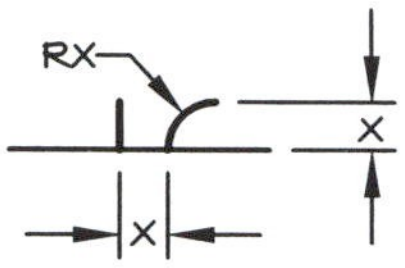 RX X X
FILLET WELD	[A] 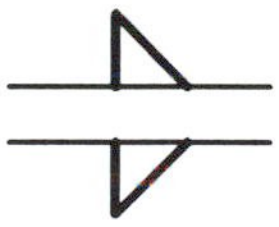	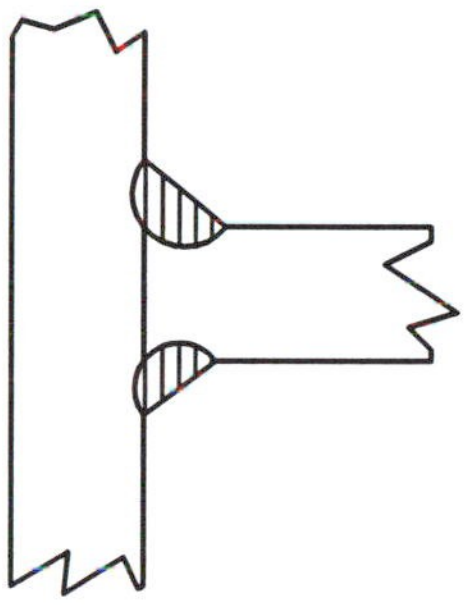	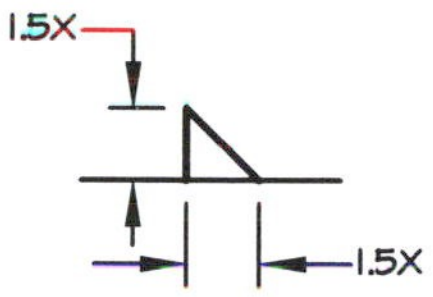 1.5X 1.5X
PLUG OR SLOT WELD	[A] 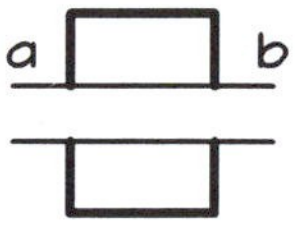 a b	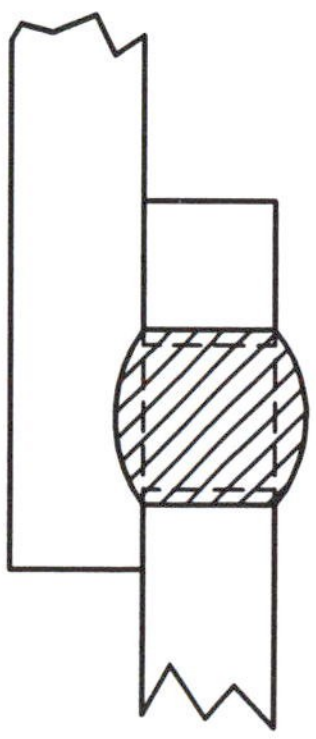 ARROW SIDE PLUG WELD	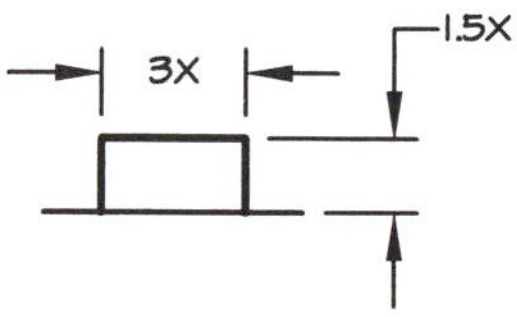 3X 1.5X PLUG & SLOT WELDS ARE SHOWN ON THE SAME SIDE OF THE REFERENCE LINE AS THE WELD SYMBOL.

WELD	SYMBOL	WELD X-SECTION	SYMBOL DIMS.
STUD WELD	[A] 	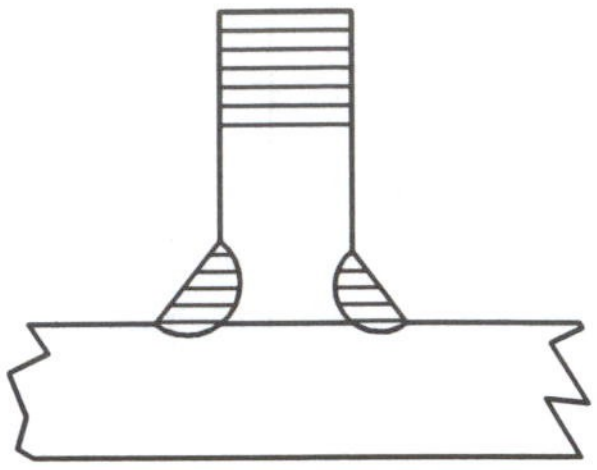	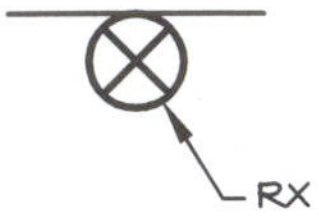 RX
SPOT OR PROJECTION WELD	[A] 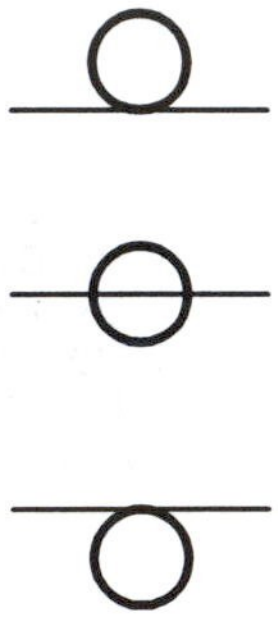	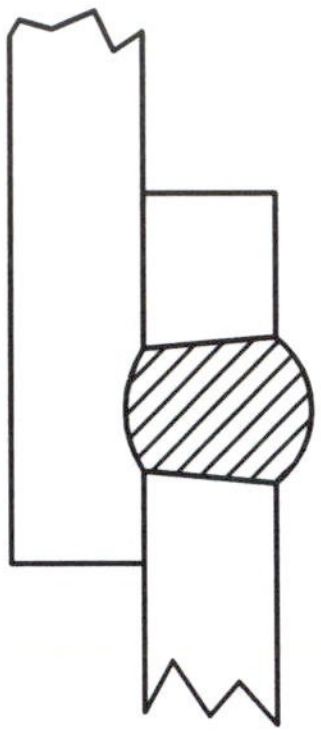 ARROW SIDE SPOT WELD	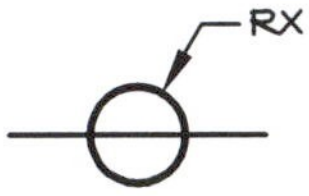 RX
SEAM WELD	[A] 	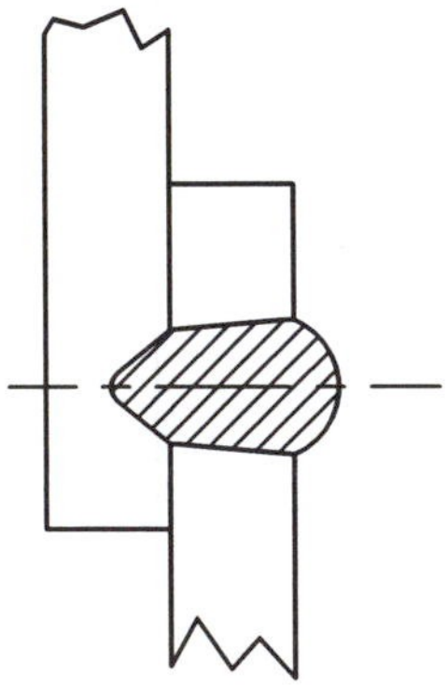 ARROW SIDE SEAM WELD	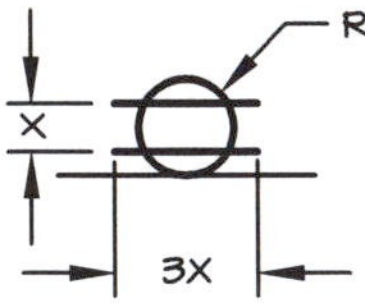 RX X 3X

WELD	SYMBOL	WELD X-SECTION	SYMBOL DIMS.
BACK OR BACKING WELD	[A] 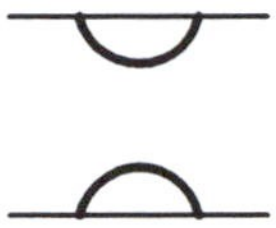	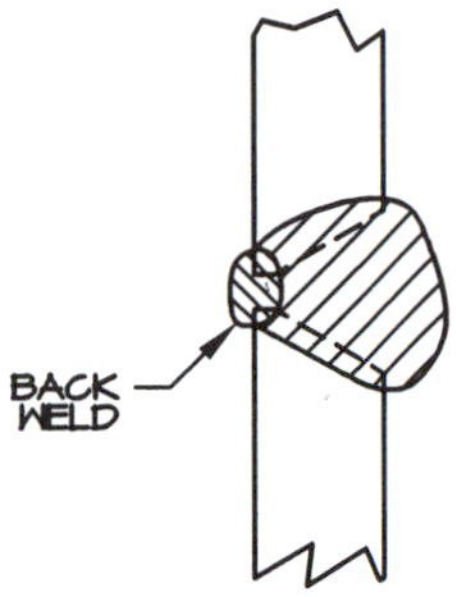 BACK WELD	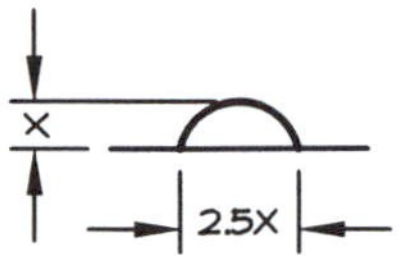 X 2.5X
SURFACING	[A] 	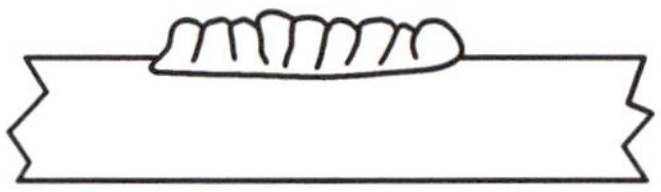	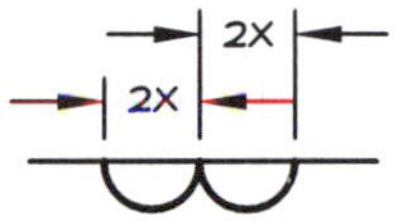 2X 2X A SURFACING WELD IS A TYPE OF WELD COMPOSED OF ONE OR MORE STRINGER OR WEAVE BEADS, DEPOSITED ON AN UNBROKEN SURFACE TO OBTAIN DESIRED PROPERTIES OR DIMENSIONS.
FLANGE WELD: EDGE	[A] 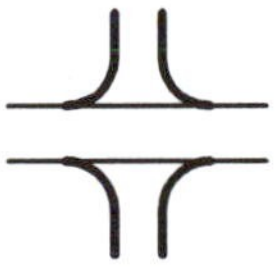	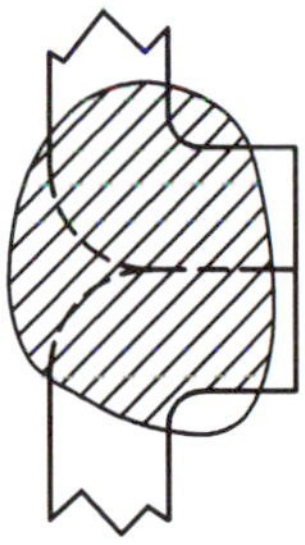	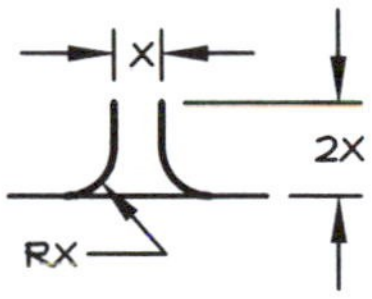 X 2X RX
CORNER	[A] 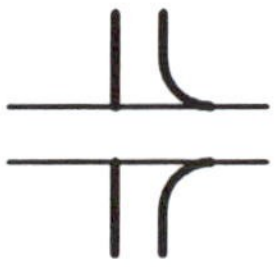	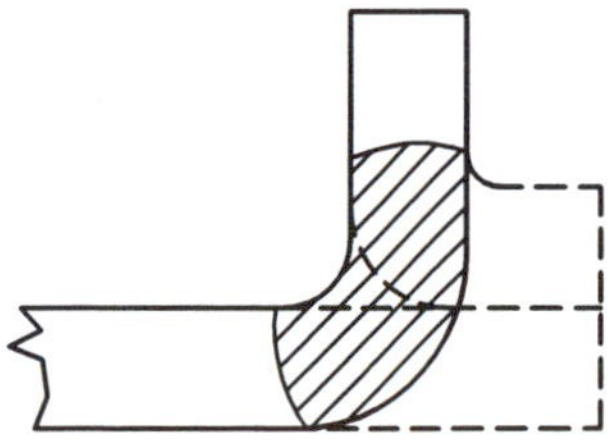	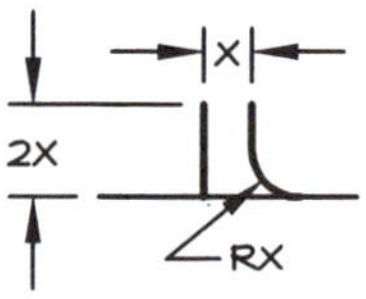 X 2X RX

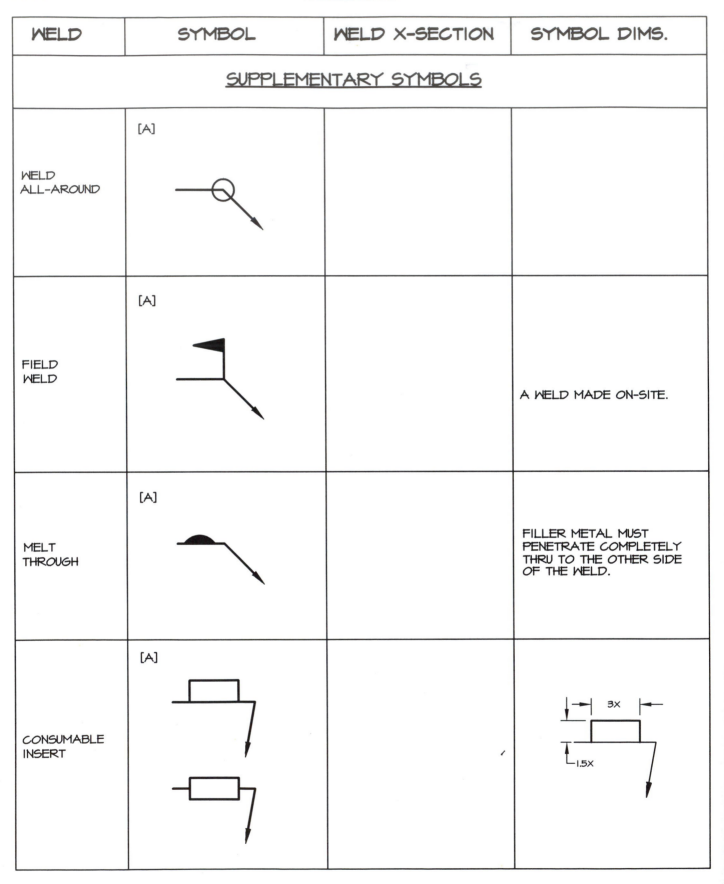

WELD	SYMBOL	WELD X-SECTION	SYMBOL DIMS.
SUPPLEMENTARY SYMBOLS			
WELD ALL-AROUND	[A] 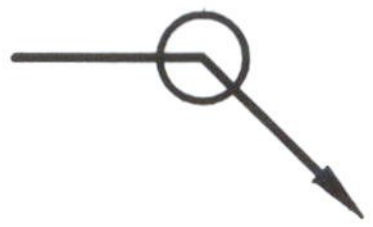		
FIELD WELD	[A] 		A WELD MADE ON-SITE.
MELT THROUGH	[A] 		FILLER METAL MUST PENETRATE COMPLETELY THRU TO THE OTHER SIDE OF THE WELD.
CONSUMABLE INSERT	[A] 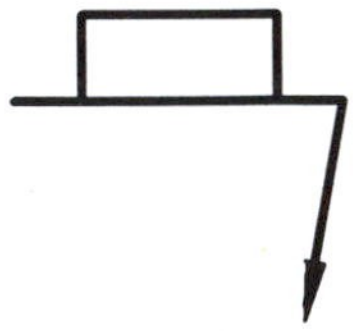 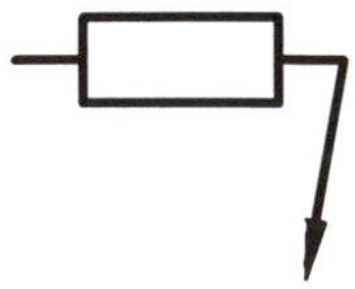		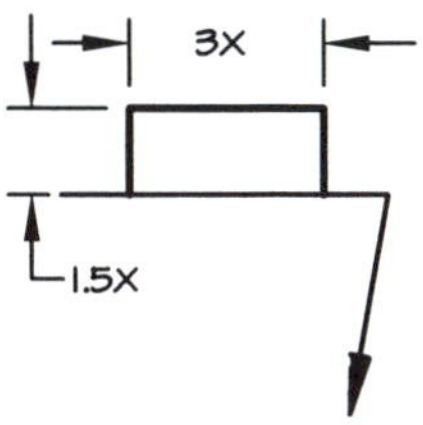 3X 1.5X

WELD	SYMBOL	WELD X-SECTION	SYMBOL DIMS.
CONSUMABLE INSERT	[A] 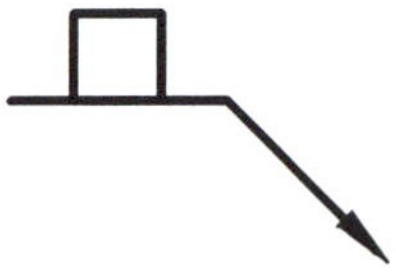	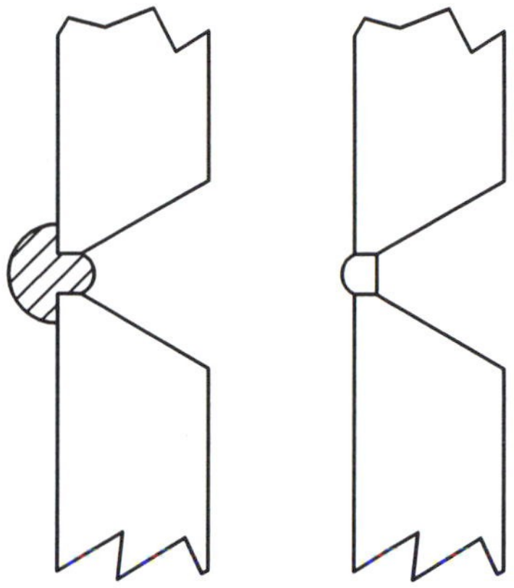 WELD WITH INSERT FINISHED WELD	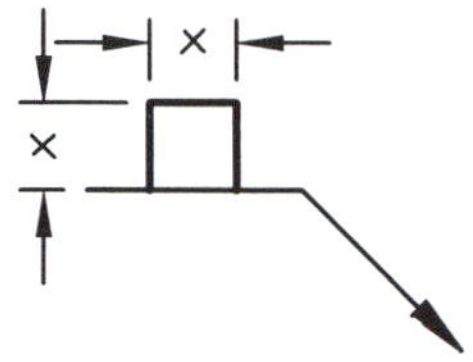 X X
CONTOUR FLUSH OR FLAT CONVEX CONCAVE	[A] [A] [A]	G MEANS	2.5X .5X .5X CONTOUR REFERS TO THE SURFACE OF THE FINISHED WELD. STANDARD FINISH SYMBOLS: C = CHIPPING M = MACHINING G = GRINDING WELDS THAT ARE TO BE WELDED FLAT WITHOUT MECHANICAL FINISHING USE NO FINISH SYMBOL.

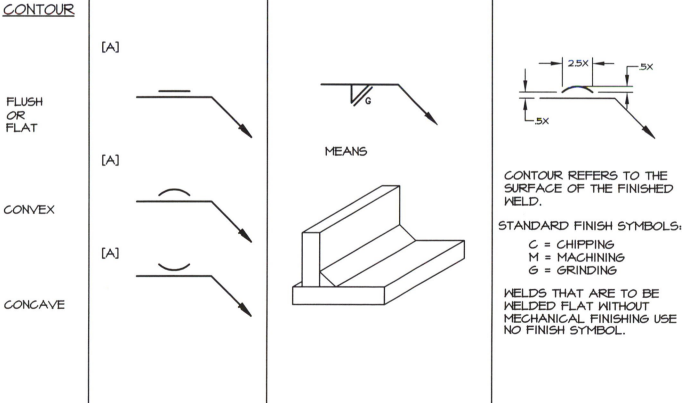

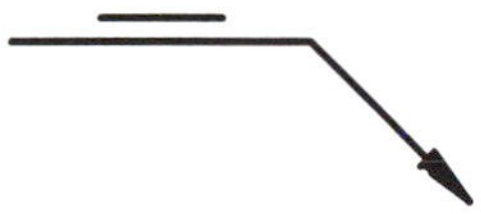

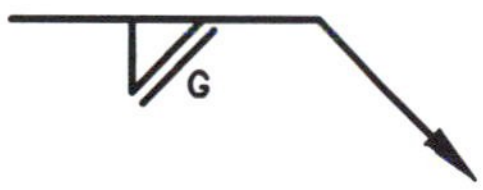

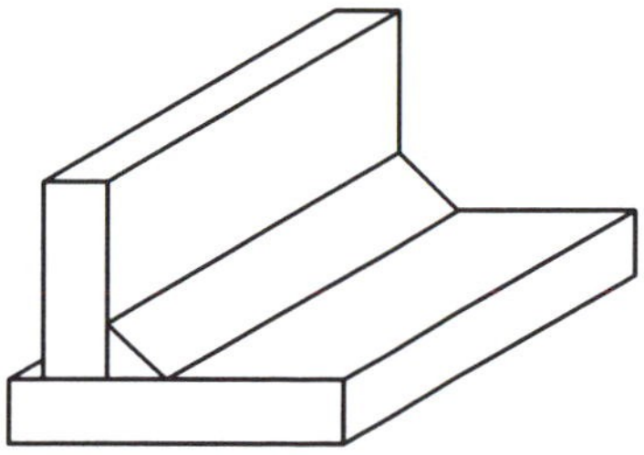

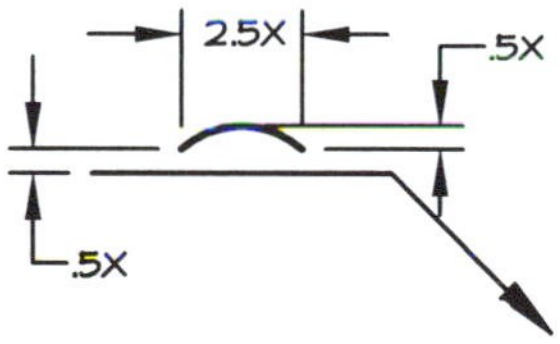

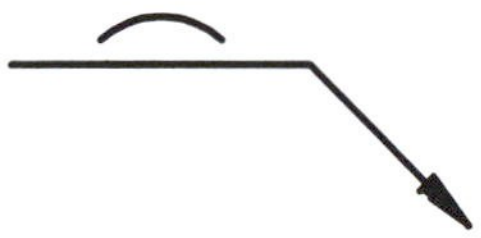

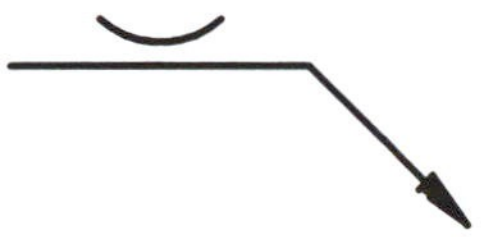

PLUMBING & HEATING SYMBOLS

NAME	SYMBOL	OBJECT	REMARKS
FLOOR SINK			
MOP SINK			
FLOOR DRAIN	   F.D. 		
OVERFLOW	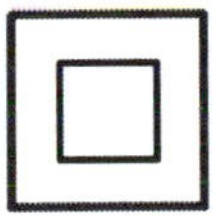		
ROOF DRAIN	 	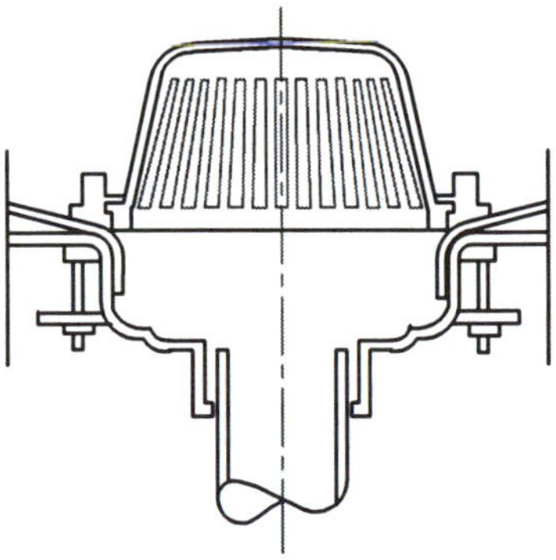	
FLOOR CLEAN-OUT			ABBREVIATED F.C.O.

NAME	SYMBOL	ELEVATION	PERSPECTIVE	REMARKS
RECTANGULAR DUCT	[A] 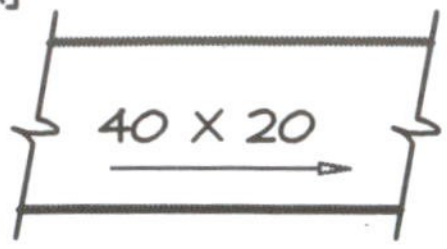 40 X 20	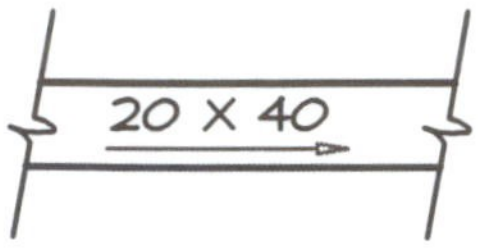 20 X 40	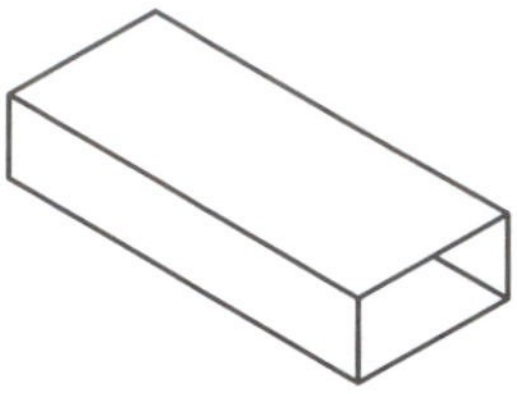	FIRST NUMBER REFERS TO WIDTH OF VISIBLE SIDE.
ROUND DUCT	 12"Φ		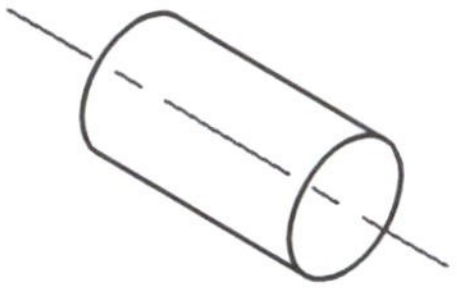	
SQUARE DUCT	 12"◫		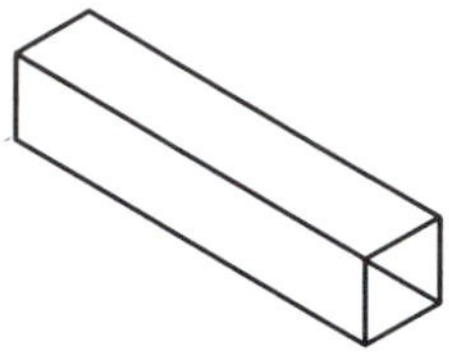	
INCLINED DROP IN RESPECT TO AIR FLOW	[A] 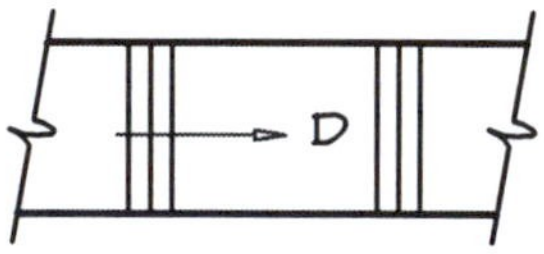 D		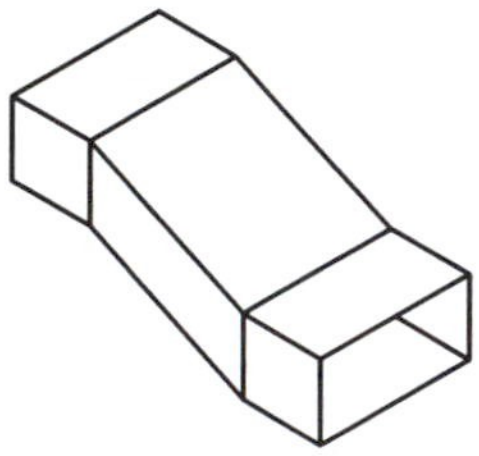	
INCLINED RISE IN RESPECT TO AIR FLOW	[A] 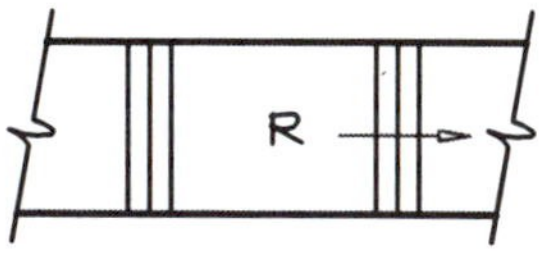 R		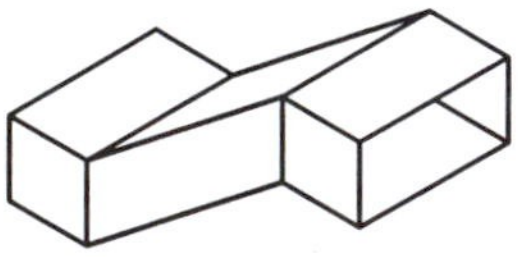	
DUCT SECTION, NEGATIVE PRESSURE (RETURN AIR)	[A] 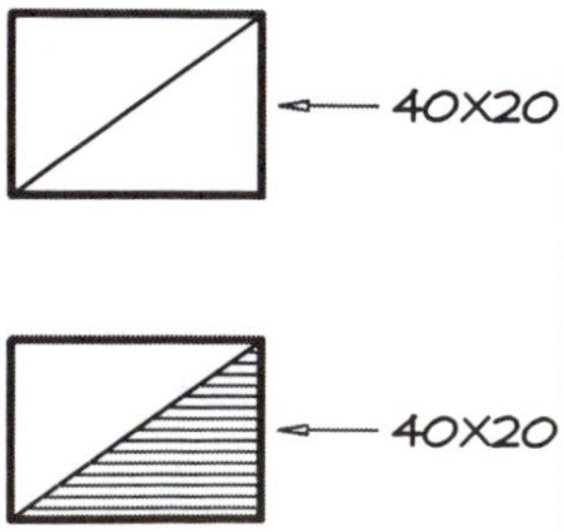 40X20 40X20			IN SECTION VIEWS, FIRST DIMENSION IS TOP PLANE.

NAME	SYMBOL	ELEVATION	PERSPECTIVE	REMARKS
DUCT SECTION, POSITIVE PRESSURE	[A] 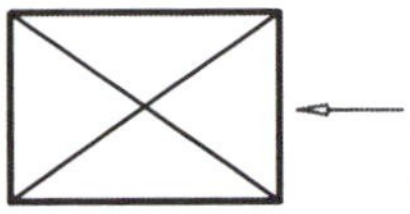 WIDTH X HEIGHT 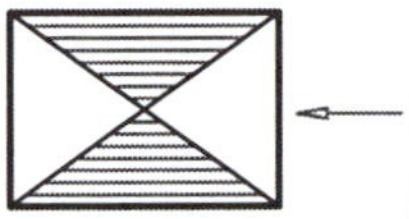 WIDTH X HEIGHT			SUPPLY DUCT SECTION
EXHAUST INLET, CEILING	[A] 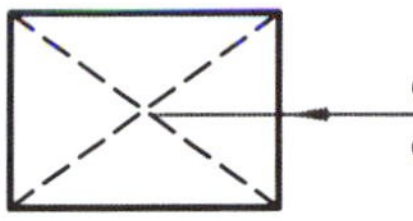 C 20X12 - 7-CFM CC 20X12-700 CFM			
ACCESS DOORS	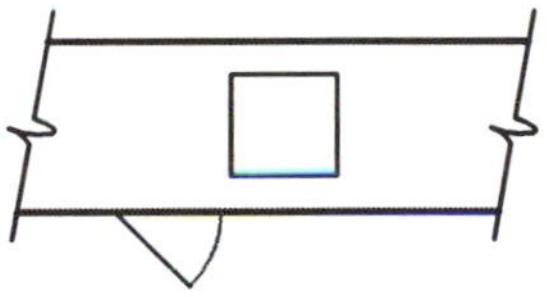 [A] 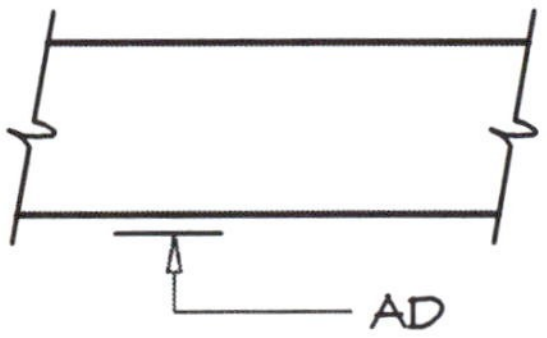 AD	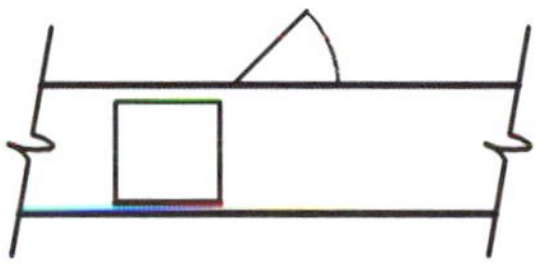	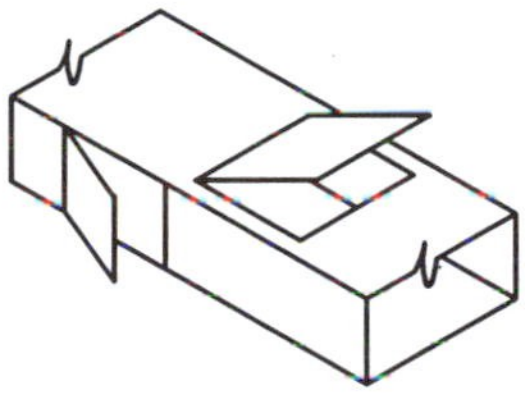	VERTICAL ACCESS DOOR SHOWN ON TOP, HORIZONTAL ACCESS DOOR SHOWN ON SIDE IN PERSPECTIVE VIEW.
ACOUSTICAL LINING (INSULATION)	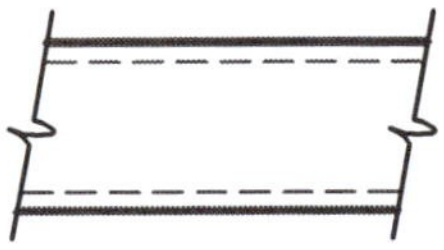	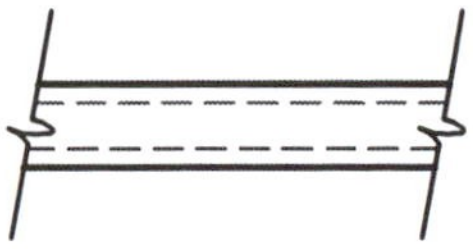	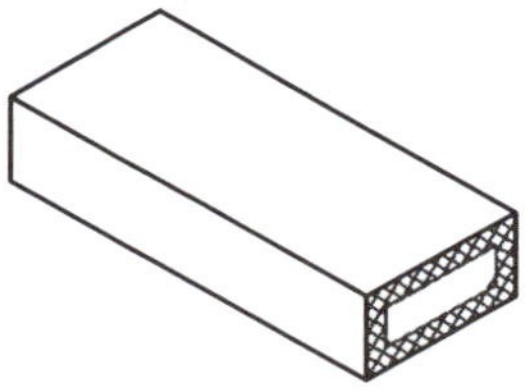	

DUCTWORK

NAME	SYMBOL	ELEVATION	PERSPECTIVE	REMARKS
DUCT TRANSITION	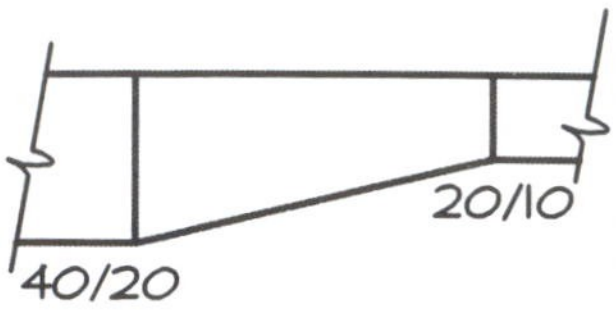 40/20 20/10	 40/20 20/10	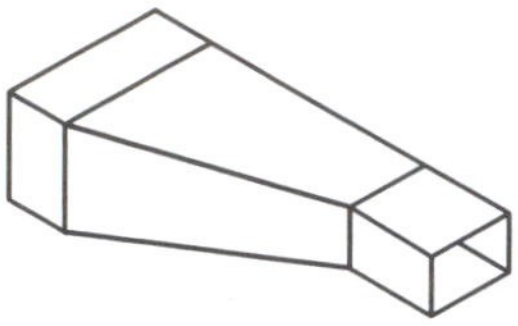	INDICATE AS FOB IF FLAT ON BOTTOM, OR FOT IF FLAT ON TOP.
TURNING VANES	[A] 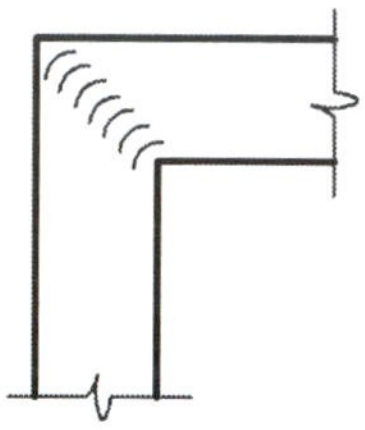 [A] 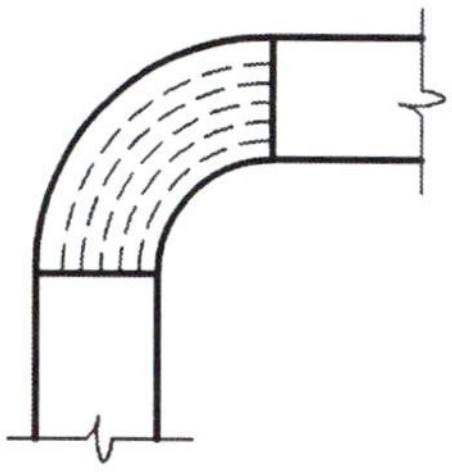	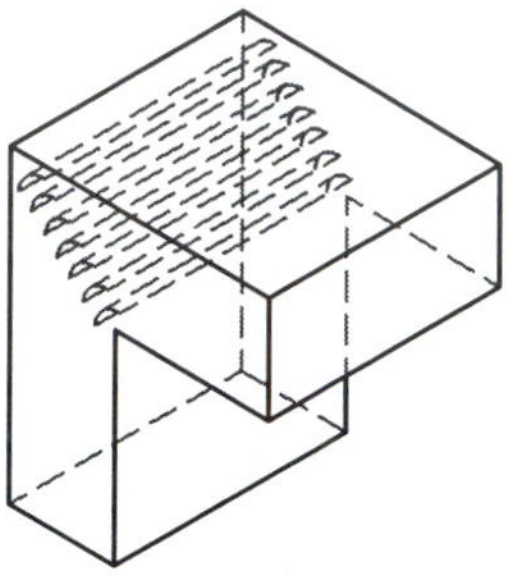 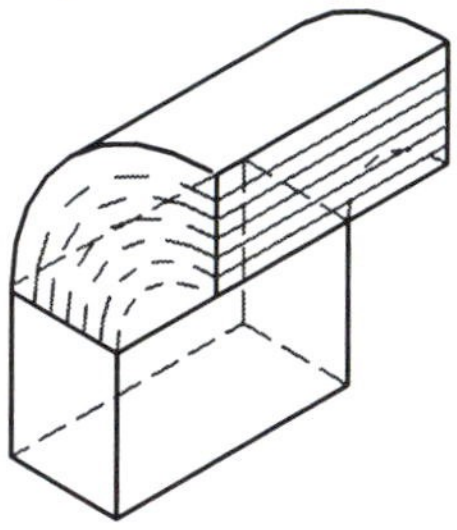		SHOWING 90° AND CURVED ELBOW. (UPPER DUCT REMOVE FOR CLARITY.)
FLEXIBLE CONNECTION	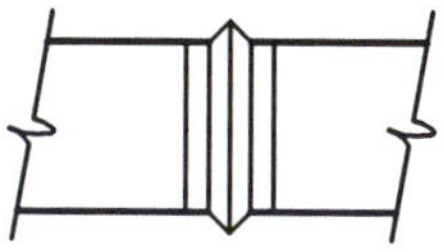 [A] 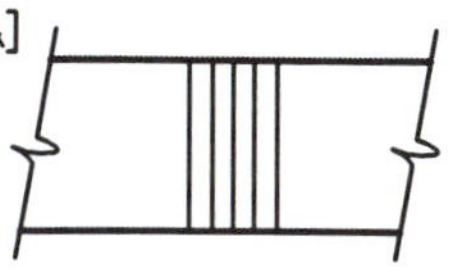	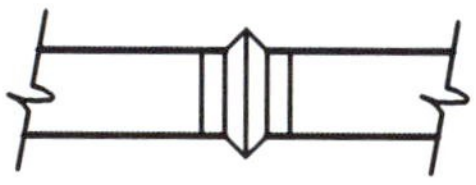 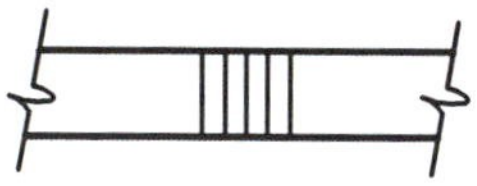	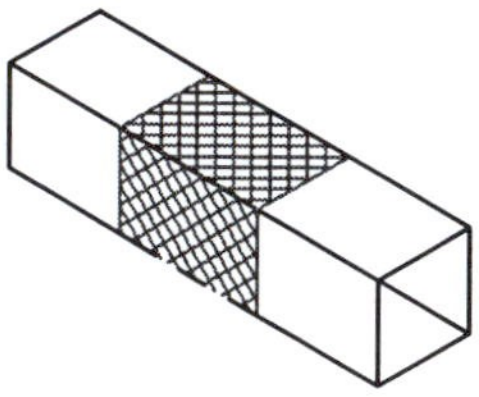	CONNECTIVE MATERIAL MAY BE CANVAS.

NAME	SYMBOL	OBJECT	REMARKS
FIRE DAMPER	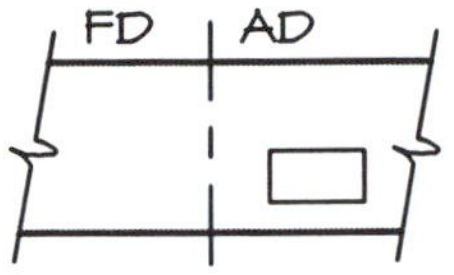 FD AD	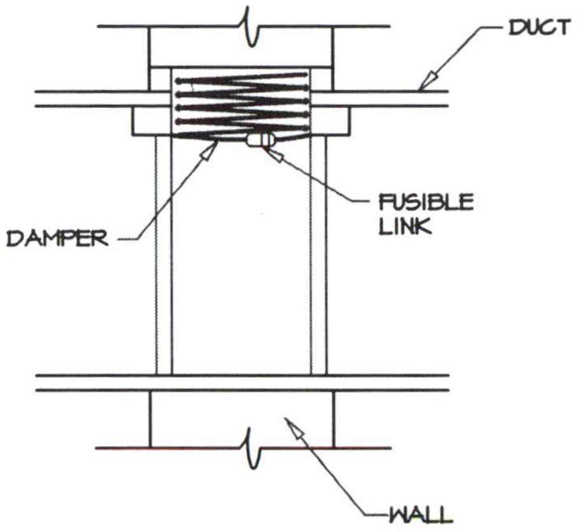 DUCT FUSIBLE LINK DAMPER WALL	INTERLOCKING BLADE TYPE ILLUSTRATED, INSTALLED AT PARTITION. REQUIRES ACCESS DOOR PER CODE.
MANUAL VOLUME DAMPER	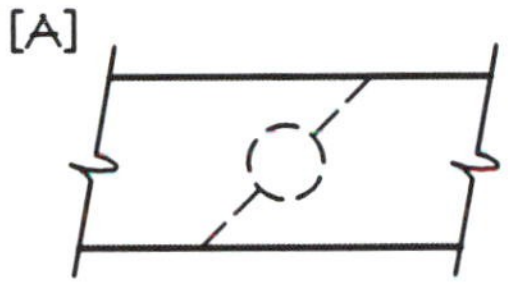 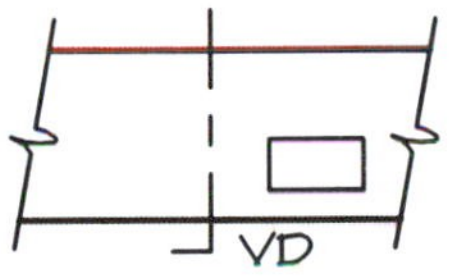 [A] VD		
AUTOMATIC VOLUME DAMPER	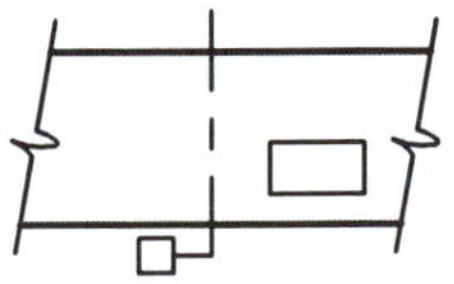		TYPICALLY ELECT. OR PNEUMATIC
AIR OUT OF REGISTER			
AIR INTO REGISTER			
AIR FLOW THRU UNDERCUT OR LOUVERED DOOR	 (A)		

NAME	SYMBOL	PLAN VIEW	REMARKS
2-WAY THROW CEILING DIFFUSER	 18" DIA CD 400 CFM	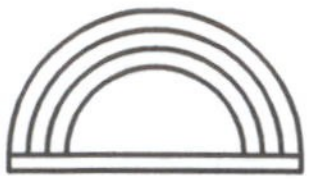	
SUPPLY OUTLET, CEILING DIFFUSER	 18" DIA CD 800 CFM		
SINGLE THROW CEILING DIFFUSER	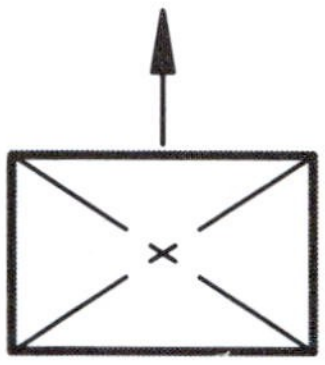 24X14 CD 300 CFM	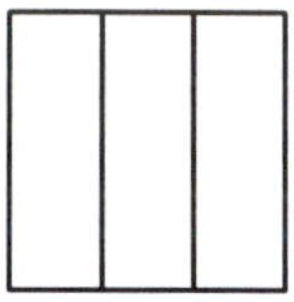	
2-WAY THROW CEILING DIFFUSER	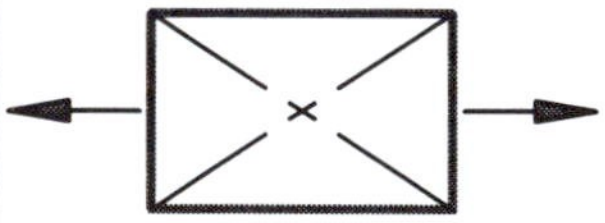 24X14 CD 600 CFM		
3-WAY THROW CEILING DIFFUSER	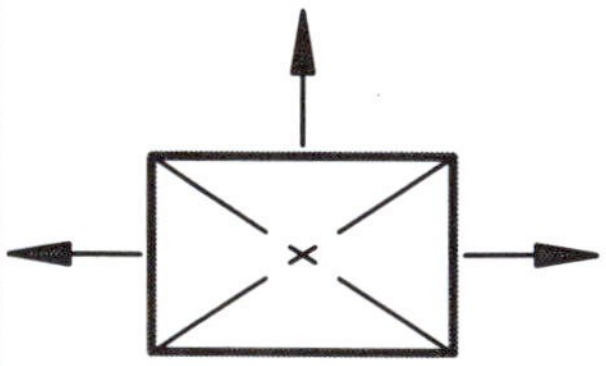 24X14 CD 600 CFM	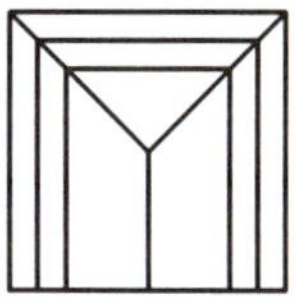	
SUPPLY OUTLET, CEILING DIFFUSER	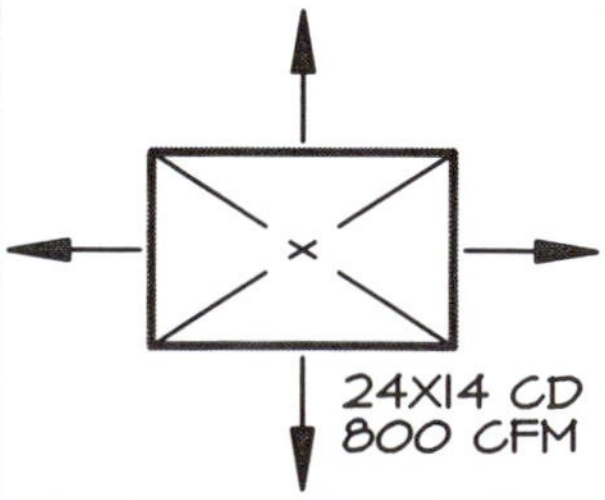 24X14 CD 800 CFM	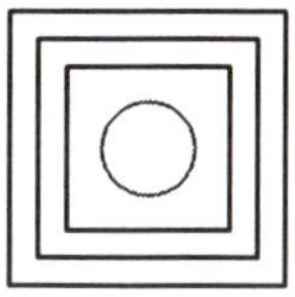	

NAME	SYMBOL	ELEVATION	PERSPECTIVE	REMARKS
VENT THRU ROOF	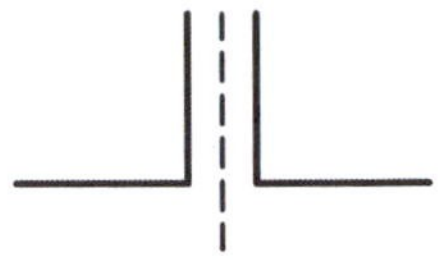			
FAN ON ROOF OR SUPPLY FAN		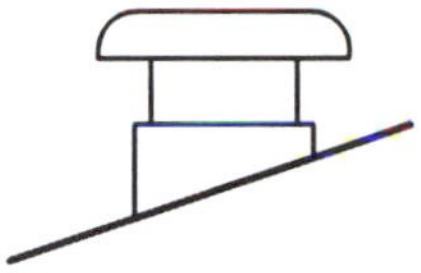 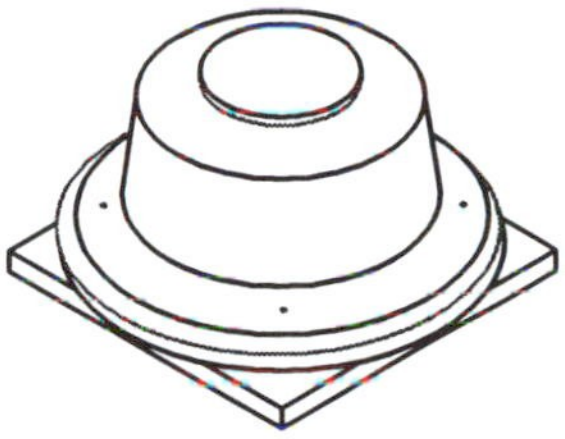	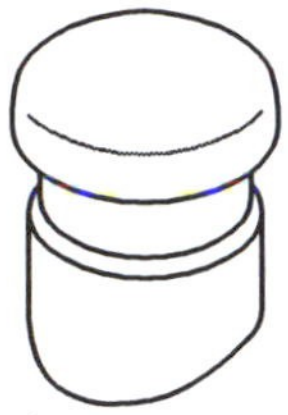	SQUARE OR ROUND DUCT DESIGNATED BY CIRCLE OR SQUARE: 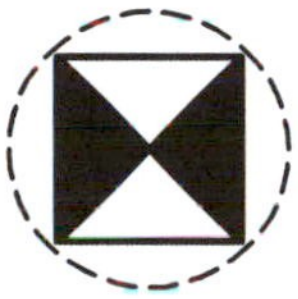 
		ROOF MOUNTED		
EXHAUST FAN		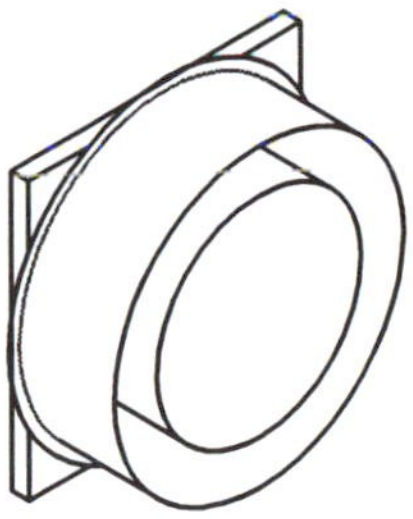 WALL MOUNTED		
LINEAR DIFFUSER	 72X6-LD 350 CFM			

DUCTWORK

NAME	SYMBOL	ELEVATION	PERSPECTIVE	REMARKS
FLOOR REGISTER	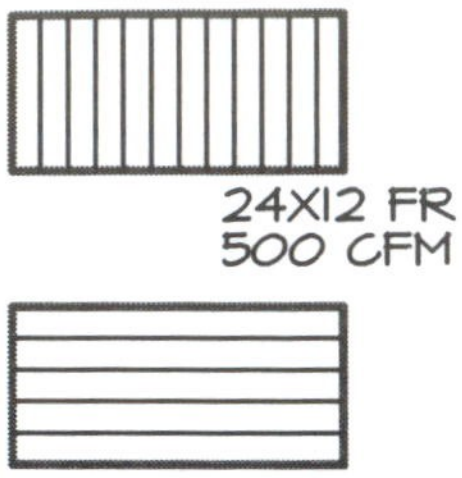 24X12 FR 500 CFM		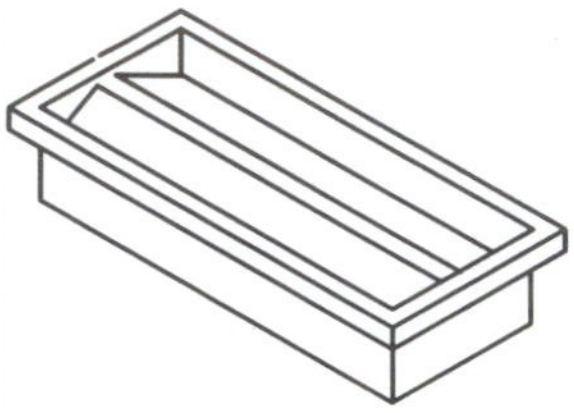	PERSPECTIVE VIEW SHOWS REGISTER WITHOUT COVER.
BASEBOARD REGISTER	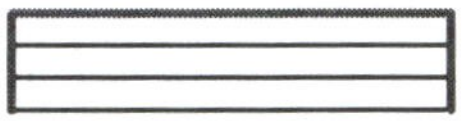	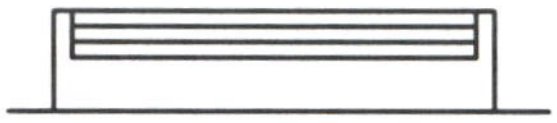	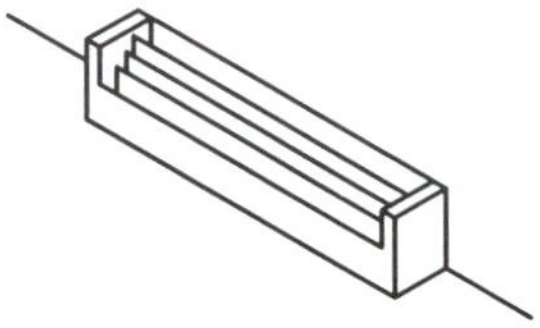	
DEFLECTING DAMPER	[A] 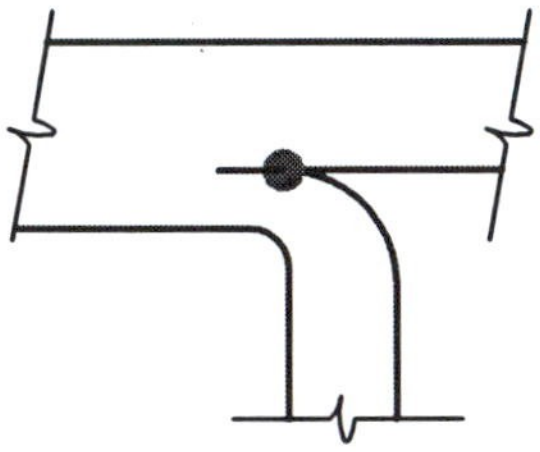	[A] 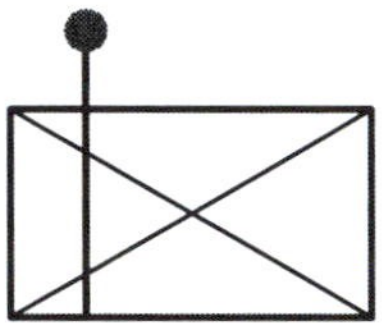		
RESIDENTIAL EXHAUST FAN		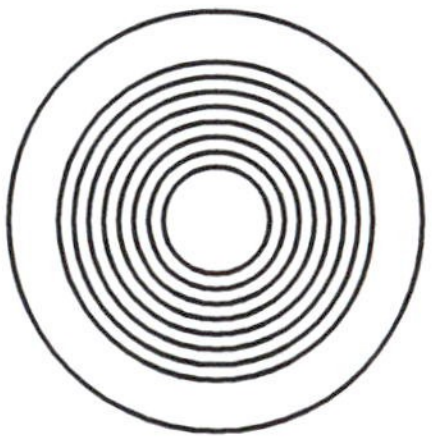	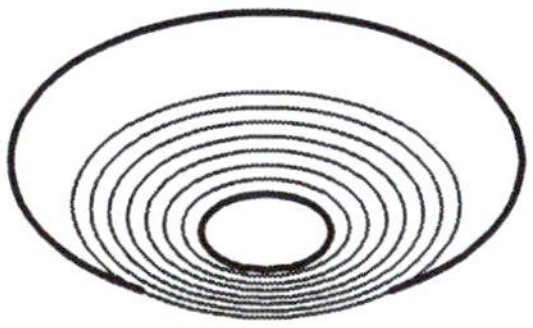	CEILING MOUNTED
UNIT HEATER - PROPELLER	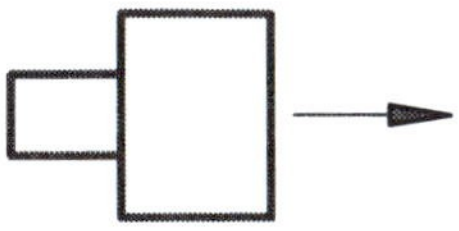			

NAME	SYMBOL	ILLUSTRATION	REMARKS
UNIT HEATER - CENTRIFUGAL			
UNIT VENTILATOR			
POINT OF CONNECTION (NEW)			INDICATES POINT OF CONN. BETWEEN NEW & EXSTG. WORK USE LEADER. 1/4" TO 3/8" Φ TYP.

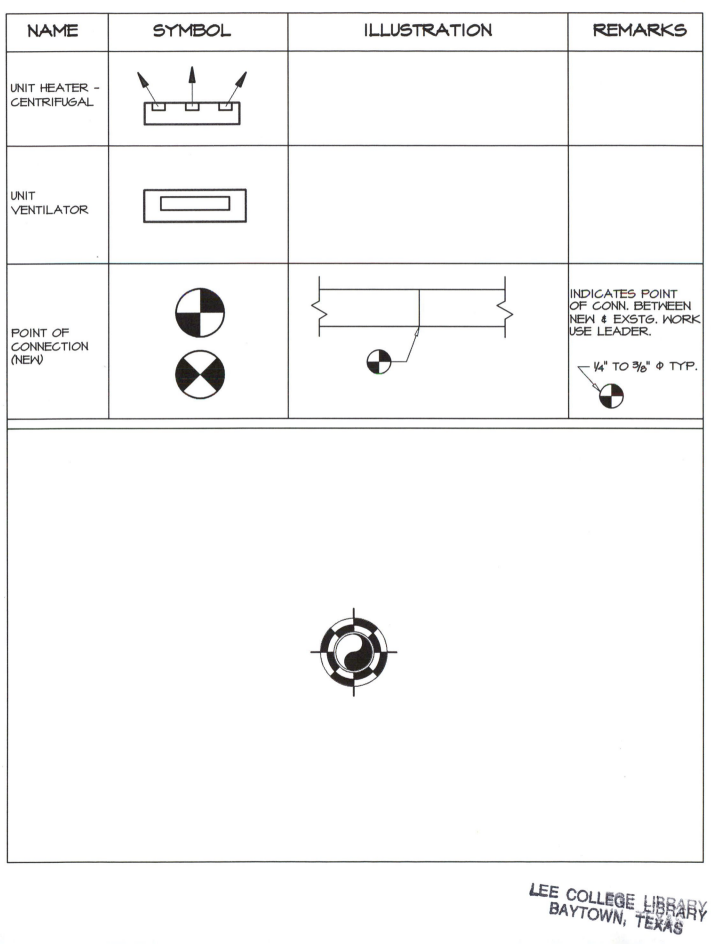

PIPING

NAME	SYMBOL	ELEVATION	PERSPECTIVE	REMARKS
PIPE	[A] 	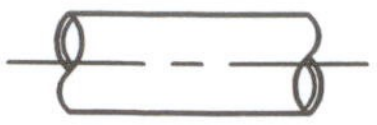 SCHEMATIC ELEVATION: 	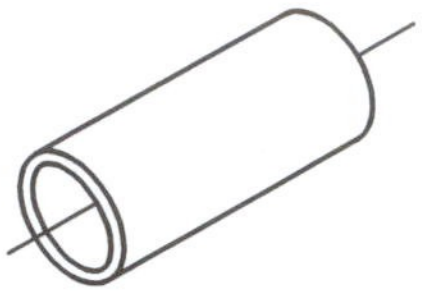	ARROW (OPTIONAL) PIPE SHOWN IN SECTION THUS: 
SLOPED PIPE	  R D	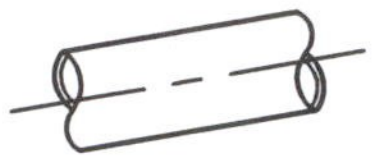 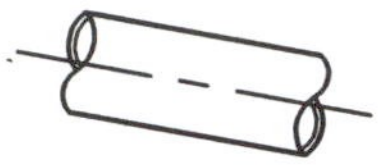		INDICATE RISE OR DROP; E.G. 3:12.
SCREWED FITTING	[A] 	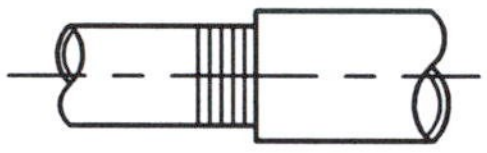	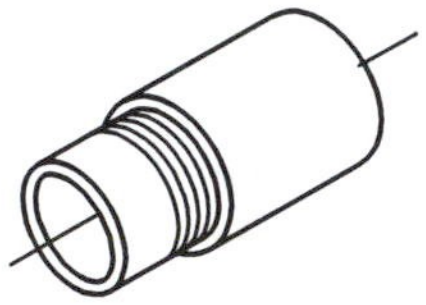	SCREW FITTING DEPICTIONTO BE USED FOR ALL FITTINGS PROVIDED TRUE CONNECTION TYPE IS SPECIFIED IN NOTES.
FLANGED FITTING	[A] 	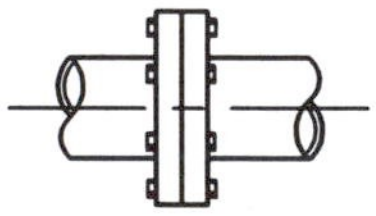	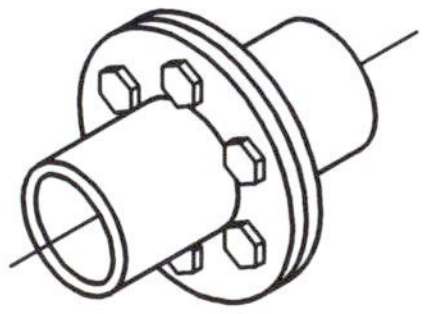	
BELL & SPIGOT FITTING	[A] 	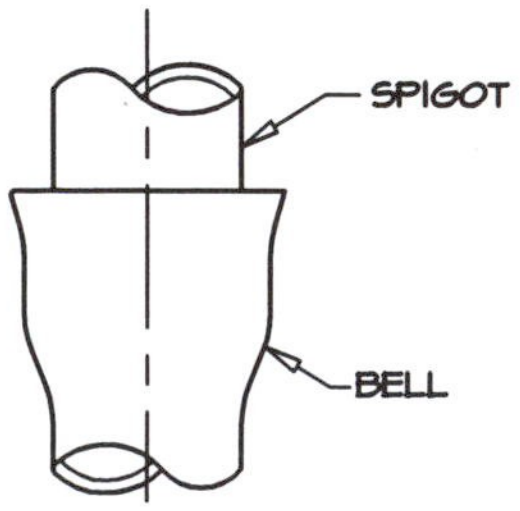 SPIGOT BELL		

NAME	SYMBOL	ELEVATION	PERSPECTIVE	REMARKS
WELDED FITTING	[A] 	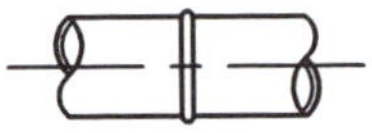	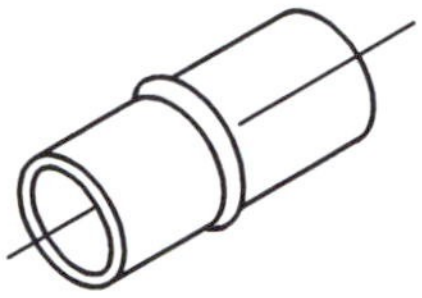	
SOLDERED FITTING	[A] 	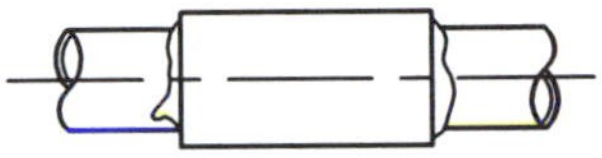	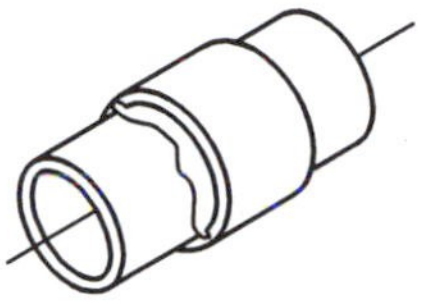	
WELDED FITTING	[A] 			
SOLDERED FITTING	[A] 			
BELL & SPIGOT FITTING	[A] 			
SOLVENT CEMENT FITTING	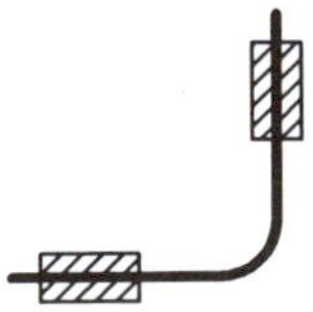			

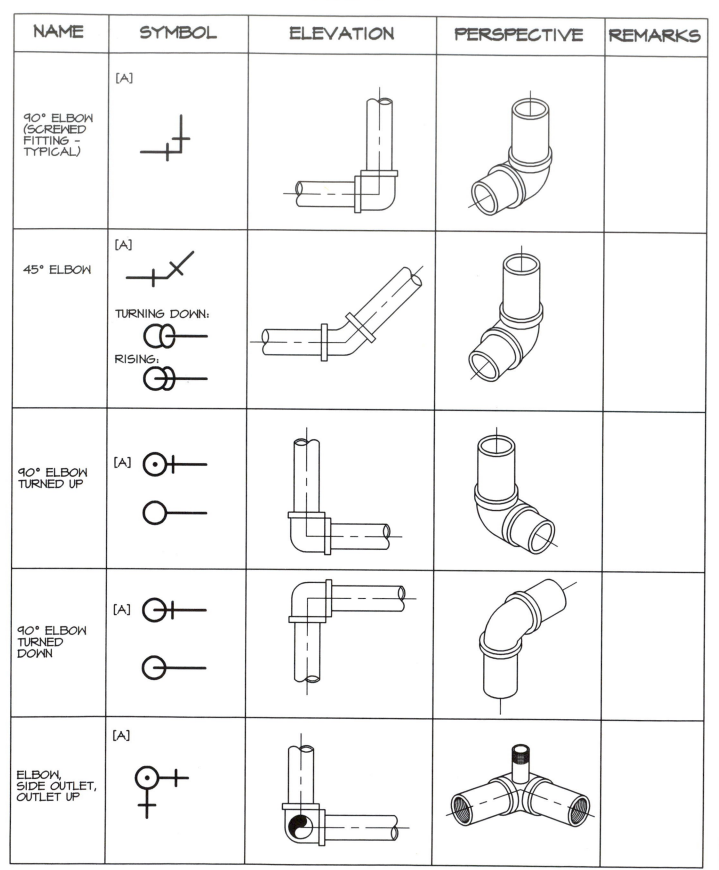

PIPING

NAME	SYMBOL	ELEVATION	PERSPECTIVE	REMARKS
90° ELBOW (SCREWED FITTING - TYPICAL)	[A] 	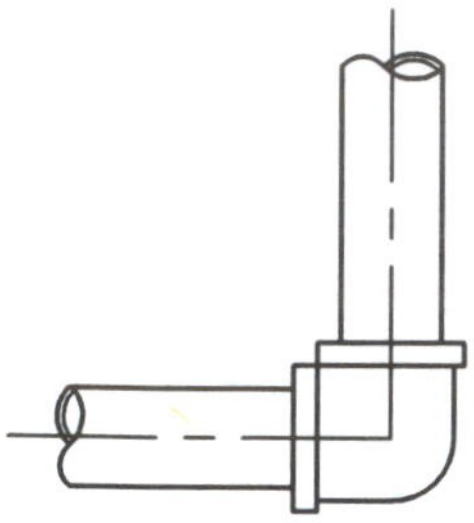	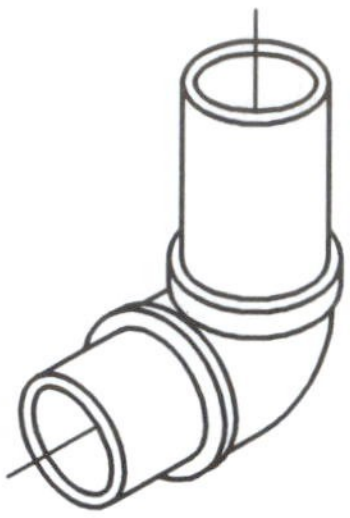	
45° ELBOW	[A]  TURNING DOWN:  RISING: 	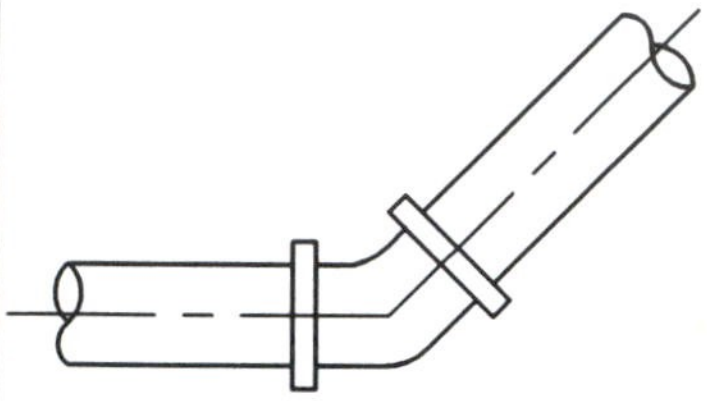	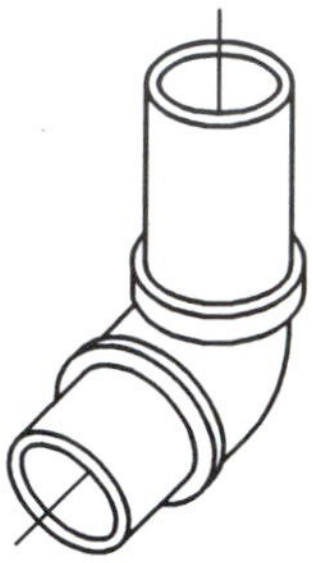	
90° ELBOW TURNED UP	[A]  	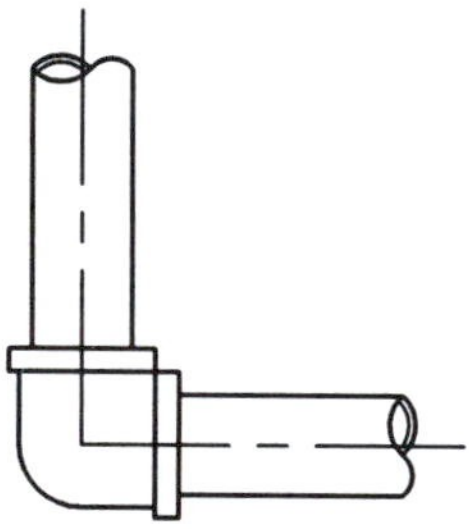	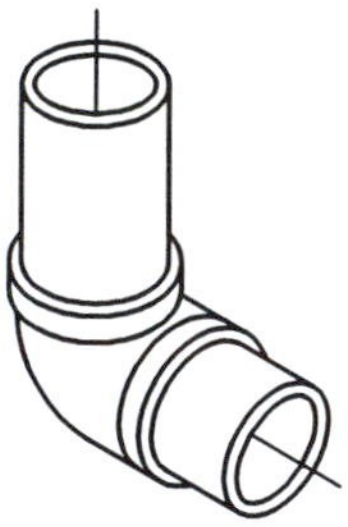	
90° ELBOW TURNED DOWN	[A]  	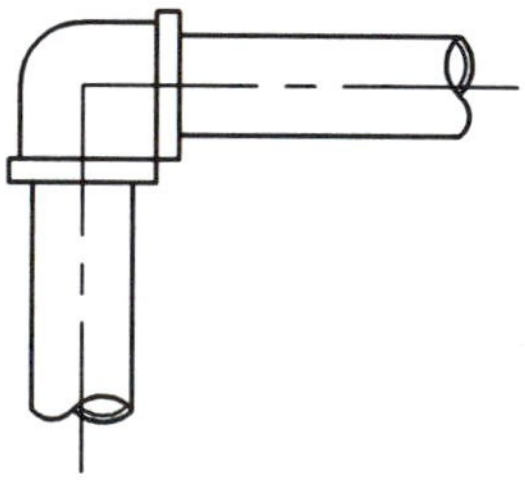	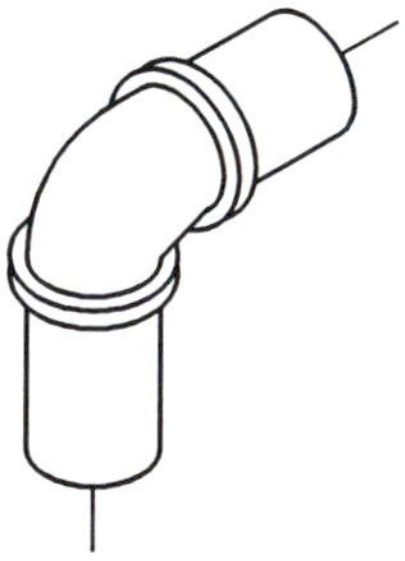	
ELBOW, SIDE OUTLET, OUTLET UP	[A] 	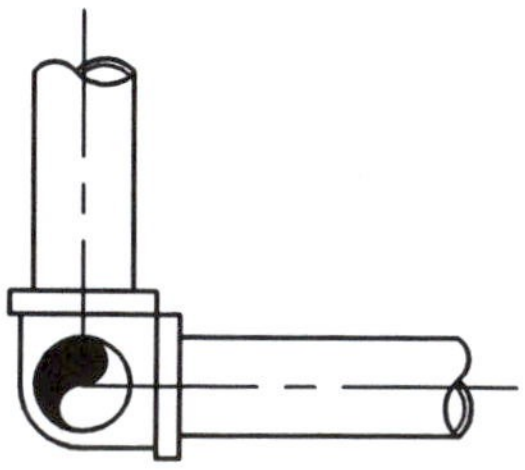	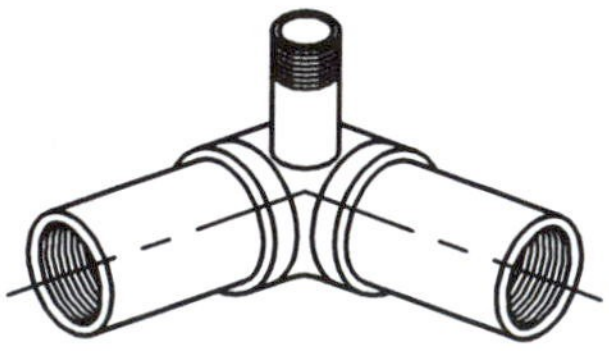	

NAME	SYMBOL	ELEVATION	PERSPECTIVE	REMARKS
ELBOW, SIDE OUTLET, OUTLET DOWN	[A] 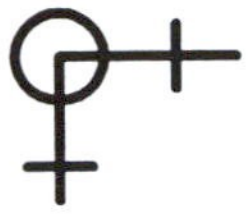	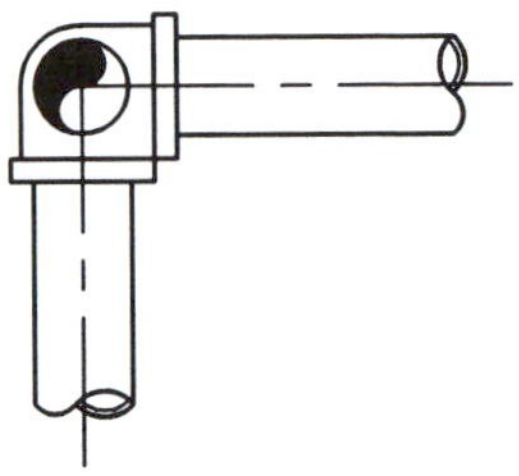	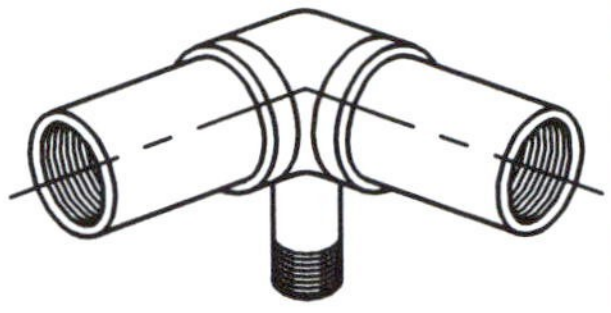	
CAP	SCREWED: [A]   BELL & SPIGOT: 	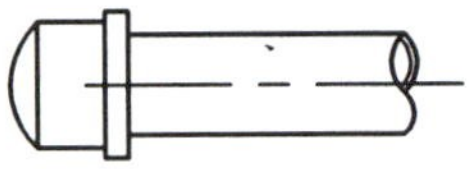	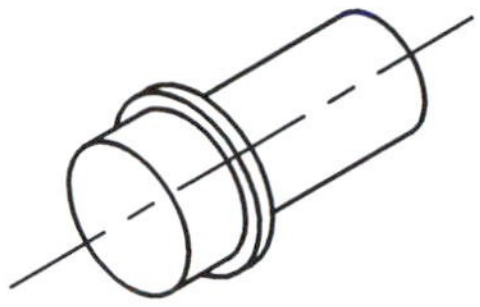	
CROSSOVER	[A] 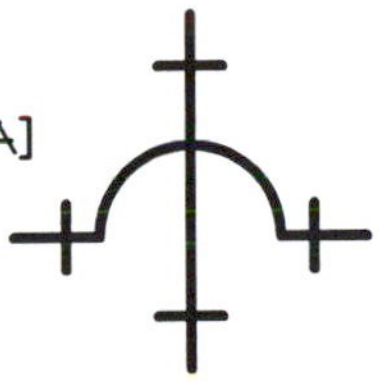 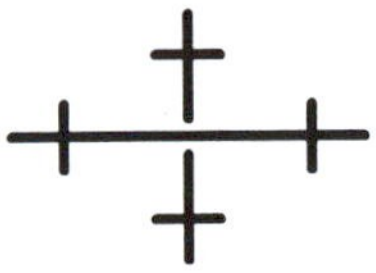	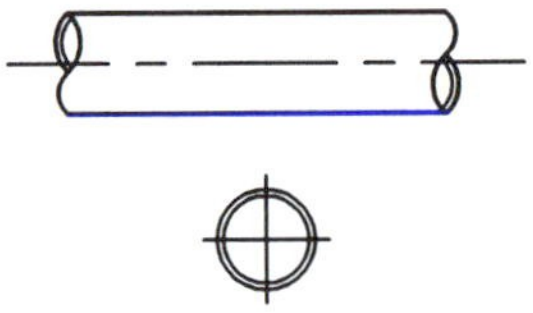	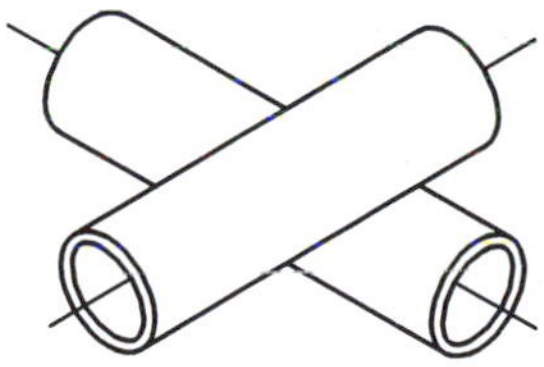	
STRAIGHT SIZE CROSS	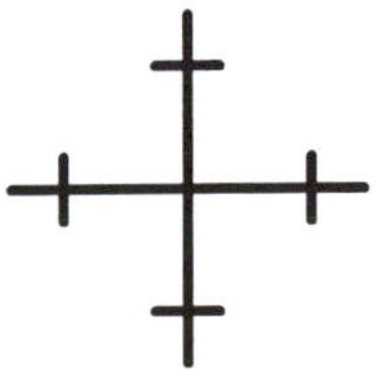	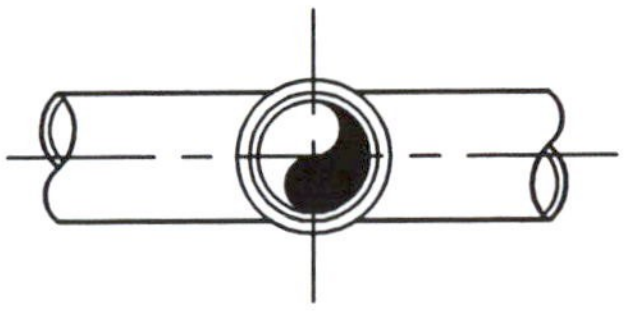	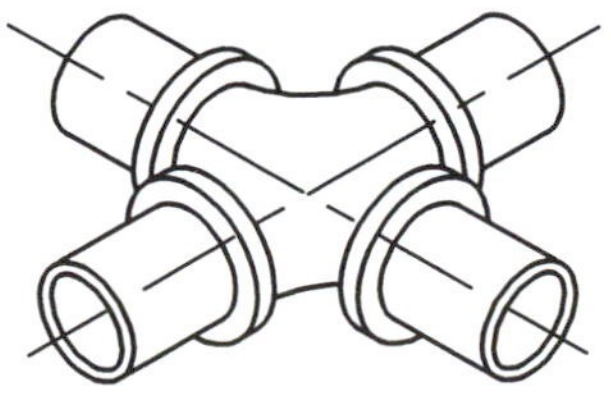	

PIPING

NAME	SYMBOL	ELEVATION	PERSPECTIVE	REMARKS
REDUCING CROSS	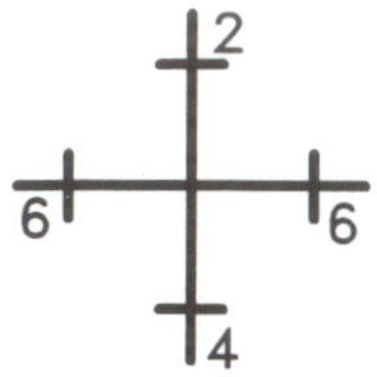 2 6 6 4	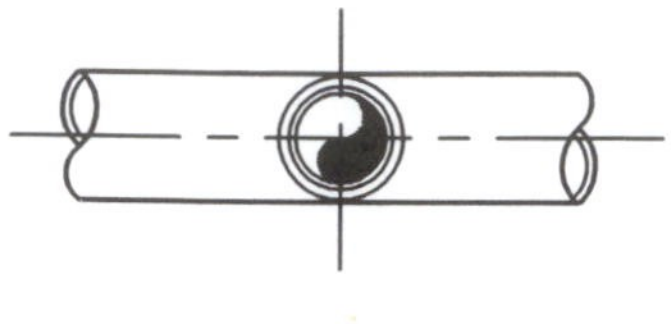	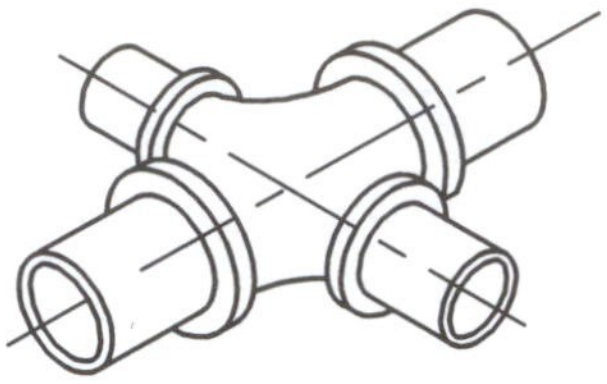	LABEL DIAMETERS
REDUCING ELBOW	[A] 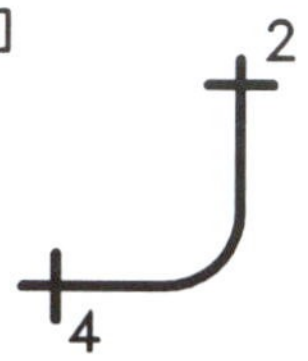 2 4	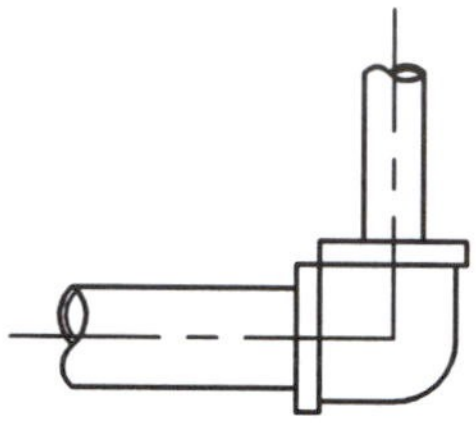	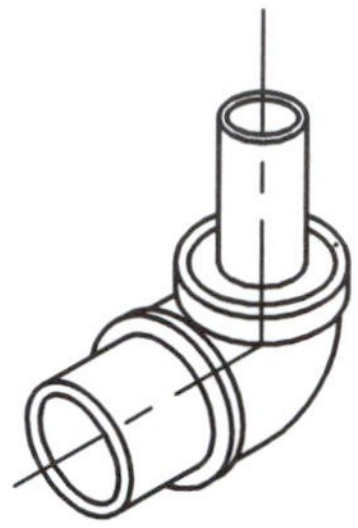	LABEL DIAMETERS
FLEXIBLE CONNECTION	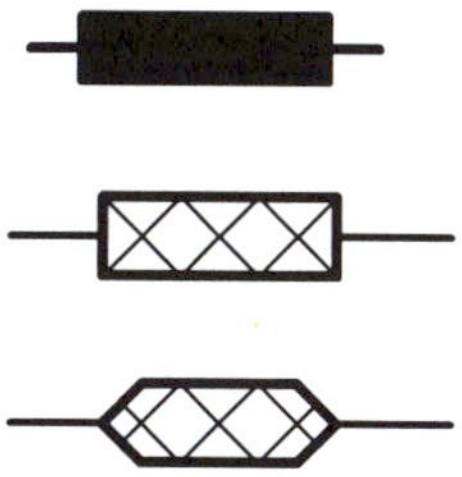 FLANGED SCHEMATIC: 	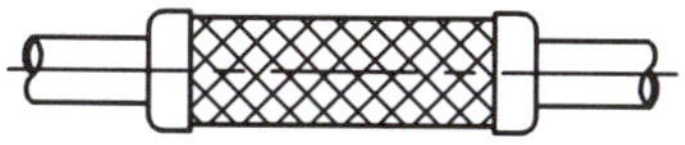	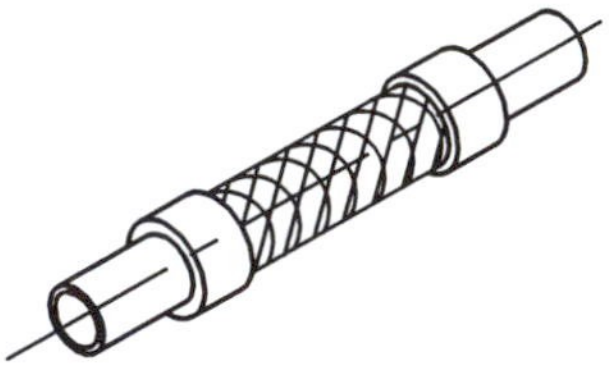	USED IN HOT WATER HEATING SYSTEMS, E.G.
EXPANSION JOINT - SLIDING	[A] 	 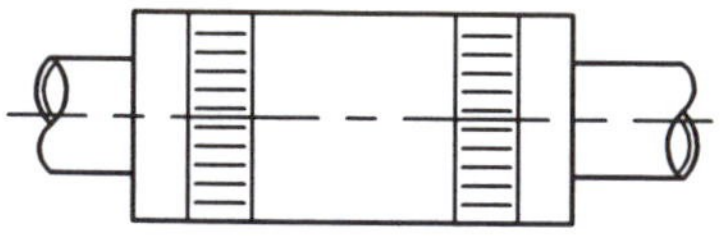	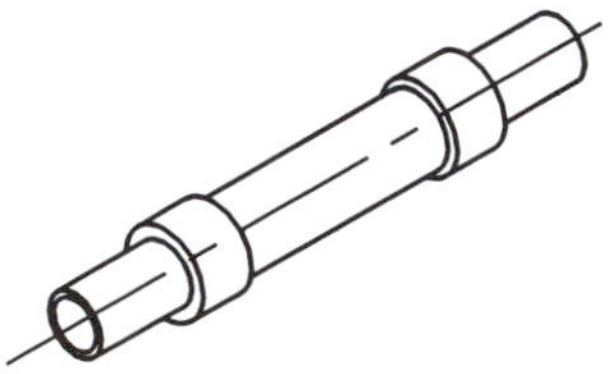 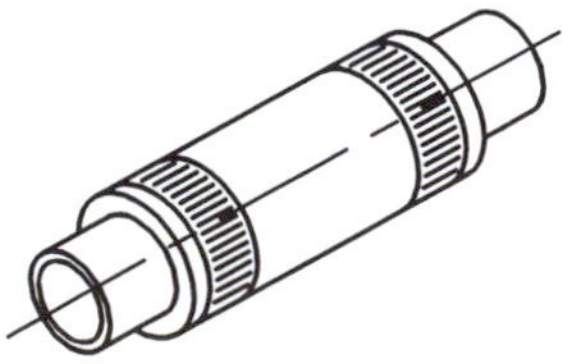	USEFUL WHERE SLIGHT MOVEMENT OR VIBRATION IS POSSIBLE BETWEEN COMPONENTS, E.G. PIPE CONNECTION TO PUMP OR COMPRESSOR. ILLUSTRATIONS: MECHANICAL SEAL ABOVE, HOSE CLAMPS BELOW.

NAME	SYMBOL	OBJECT	REMARKS
EXPANSION JOINT - BELLOWS TYPE *OR* LINE VIBRATION ABSORBER	 [A]  ELEVATION SCHEMATIC: 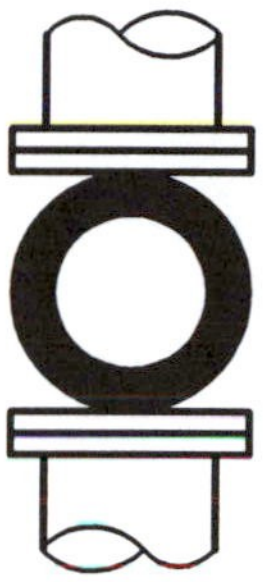	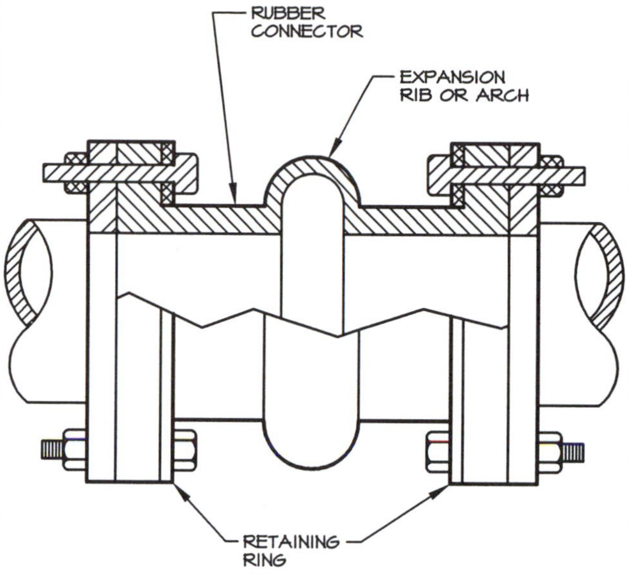 RUBBER CONNECTOR EXPANSION RIB OR ARCH RETAINING RING	EXPANSION JOINTS ALLEVIATE STRESS CAUSED BY EXPANSION AND CONTRACTION OF PIPES, & REDUCE NOISE & VIBRATION.
BULKHEAD FLANGE	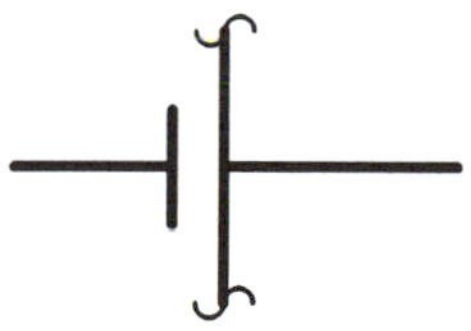	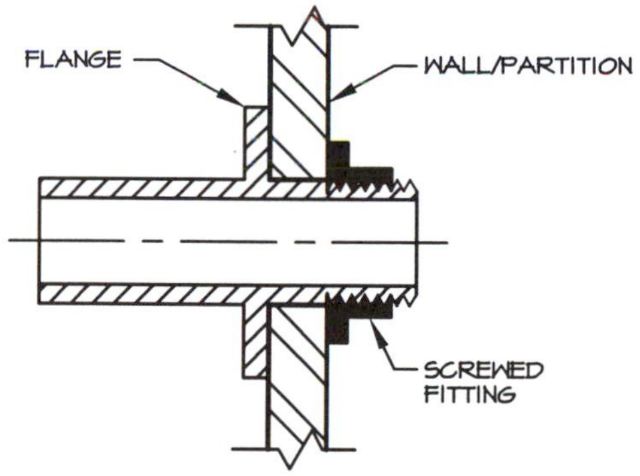 FLANGE WALL/PARTITION SCREWED FITTING	USED WHERE PIPE PASSES THRU A PARTITION - FOR EXAMPLE A DRAIN IN A TANK.
CONCENTRIC REDUCER	[A] 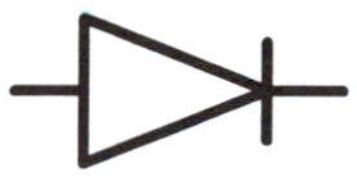 RISING & DESCENDING: 	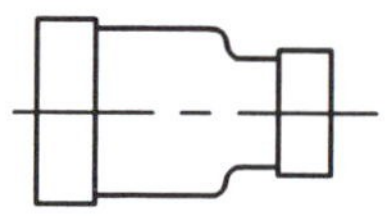	

NAME	SYMBOL	ELEVATION	PERSPECTIVE	REMARKS
ECCENTRIC REDUCER-STRAIGHT INVERT	[A] 	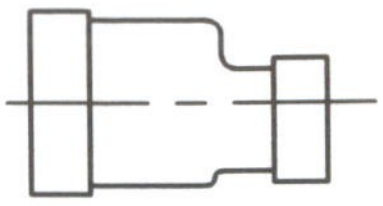		ELIMINATES STANDING FLUID IN HORIZONTAL APPLICATIONS.
ECCENTRIC REDUCER-STRAIGHT CROWN	[A] 	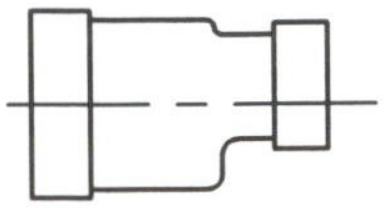		
SLEEVE	[A] 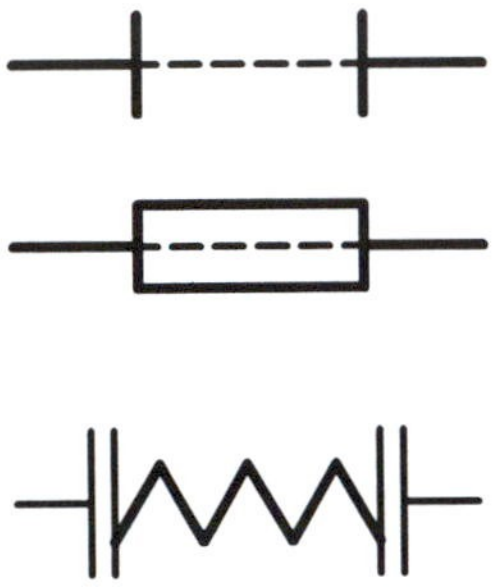	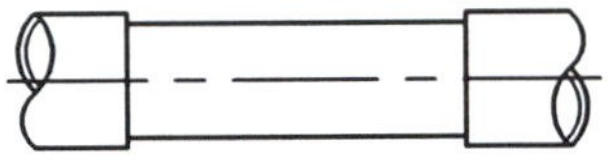	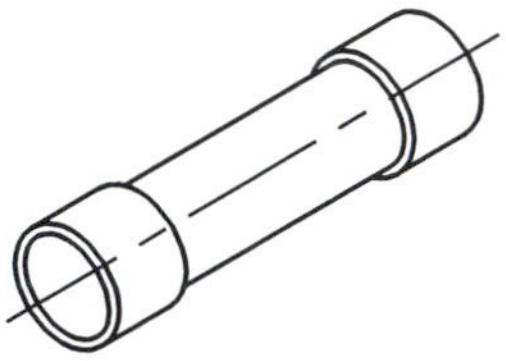	
STRAIGHT SIZE TEE	[A] 	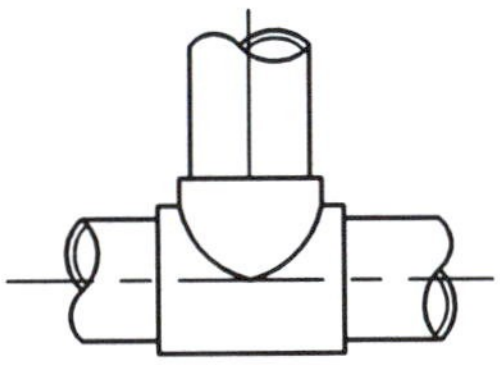		
TEE-UP	[A] 	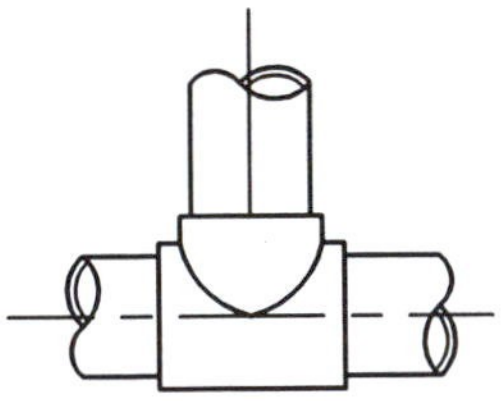		FREQUENTLY SEEN WITHOUT THE DOT.
TEE-DOWN		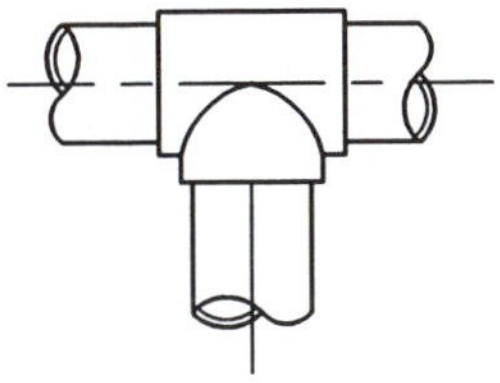		

NAME	SYMBOL	ELEVATION	PERSPECTIVE	REMARKS
REDUCING TEE	[A] 	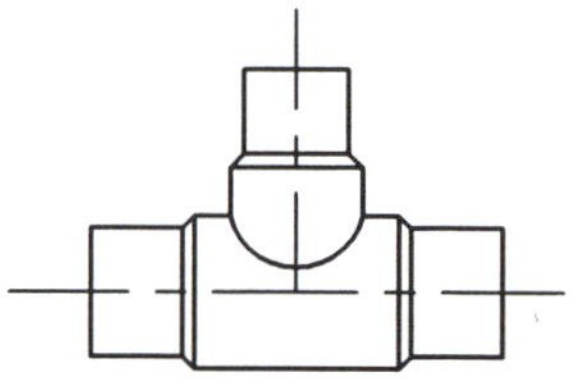		LABEL DIAMETERS AS SHOWN.
TEE, SIDE OUTLET, OUTLET UP	[A] 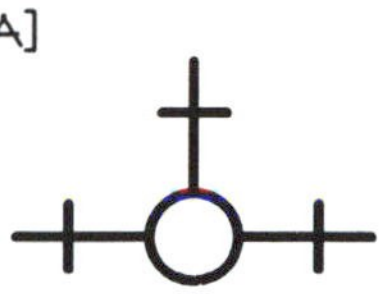	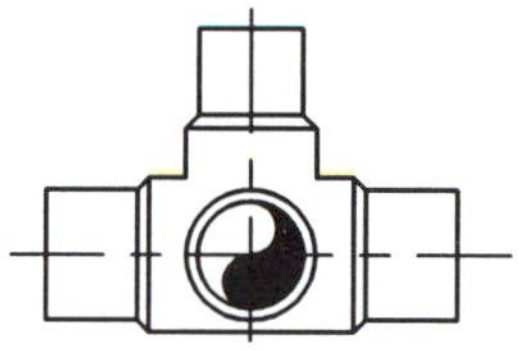		
TEE, SIDE OUTLET, OUTLET DOWN	[A] 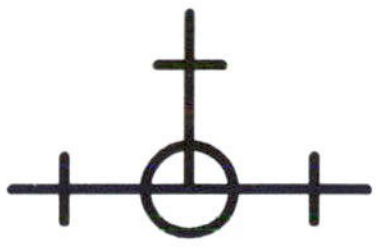	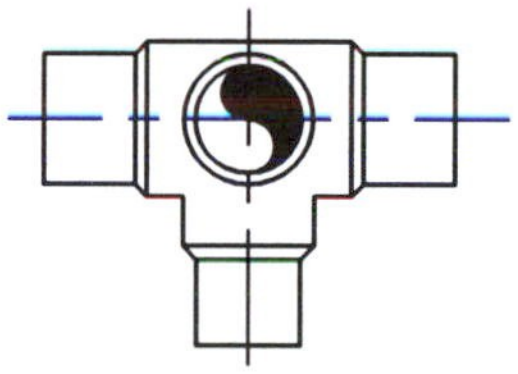		
SCREWED UNION	[A] 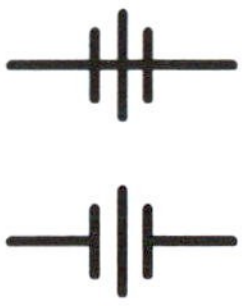	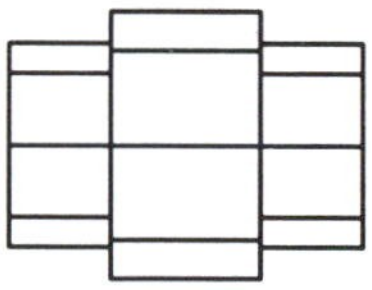	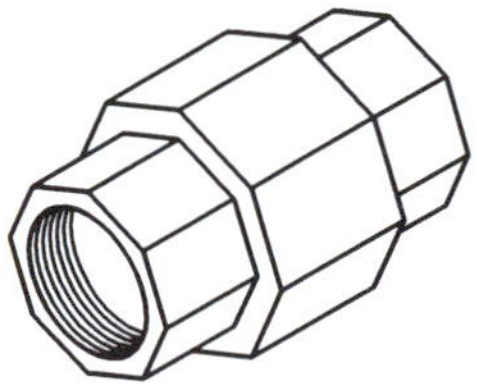	
FLANGED UNION	[A] 	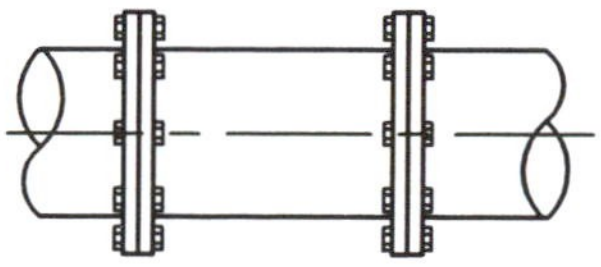	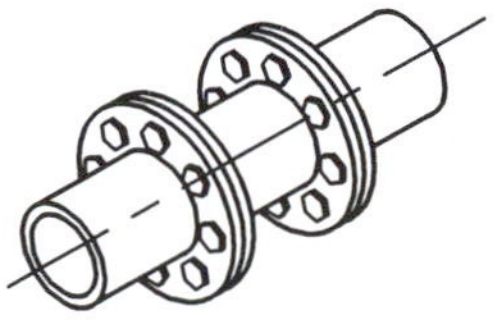	

NAME	SYMBOL	OBJECT	REMARKS
FLOW STRAIGHTENER	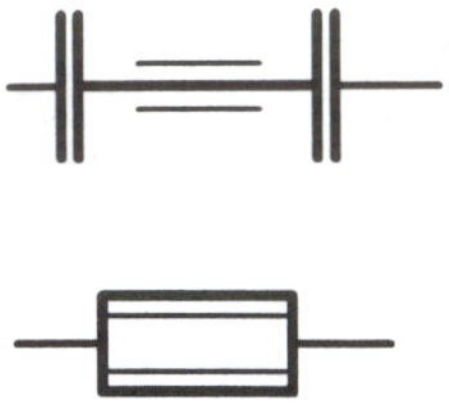	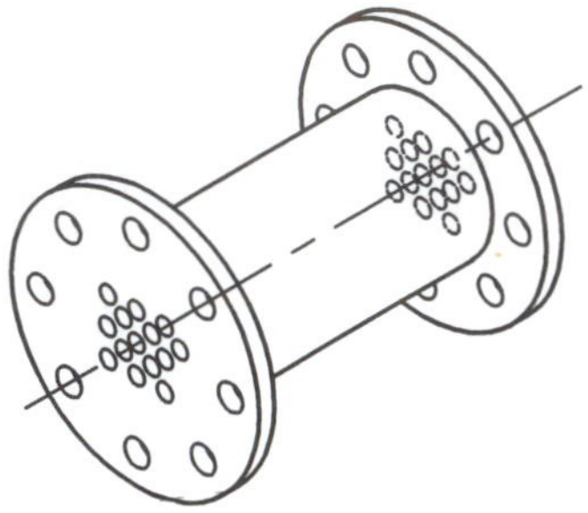	SHOWN FLANGED. COMMONLY OF GREATER LENGTH THAN SHOWN HERE.
QUICK COUPLE HOSE CONNECTOR		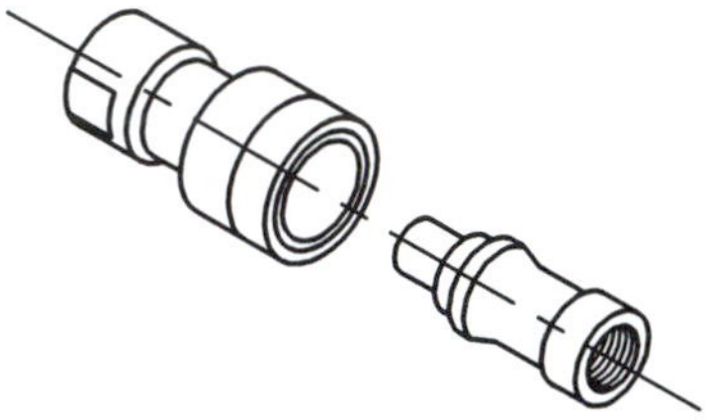	
PETE'S PLUG	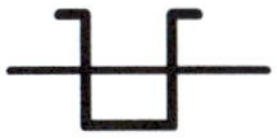		REMOVEABLE PLUG IN TOP PROVIDES ACCESS FOR PRESSURE OR TEMPERATURE, OR PRESSURE AND TEMPERATURE SENSORS.
BULL PLUG	FLANGED: [A] 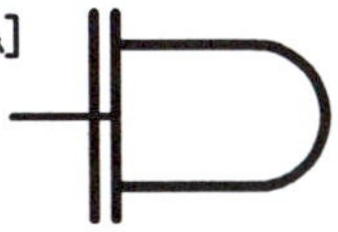 SCREWED: 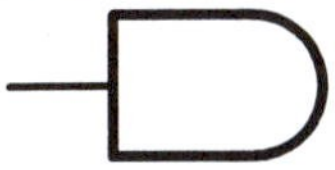 BELL & SPIGOT: [A] 	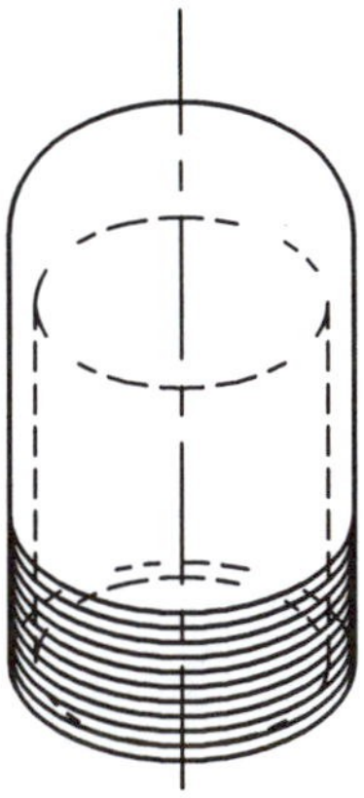	PLUG ILLUSTRATED IS HOLLOW. SOME ARE SOLID METAL.

NAME	SYMBOL	ILLUSTRATION	REMARKS
PIPE PLUG OR CLEANOUT PLUG	SCREWED: [A]		ILLUSTRATIONS SHOW SQUAREHEAD AND SLOTTED HEAD PLUGS.
	BELL & SPIGOT: [A]		ILLUSTRATION SHOWS SHOWS GASKETED BELL WITH A MINNEAPOLIS CODE CAP.
BUSHING			
POINT OF CONNECTION		(OLD WORK) (NEW WORK) 1/4" TO 3/8" Φ TYP.	INDICATES POINT OF CONNECTION BETWEEN NEW & EXISTING WORK. USE WITH ARROW. SHOW NEW WORK WITH THICKER LINE.

NAME	SYMBOL	OBJECT	REMARKS
SWING TYPE JOINT OR BALL JOINT			USED ON FUEL TRUCKS AND IN CAR WASHES, FOR EXAMPLE.
DOUBLE PLANE SWING TYPE JOINT			

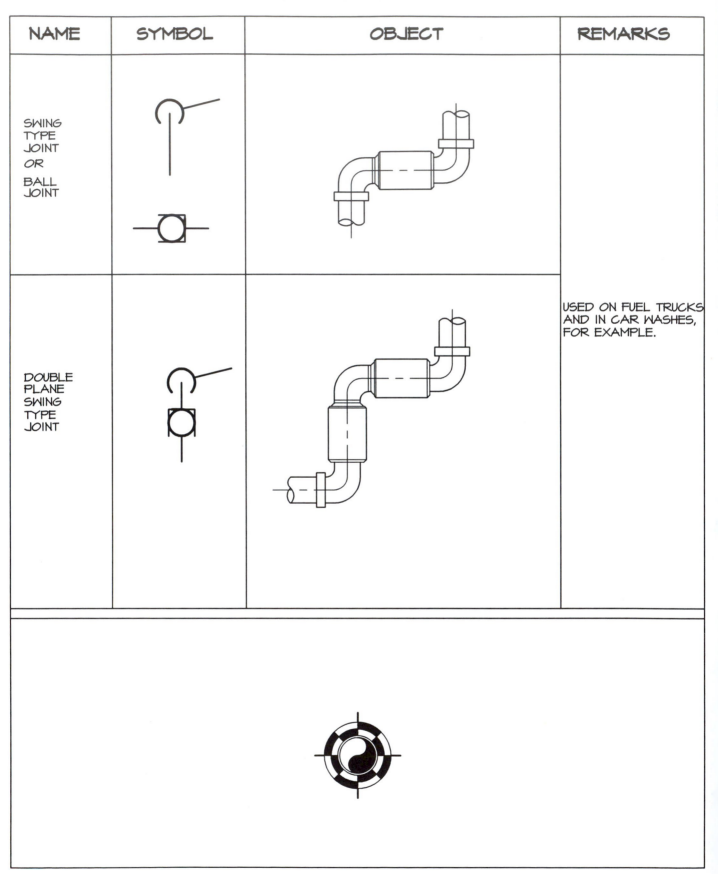

NAME	SYMBOL	OBJECT	REMARKS
CHOKE NIPPLE			
PIPE INSULATION:		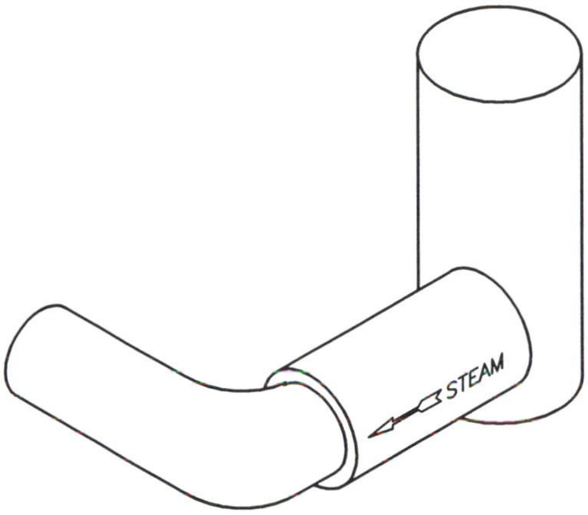	
INSULATION WITH THICKNESS	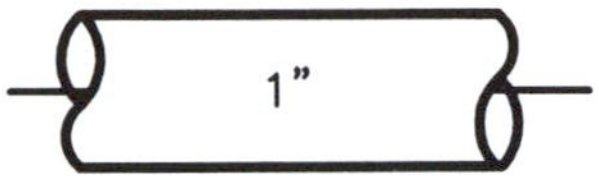 1"		
INSULATION WITH STEAM TRACE	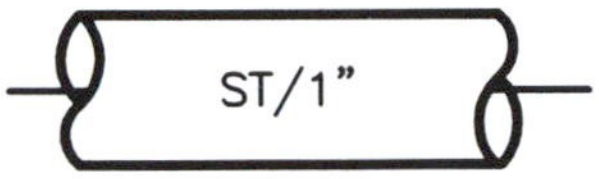 ST/1"	STEAM	
INSULATION WITH ELECTRIC TRACE	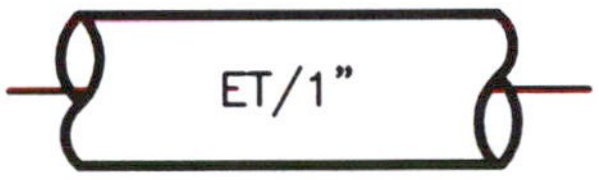 ET/1"		
PERSONNEL PROTECTION	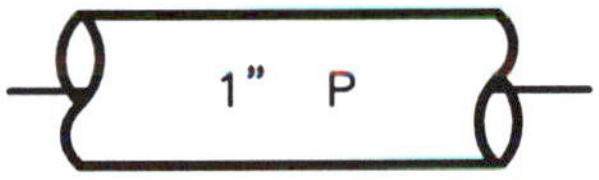 1" P		
CLEVIS TYPE PIPE HANGER	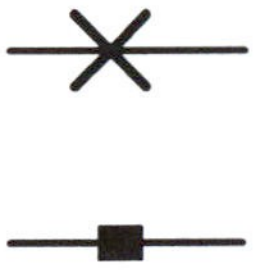	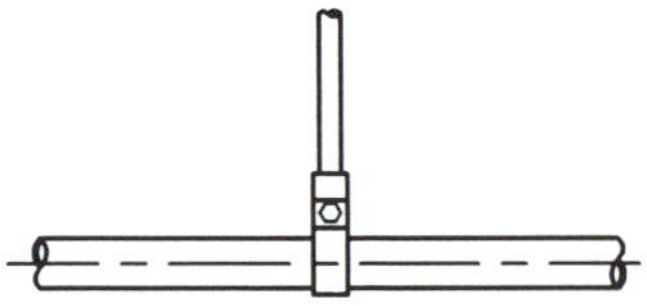 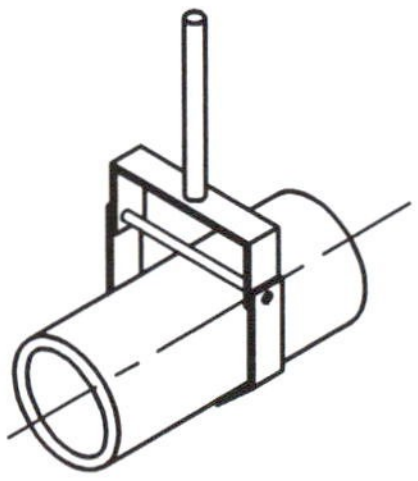	
ROLLER TYPE HANGER	 RH	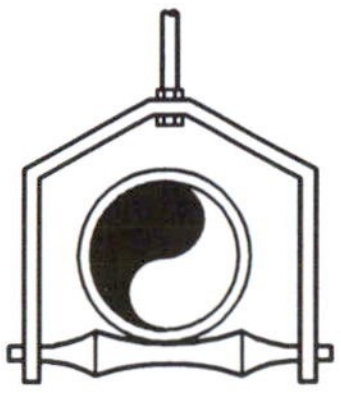	

PIPE EQUIPMENT

NAME	SYMBOL	OBJECT	REMARKS
VARIABLE SPRING TYPE HANGER	 SH	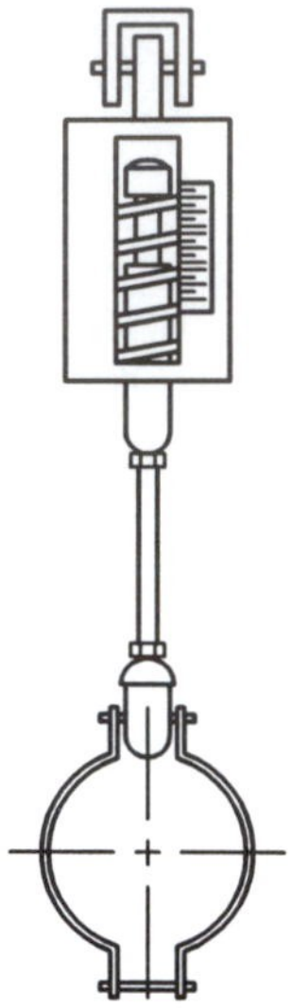	SPRING HANGERS ARE USED WHERE PIPES MOVE FROM EXPANSION AND CONTRACTION AND WHERE PIPE MOVEMENT RESULTS FROM EQUIPMENT VIBRATION.
SPRING CUSHION TYPE HANGER	 SCH	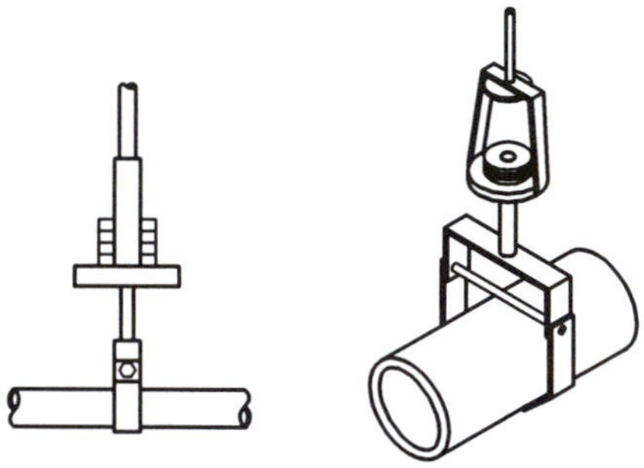	
TRAPEZE HANGER	 TH	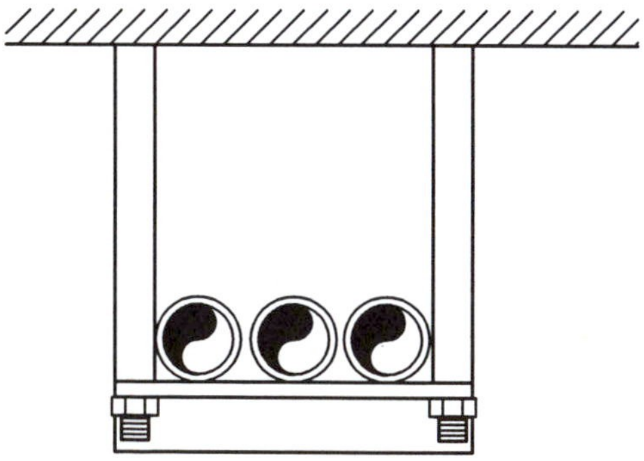	
ALIGNMENT GUIDE			

NAME	SYMBOL	OBJECT	REMARKS
FLOOR-SUPPORTED PIPE STAND	PS	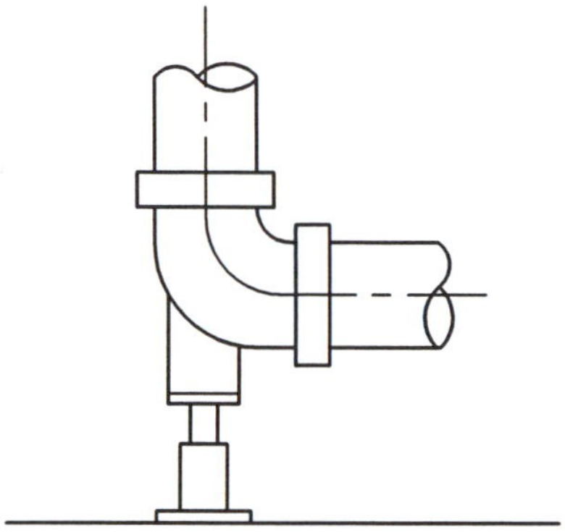	
WALL BRACKET	WB	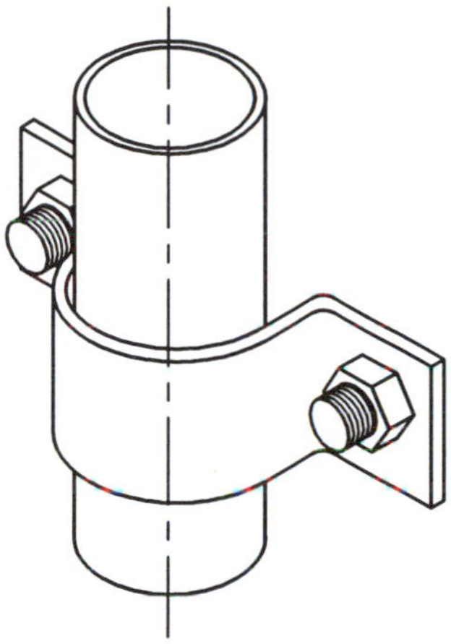	
CONSTANT SUPPORT HANGER	CSH		
SLIDING SUPPORTS	SS	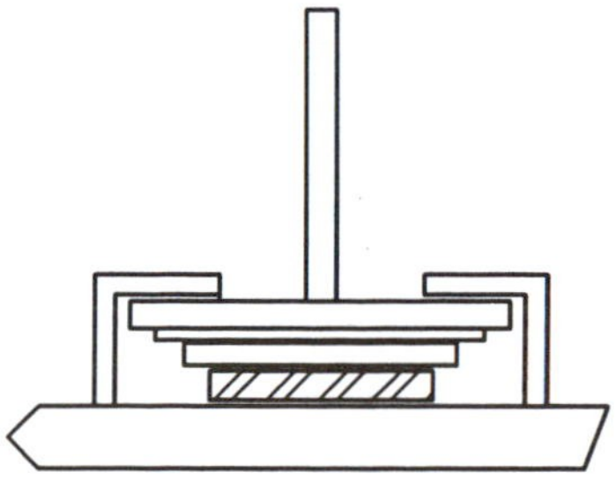	

NAME	SYMBOL	OBJECT	REMARKS
RISER CLAMP	RC	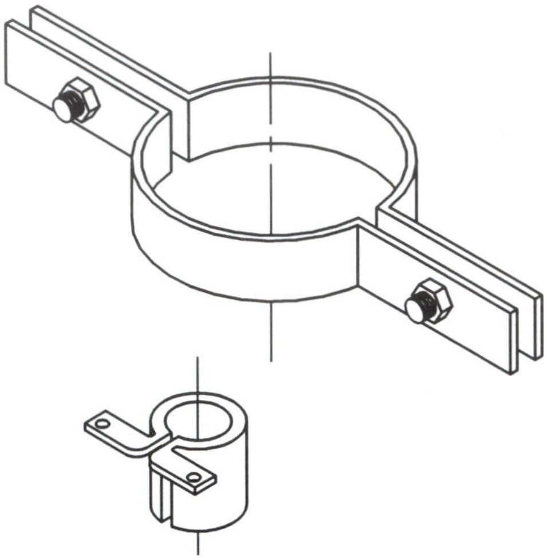	ILLUSTRATIONS SHOW HEAVY COMMERCIAL CLAMP ABOVE, LIGHT COPPER TUBING TYPE BELOW.
FOR OTHER TYPES OF HANGERS (SPLIT PIPE, ADJUSTABLE RING, ETC.) USE APPROPRIATE LETTERS & KEY.	X		
FLOW SWITCH	FS		
PRESSURE SWITCH	PS		

NAME	SYMBOL	OBJECT	REMARKS
EXPANSION TANK			COMPENSATES FOR CHANGE IN WATER VOLUME IN HOT WATER SYSTEMS. MAY CONTAIN A BLADDER, AND FUNCTION IN NOISE REDUCTION.
SAMPLE CONNECTION	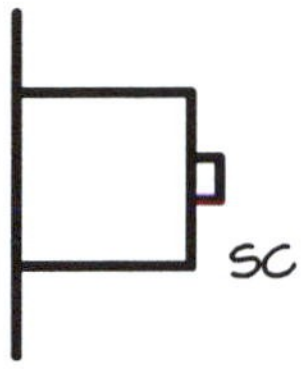 SC		
FLEXIBLE CONNECTION HOSE	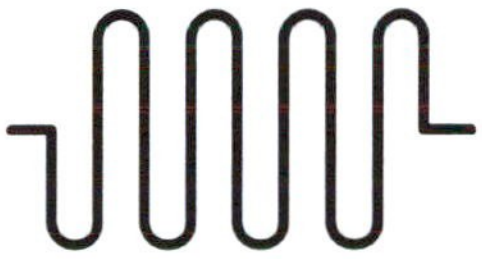		
FLOW SIGHT GLASS OR PLAIN SIGHT INDICATOR		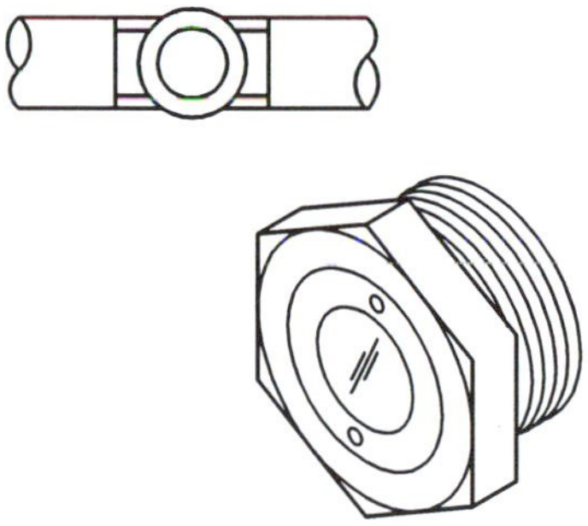	SHOWS FLOW AND CONDITION OF FLUID.
SIGHT FLOW INDICATOR			FLAPPER TYPE SHOWN IN PERSPETIVE,ROTATOR SHOWN IN ELEVATION. OTHER TYPES OF INDICATORS ARE: SIDE FLAPPER, MID FLAPPER & DRIP TUBE.

NAME	SYMBOL	OBJECT	REMARKS
LEVEL GAUGE		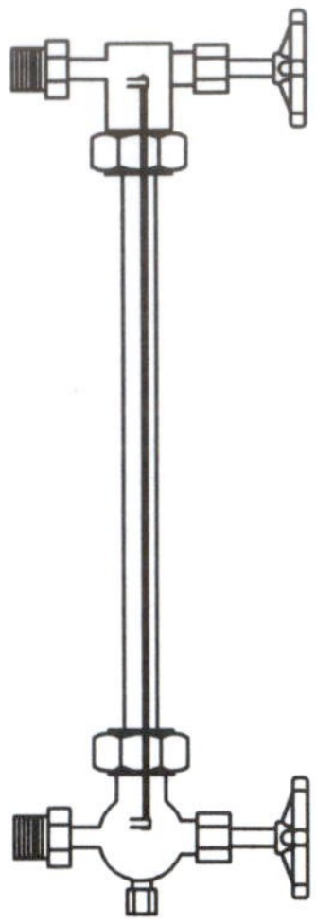	
EXPOSED RADIATOR			
RECESSED RADIATOR	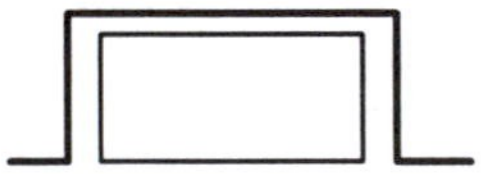		
ENCLOSED RADIATOR - FLUSH			
ENCLOSED RADIATOR - PROJECTING	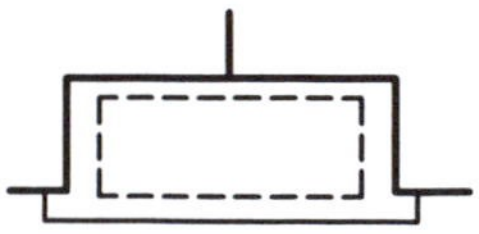		

NAME	SYMBOL	OBJECT	REMARKS
RUPTURE DISK			PROVIDES QUICK PRESSURE RELIEF AT A SPECIFIC PSI.
PUMP		TRIANGLE IN SYMBOL INDICATES DIRECTION OF FLOW. ILLUSTRATION SHOWS TWO PUMPS ON A BOILER.	

SEWER PIPE

NAME	SYMBOL	OBJECT	REMARKS
LONG RADIUS ELBOW	[A] 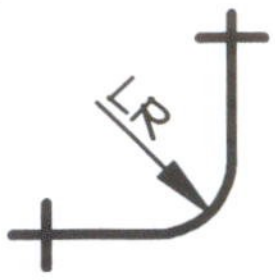 LR		BROADER SWEEP MAKES FOR SMOOTHER FLOW.
STREET ELBOW	[A] 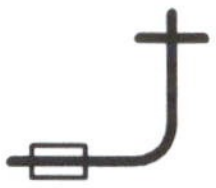		FEMALE FITTING SHOWN AT TOP, MALE FITTING SHOWN BELOW ON THE SYMBOL.
90° BASE ELBOW	[A] 	[A] 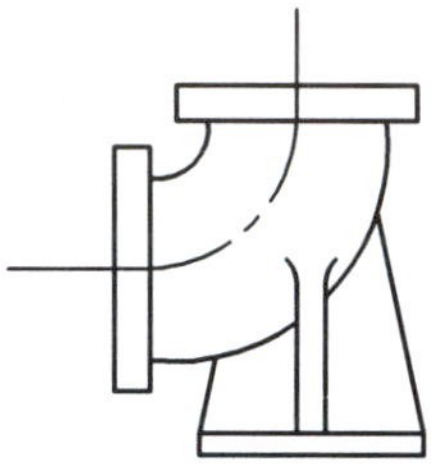	SUPPORT AT BOTTOM IS STEEL OR IRON PROTU-BERANCE.
LATERAL CONNECTION	[A]  RISING:  TURNING DOWN: 	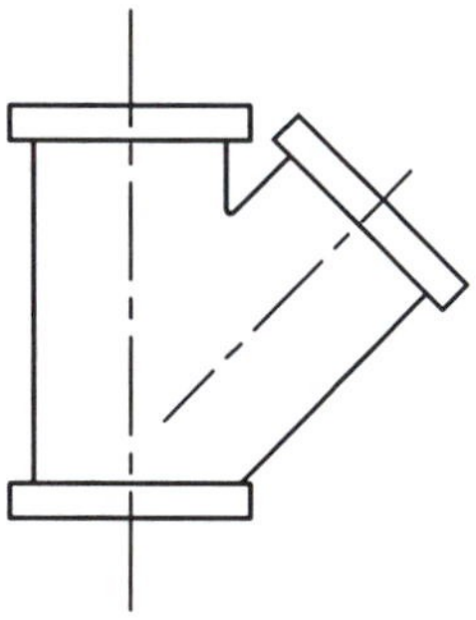	
TEE, SIDE OUTLET, OUTLET UP	[A] 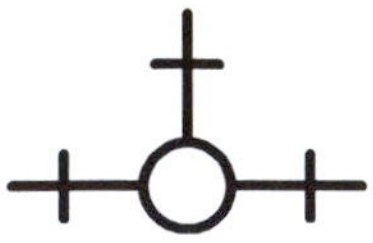	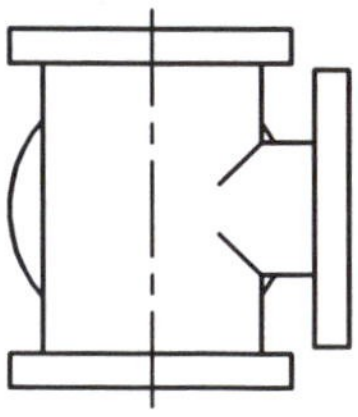	

NAME	SYMBOL	OBJECT	REMARKS
DOUBLE SWEEP	[A] 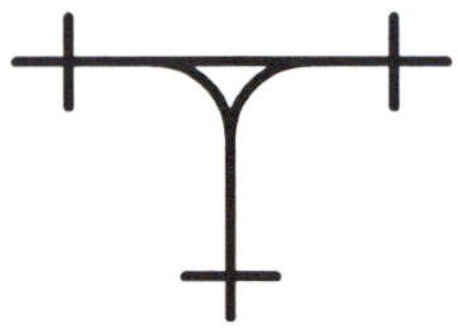	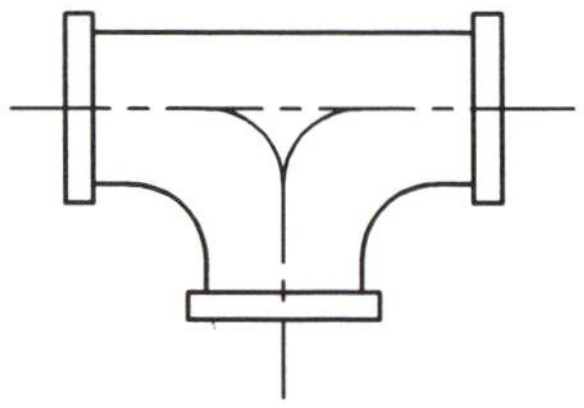	
SINGLE SWEEP	[A] 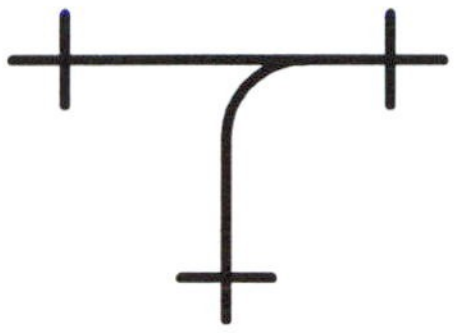	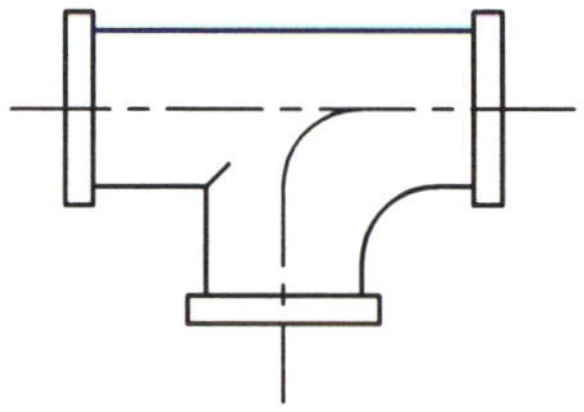	
DOUBLE BRANCH ELBOW	[A] 	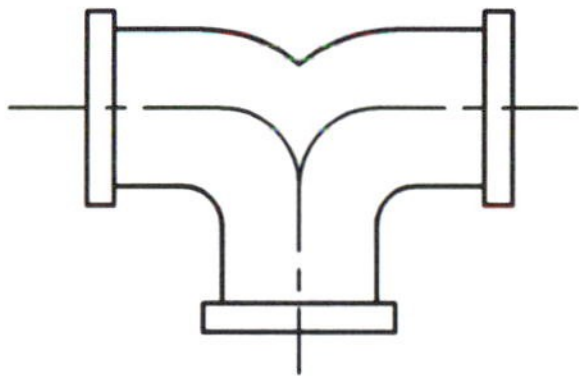	
THROUGH DOUBLE T-Y	[A] 	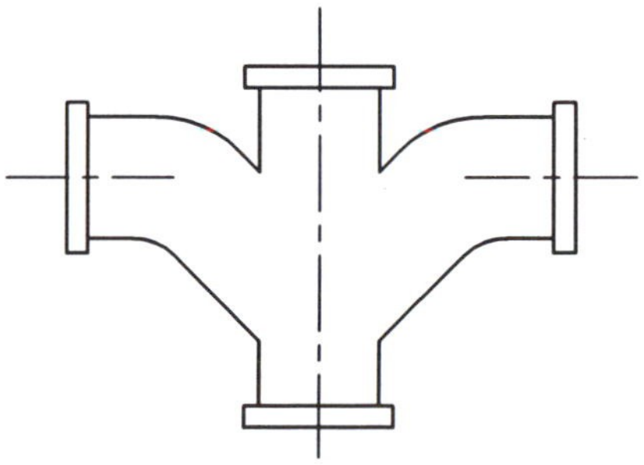	

NAME	SYMBOL	OBJECT	REMARKS
THROUGH DOUBLE Y	[A] 	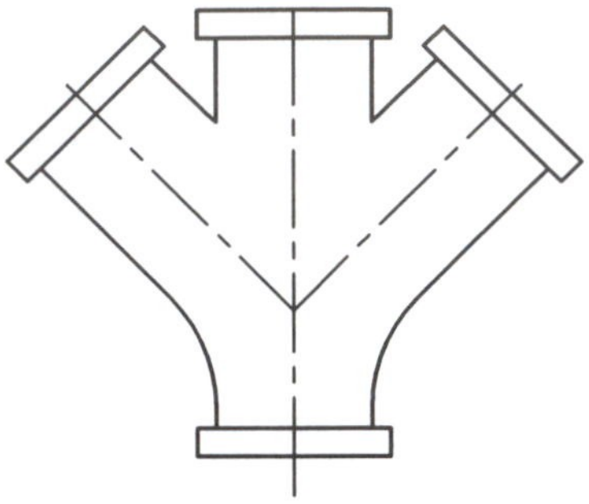	
TRUE Y	[A] 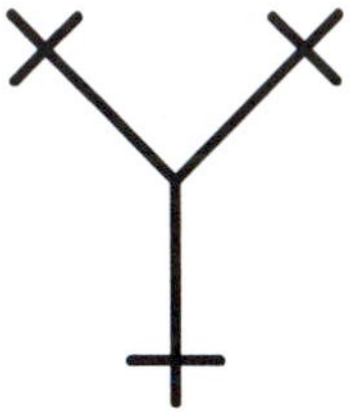	[A] 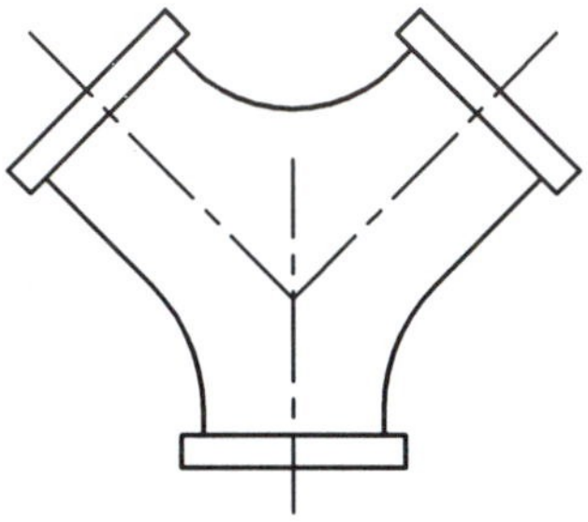	
RETURN BEND	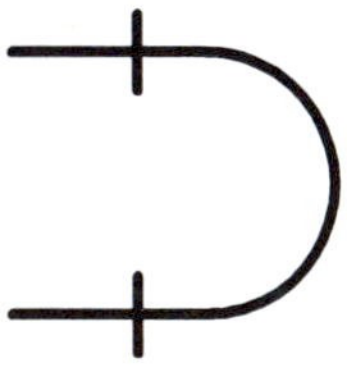	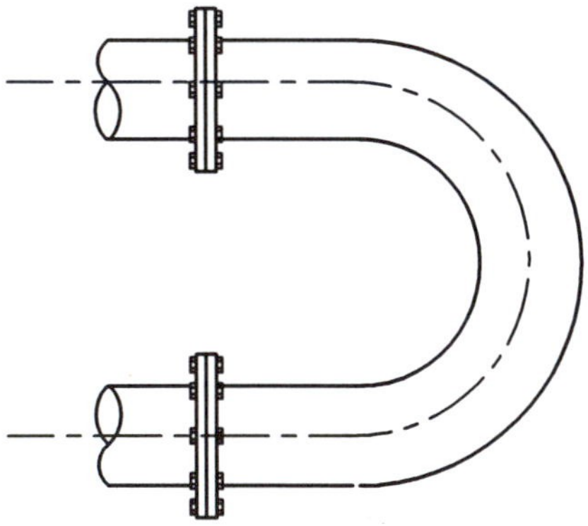	
BLANK FLANGE	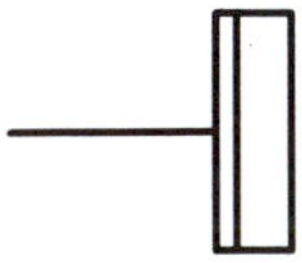	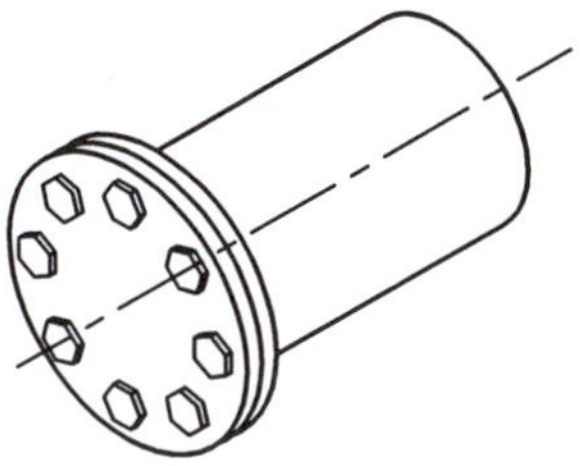	

NAME	PLAN	SIDE	FRONT
TOILETS:			
FLOOR OUTLET TOILET	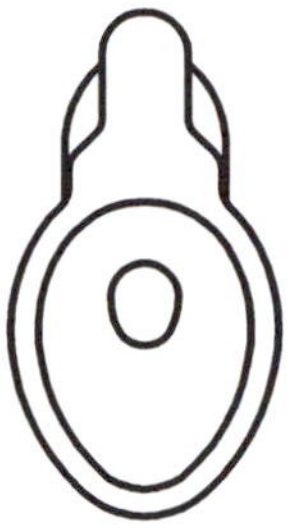	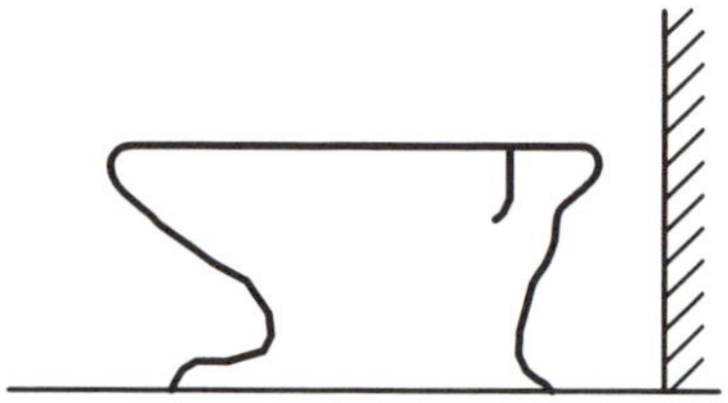	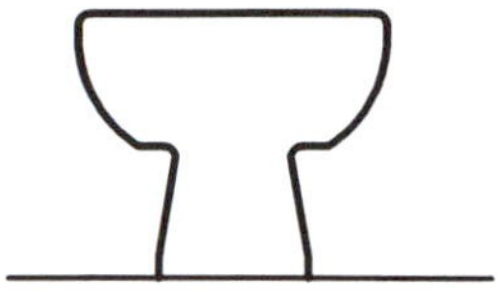
	[A] 	[A] 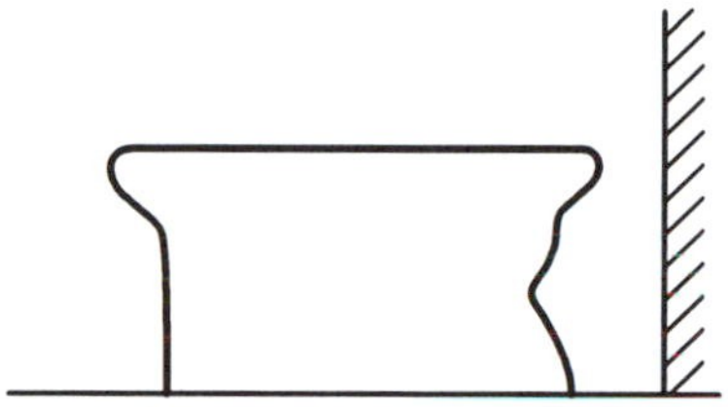	[A] 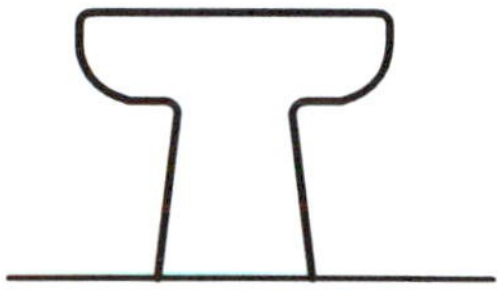
WALL HUNG TOILET		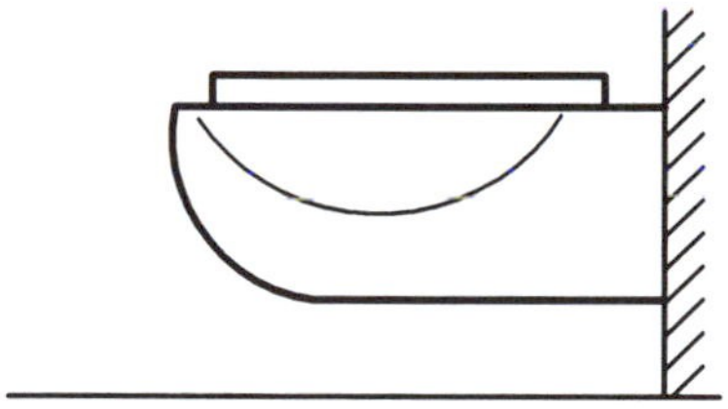	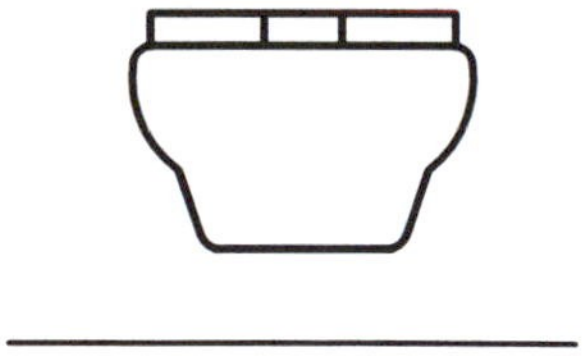
	[A] 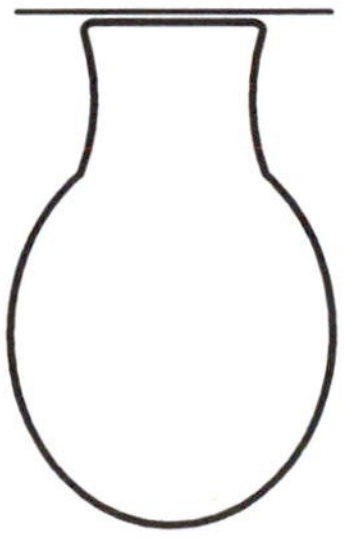	[A] 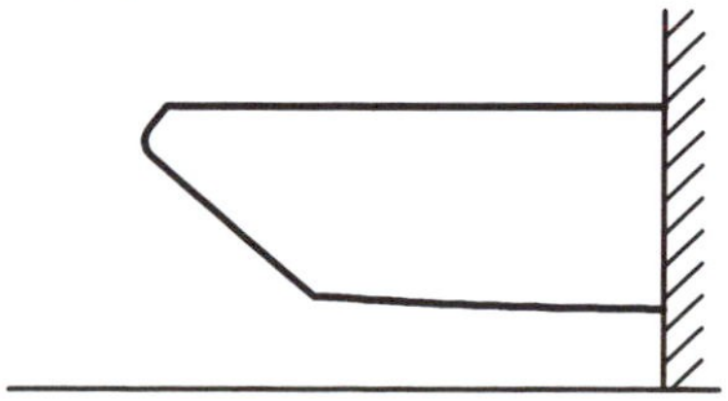	[A] 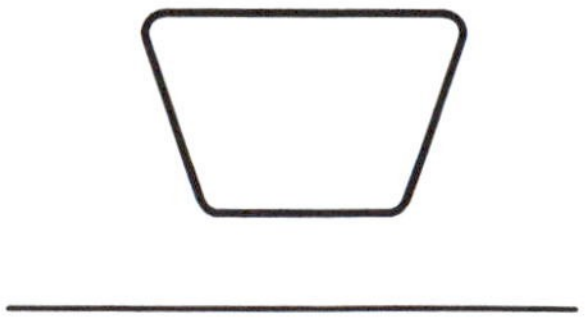

NAME	PLAN	SIDE	FRONT
TOILETS: (CONTD.) TANK TYPE TOILET	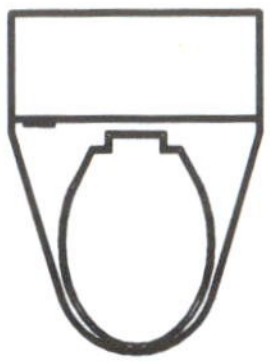	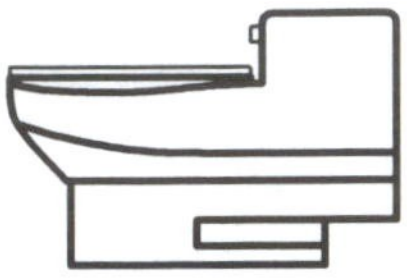	

	[A] 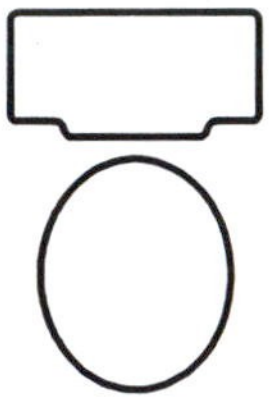	[A] 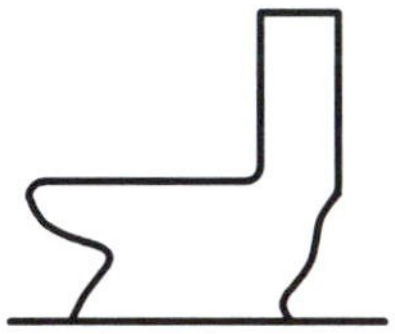	[A] 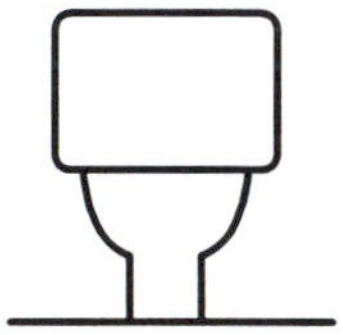
BIDET	[A] 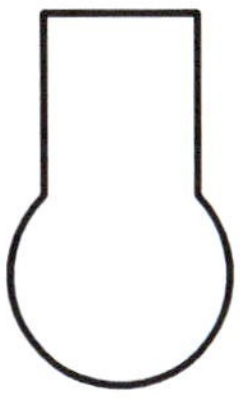	[A] 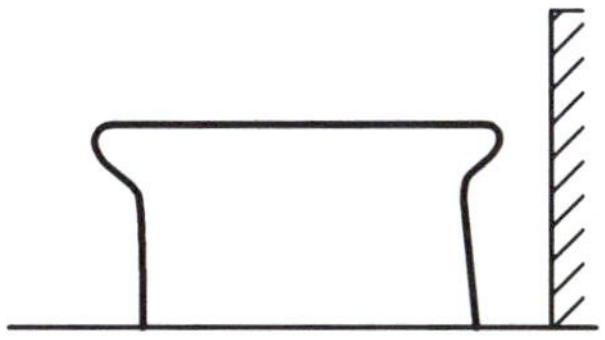	[A] 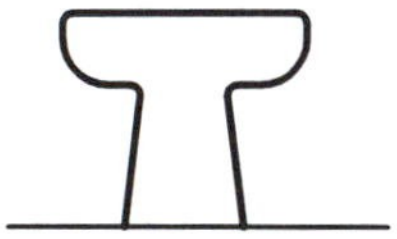
	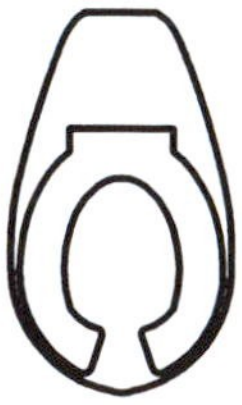	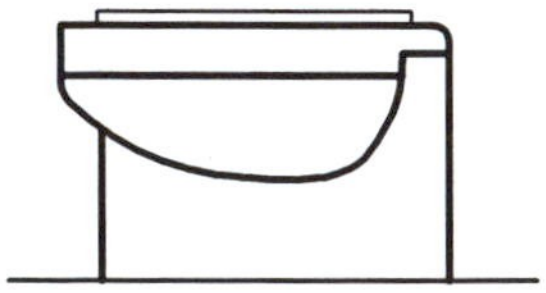	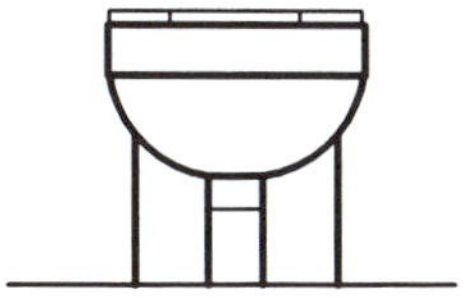

NAME	PLAN	FRONT	SIDE
WALL HUNG URINAL	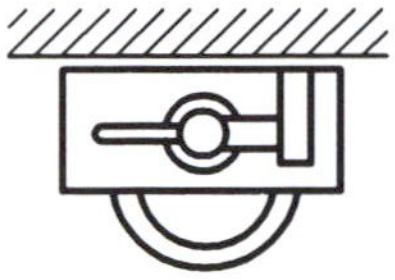 [A] 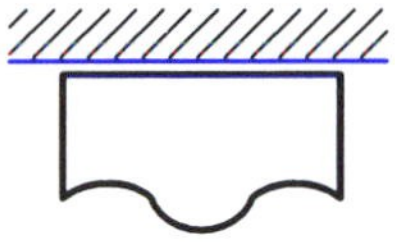	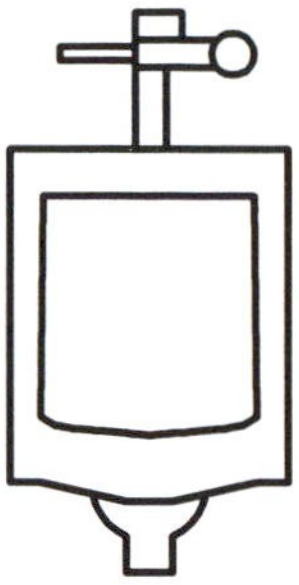 [A] 	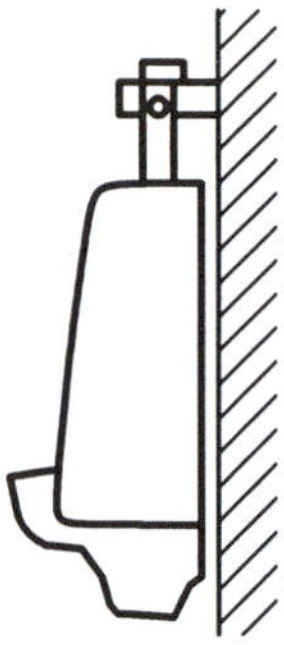 [A] 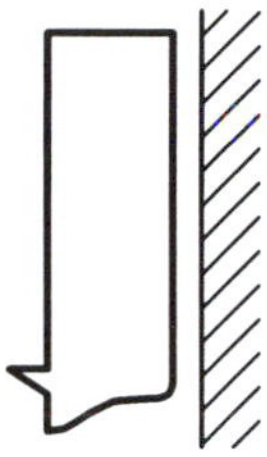
SINK - GENERAL	[A]		
KITCHEN SINK	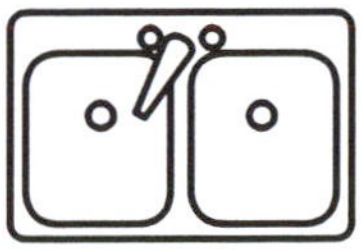	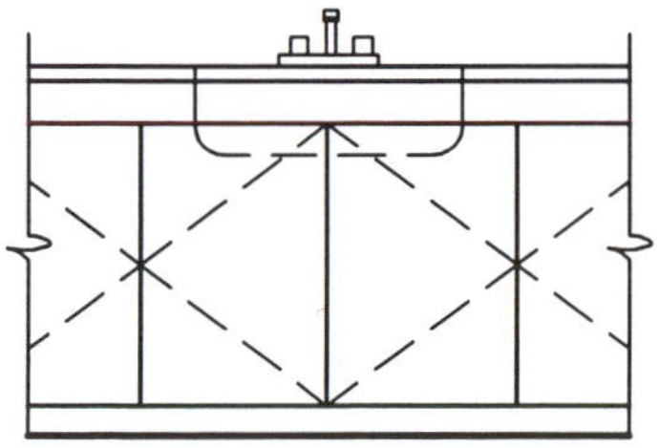	

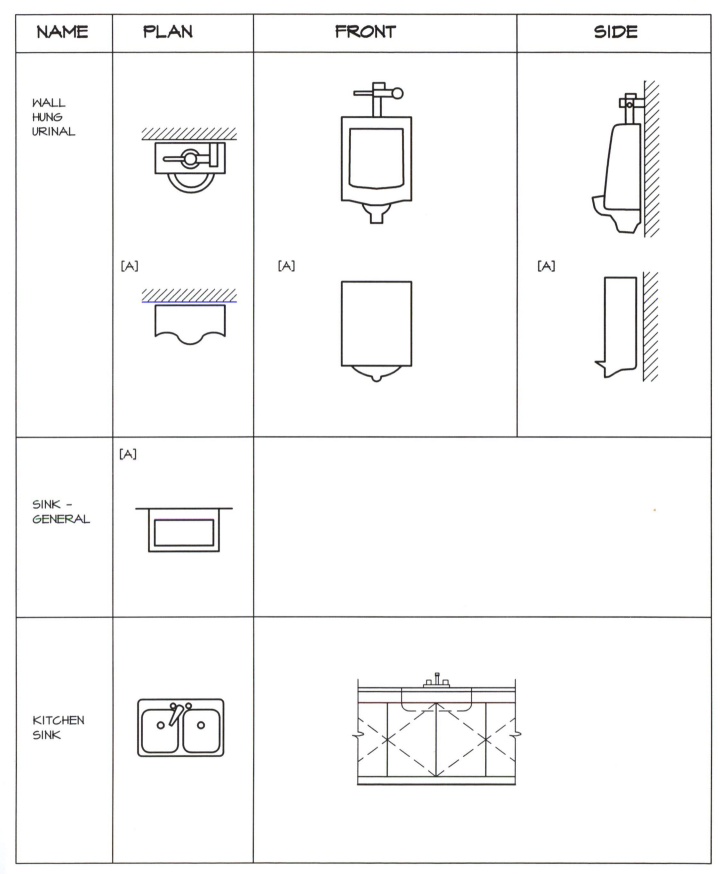

NAME	PLAN	FRONT	SIDE
UTILITY SINK			
BATHTUB	[A] RECESSED STYLE - PLAN PLAN ELEVATION		[A]

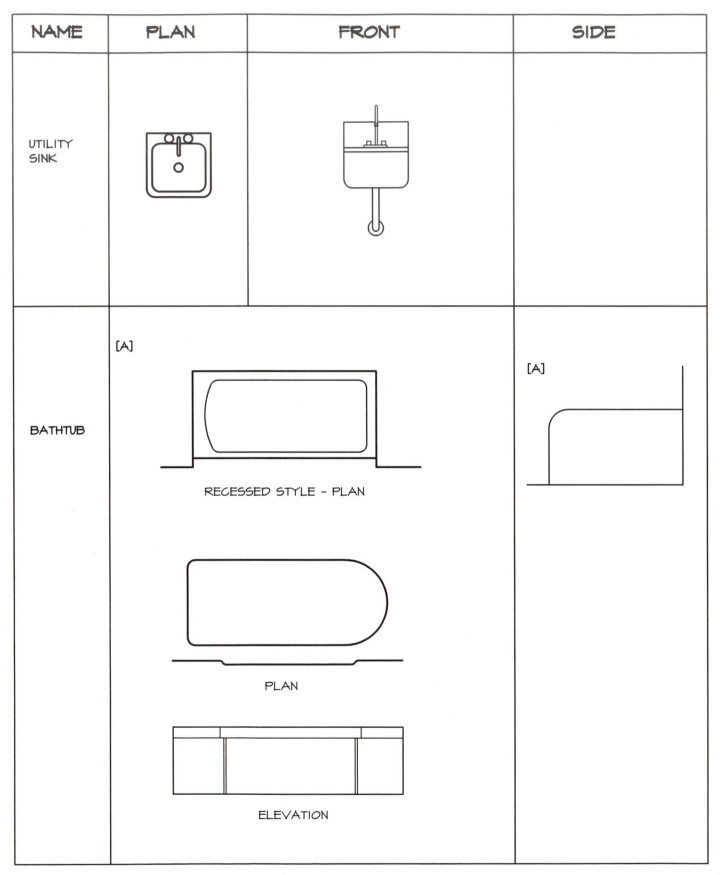

NAME	PLAN	FRONT
WATER FOUNTAIN - FLOOR MOUNTED	[A]	[A]
WATER FOUNTAIN - WALL MOUNTED (HANDICAPPED)		
SHOWER (STALL TYPE)	[A]	
SHOWER (HANDICAPPED)	FOLD-DOWN SEAT	

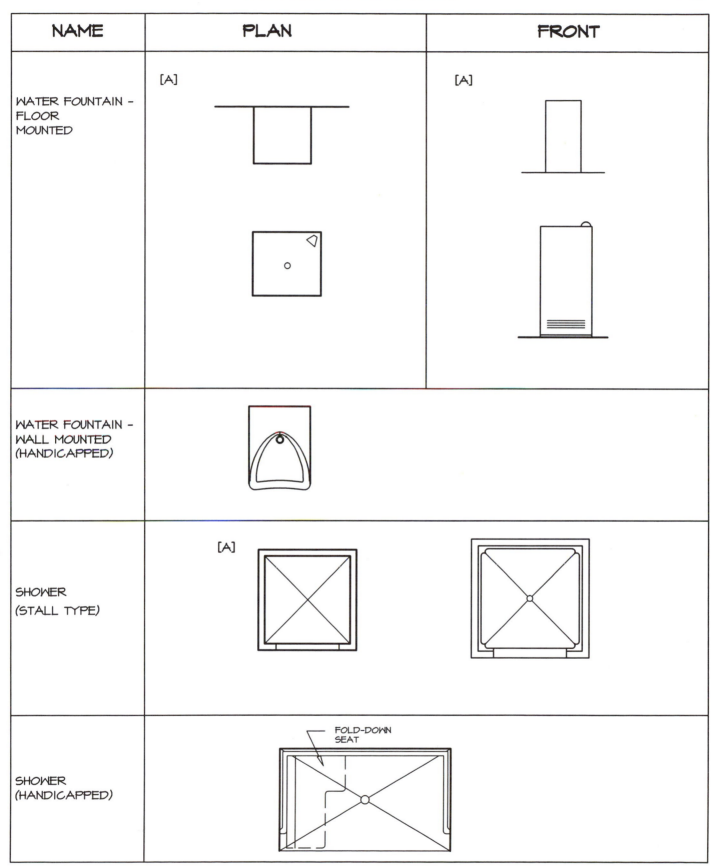

NAME	PLAN	FRONT
JACUZZI		
ADA TOILET		

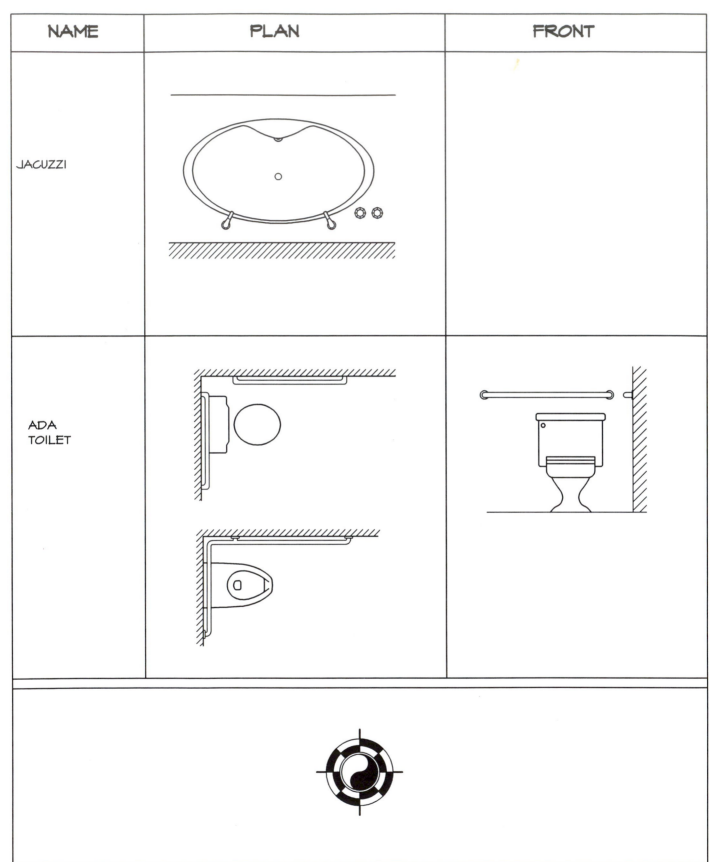

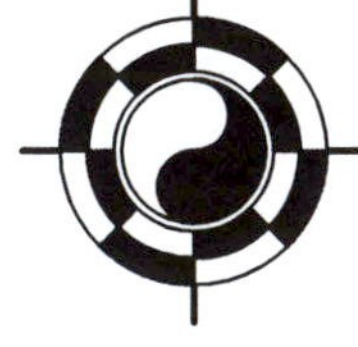

NAME	SYMBOL	OBJECT	REMARKS
SEDIMENT STRAINER *OR* Y STRAINER	[A]  	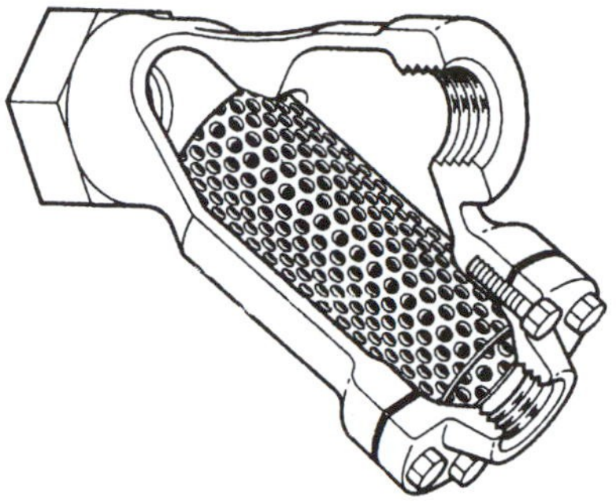	ILLUSTRATION SHOWS STRAINER WITH DRAIN PLUG (REMOVED).
BLOW OFF STRAINER			STRAINER WITH VALVE.
"T" TYPE STRAINER *OR* BASKET STRAINER	  SCHEMATIC PLAN: 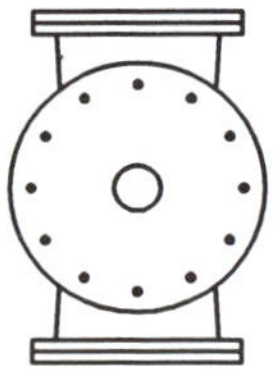	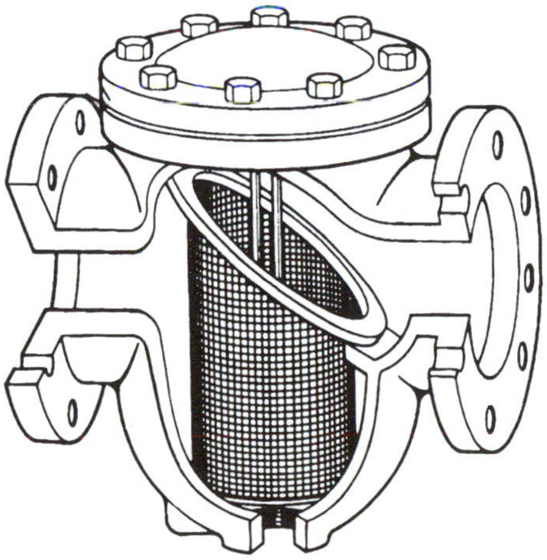	

STRAINERS

NAME	SYMBOL	OBJECT	REMARKS
CONTINUOUS STRAINER			
TEMPORARY STRAINER		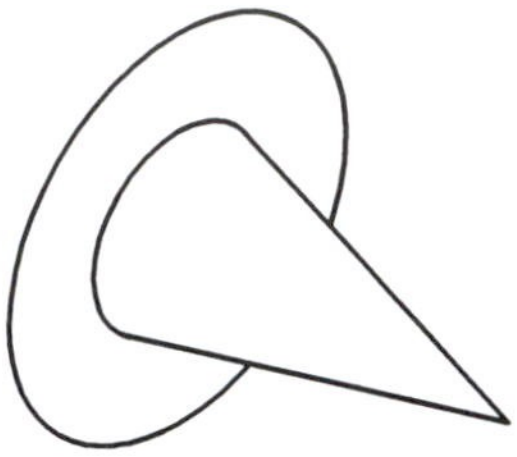	DISPOSABLE INTERNAL STRAINING ELEMENT. AVAILABLE IN PLATE, BASKET AND CONICAL STYLES. CONICAL STYLE ILLUSTRATED.
CONE STRAINER		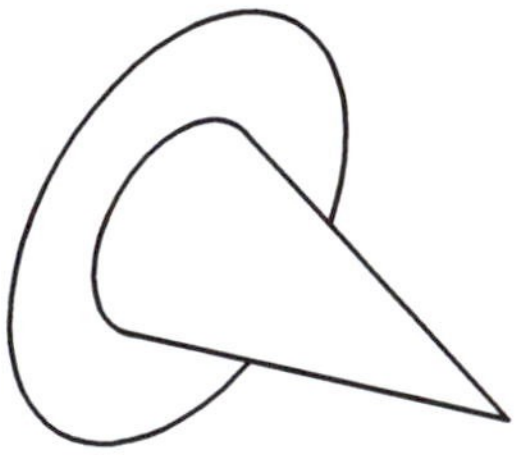	STEEL OR STAINLESS STEEL, TYP.
PURITAN HAT STRAINER			FILTER ELEMENT RESEMBLES PURITAN HAT.
METAL EDGE STRAINER	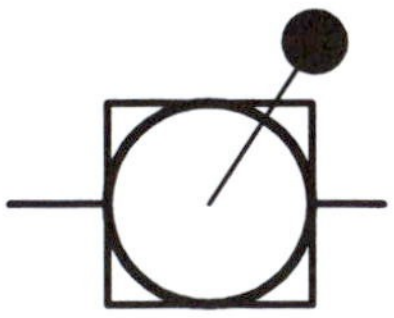		

NAME	SYMBOL	OBJECT	REMARKS
PLATE STRAINER			
SELF-CLEANING STRAINER			
STEAM (BASKET) STRAINER	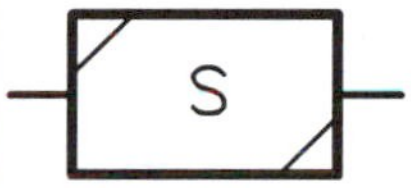 S		
SIMPLEX STRAINER			GENERIC SINGLE ELEMENT STRAINER
BOX STRAINER			ARCHAIC DEVICE. MAY HAVE STRAINER ELEMENT OR NOT.

NAME	SYMBOL	OBJECT	REMARKS
DUPLEX STRAINER		DIVERTER HANDLE INLET BASKET STRAINERS	
DUPLEX OIL STRAINER			DUAL STRAINER UNIT, WITH HANDLE FOR DIVERTING FLOW; ONE SIDE MAY BE ISOLATEDTO CHANGE OR CLEAN ITS ELEMENT.
MAGNETIC STRAINER			SIMILAR TO DUPLEX STRAINER ABOVE.

NAME	SYMBOL	OBJECT	REMARKS
TRAP	 T		GENERIC TRAP SYMBOL.
FLOAT TRAP	[A] 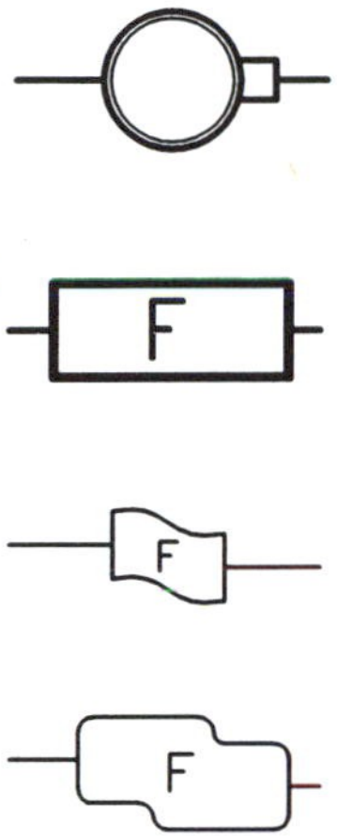 F F F	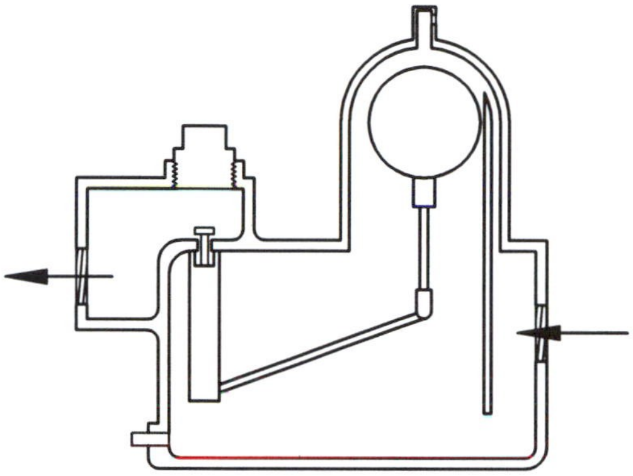	
VACUUM TRAP	 V		
BOILER RETURN TRAP	[A] 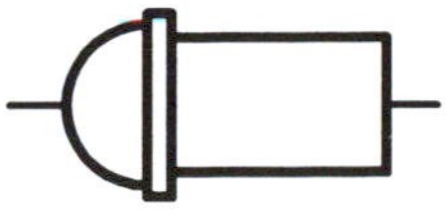	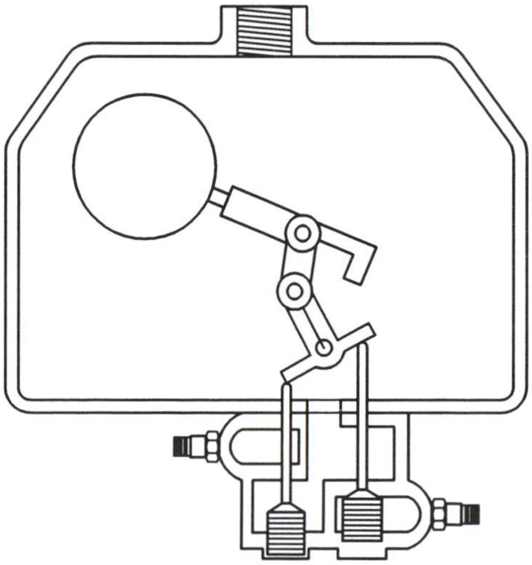	

NAME	SYMBOL	OBJECT	REMARKS
THERMOSTATIC BLAST TRAP	[A] 		
FLOAT & THERMOSTATIC STEAM TRAP	      F&T F&T F&T F [A] 	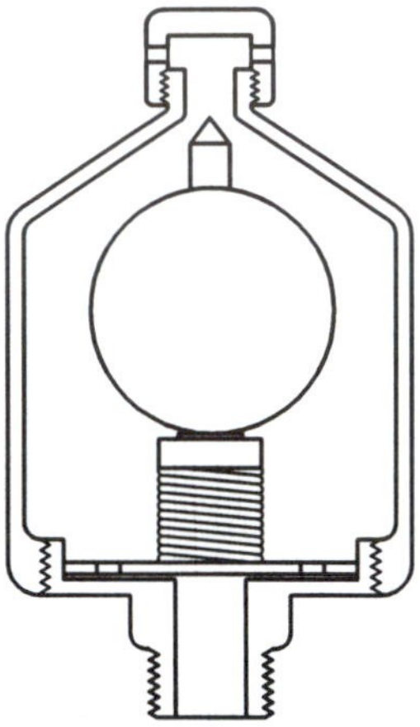 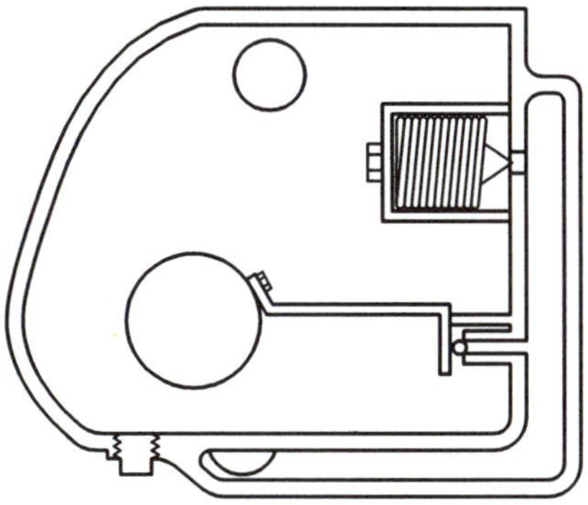	THERMOSTATIC ELEMENT RELEASES AIR & FLOAT ELEMENT RELEASES CONDENSATION. TWO OF MANY STYLES ARE ILLUSTRATED.
VACUUM TRAP	 V		
CLOSED-FLOAT THERMOSTATIC STEAM TRAP ASSEMBLY			

NAME	SYMBOL	OBJECT	REMARKS
AUTOMATIC AIR VENT OR AIR ELIMINATOR	AV [A]		REMOVES AIR BUBBLES FROM HEATED WATER. COMMONLY MOUNTED ON A PURGE UNIT.
AIR PURGE	P		FREQUENTLY MOUNTED ON A BOILER WITH AIR VENT ON TOP.
AIR VENT - MANUAL	MV		
SCALE TRAP			

NAME	SYMBOL	OBJECT	REMARKS
THERMOSTATIC STEAM TRAP	[A] 	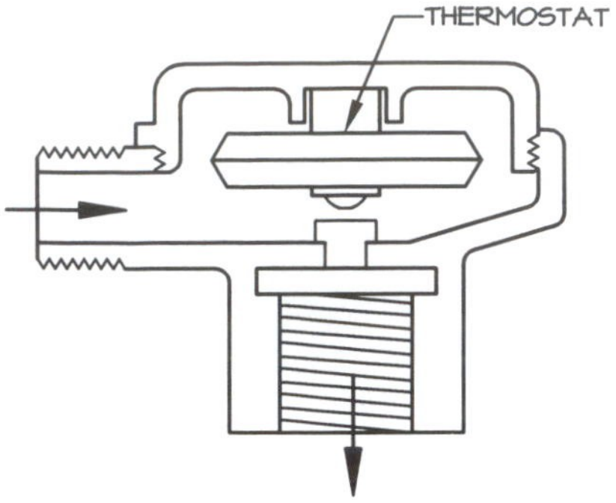 THERMOSTAT	
THERMODYNAMIC STEAM TRAP	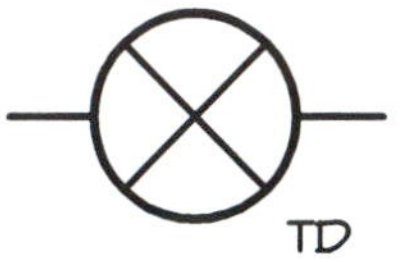 TD	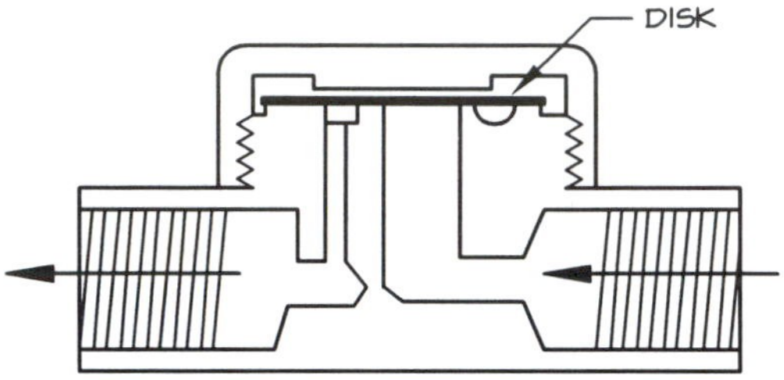 DISK	DISK MOVES UP AND DOWN. STEAM RETAINED, WATER RELEASED.
(UPRIGHT) BUCKET TRAP	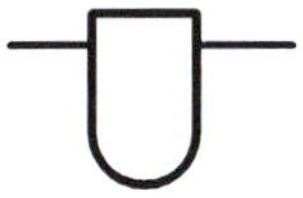	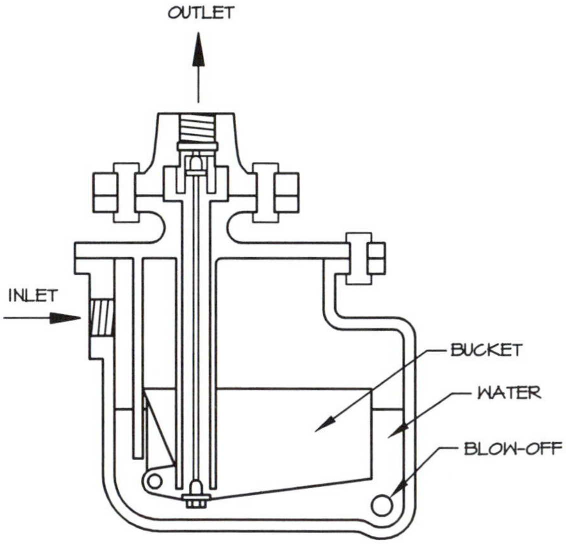 OUTLET INLET BUCKET WATER BLOW-OFF	
IMPULSE TRAP	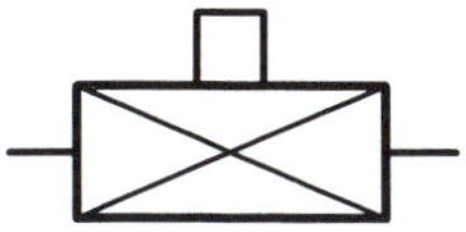		

NAME	SYMBOL	OBJECT	REMARKS
INVERTED BUCKET STEAM TRAP ASSEMBLY	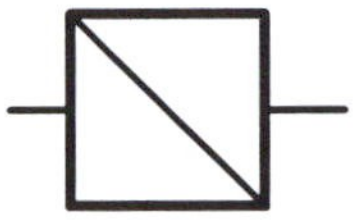	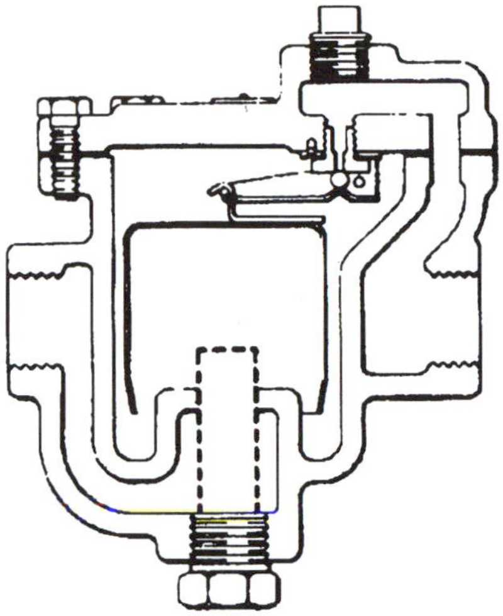	
P TRAP	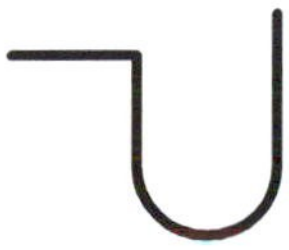	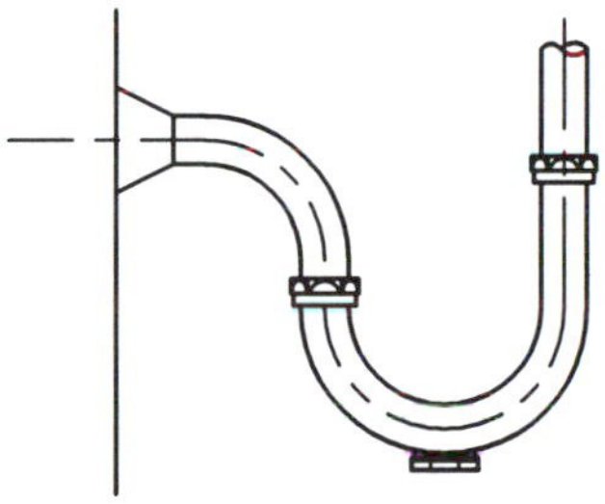	COMMONLY USED UNDER SINKS. SHOWN WITH DRAIN\CLEAN- OUT PLUG. PREVENTS GASES FROM ESCAPING FROM DRAIN.
RUNNING TRAP	ELEVATION:  PLAN: 	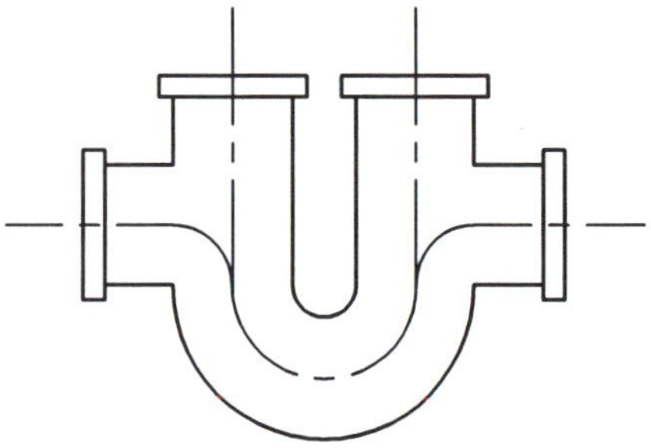	SEWER TYPE RUNNING TRAP WITH TWO VENTS ILLUSTRATED.
EXPOSED CLEAN-OUT			

NAME	SYMBOL	OBJECT	REMARKS
CLEAN-OUT IN FLOOR	O——		

SYMBOL	NAME	ELEVATION/PLAN	SECTION	REMARKS
[A] 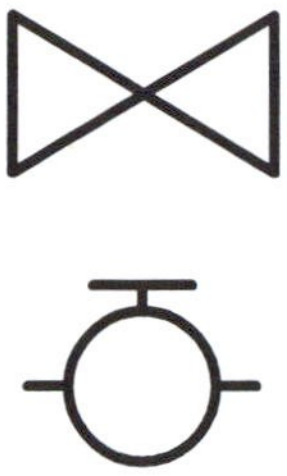	GATE VALVE NORMALLY OPEN	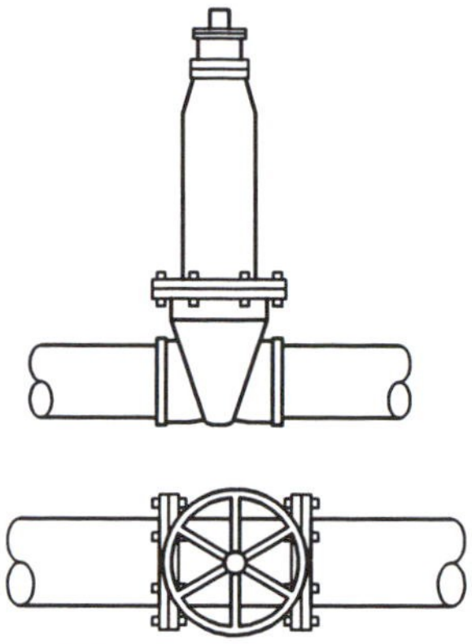	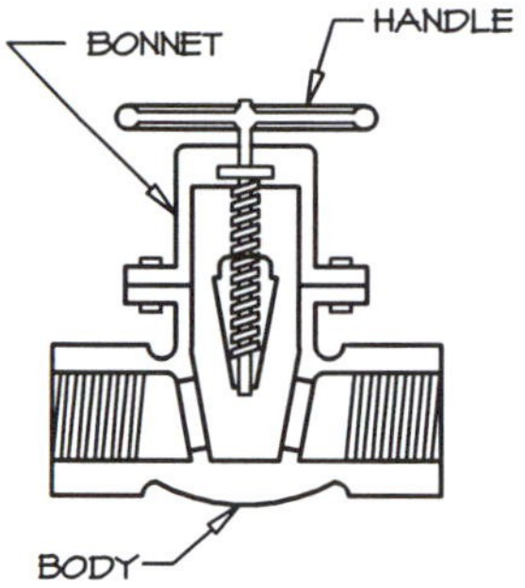 BONNET HANDLE BODY	ELEVATION SHOWS UTILITY STYLE GATE VALVE. NOTE LACK OF HANDLE - TYP. HOUSED IN VALVE BOX AND TURNED WITH KEY. THESE SYMBOLS USED FOR VALVES OF UNSPECIFIED TYPE.
	GATE VALVE NORMALLY CLOSED			FOR EXAMPLE, A BLEED VALVE OR DRAIN VALVE.
 X V-X	VALVE DESIGNATED BY ABBREVIATION OR KEY			FOR EXAMPLE: X= TSO, FOR TIGHT SHUTOFF VALVE.
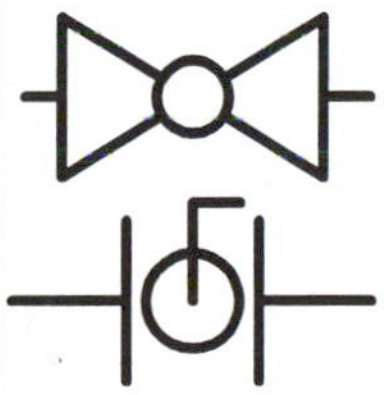	BALL VALVE		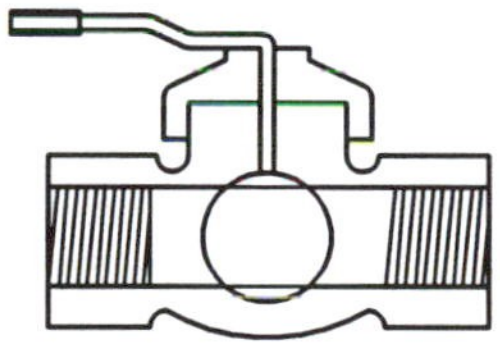	HANDLE SHOWS ALIGNMENT OF BALL.
	VEE BALL VALVE		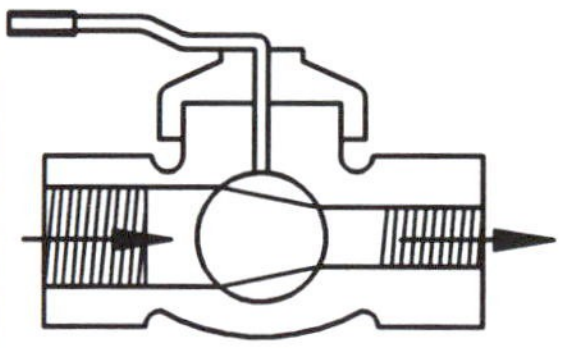	PASSAGE THRU BALL IS TAPERED.
	PIG BALL VALVE			USED IN OIL FIELDS. VALVE HAS OPENING FOR INSERTION OF CLEANING DEVICE.

SYMBOL	NAME	ELEVATION	SECTION	REMARKS
[A] 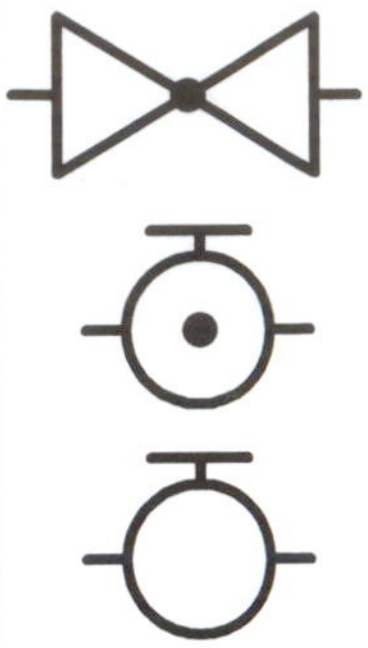	GLOBE VALVE	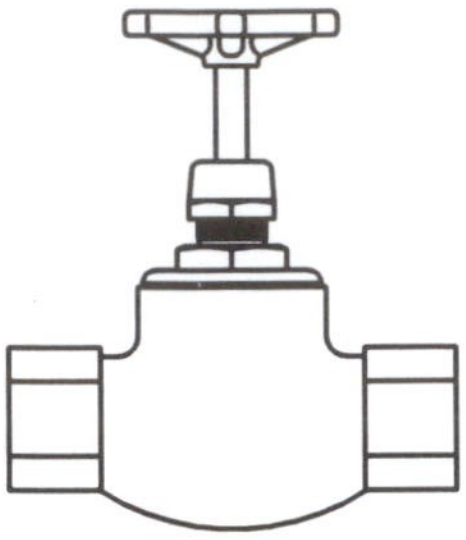	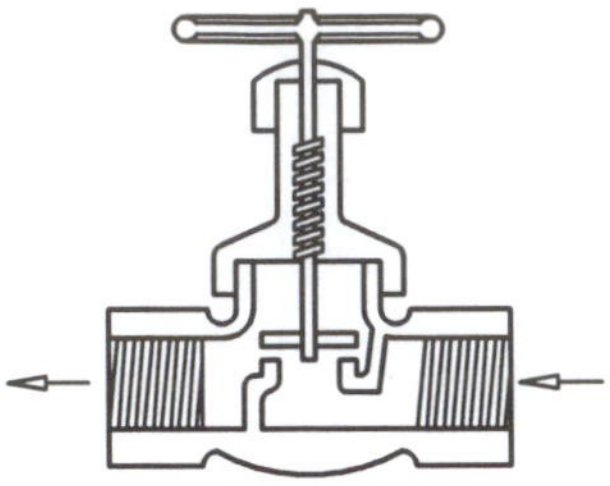	
[A] [I] 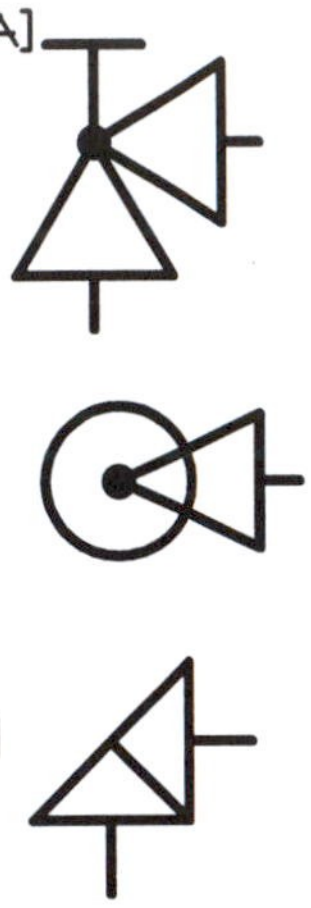	ANGLED GLOBE VALVE ANGLED GATE VALVE (PLAN)		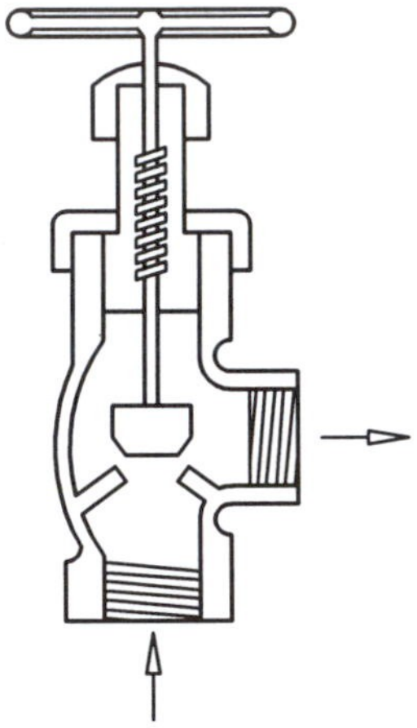	BOTTOM SYMBOL MORE TYPICALOF PROCESS DRAWINGS.
[I] 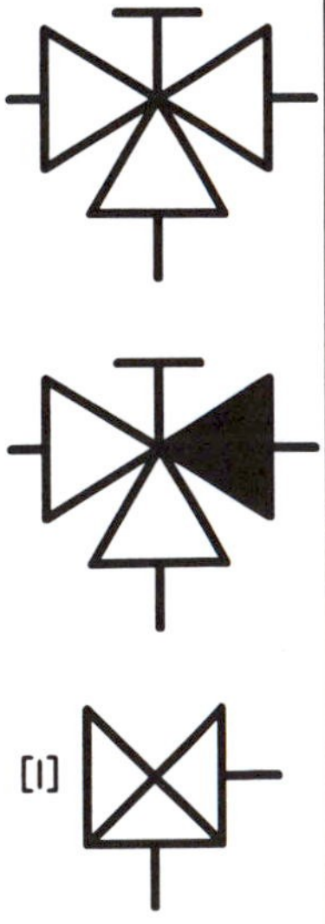	3-WAY VALVE		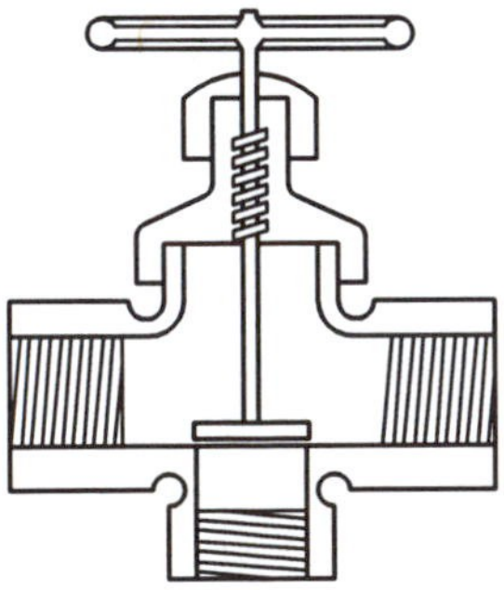	HORIZONTAL IS RUN, VERTICAL IS BRANCH. SHADING INDICATES CLOSED PORT OF VALVE. BOTTOM SYMBOL MORE TYPICAL OF PROCESS DRAWINGS.

SYMBOL	NAME	ELEVATION	SECTION	REMARKS
 [I] 	4-WAY VALVE			BOTTOM SYMBOL MORE TYPICAL OF PROCESS DRAWINGS.
[A] 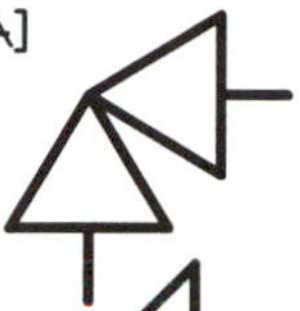  	ANGLED GATE VALVE (ELEVATION) ANGLED GATE VALVE (PLAN)			MIDDLE SYMBOL MORE TYP. OF PROCESS DRAWINGS.
[I]    	BUTTERFLY VALVE		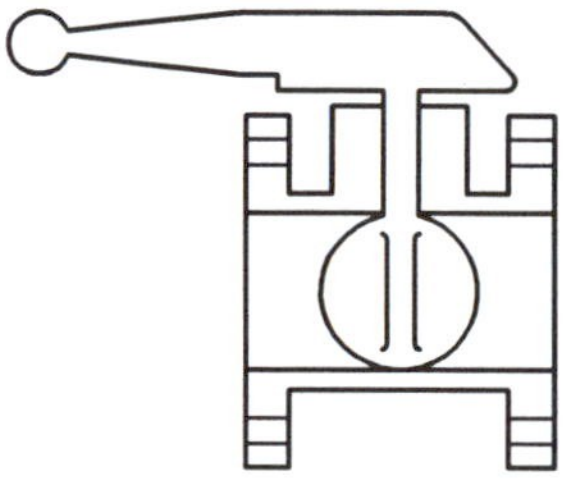	SHOWN FLANGED WITH MANUAL CONTROL. DIRECTION OF HANDLE INDICATES POSITION OF VALVE.
	INDICATING BUTTERFLY VALVE			180° DIAL ON TOP OF VALVE INDICATES DEGREE OF OPENNESS OF VALVE.

SYMBOL	NAME	ELEVATION/SECTION	REMARKS
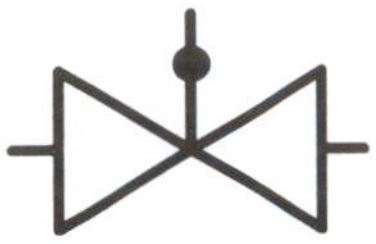	POST-INDICATOR VALVE	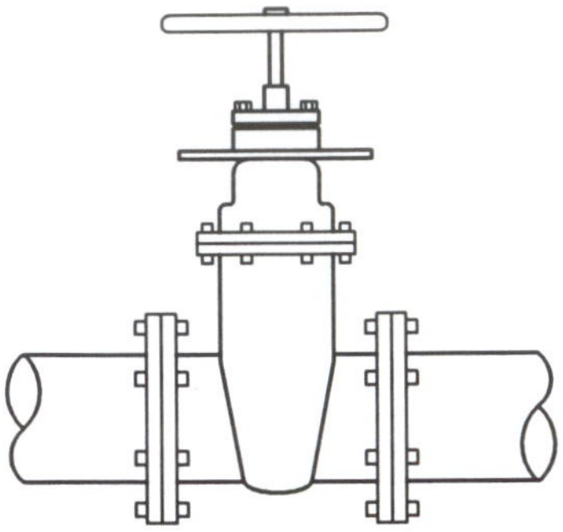	DIAL ON TOP OF VALVE INDICATES DEGREE OF OPENNESS OF VALVE.
	LISTED INDICATING VALVE		DIAL ON TOP OF VALVE INDICATES CORRECT POSITION(S) OF VALVE.
	DIAPHRAGM VALVE	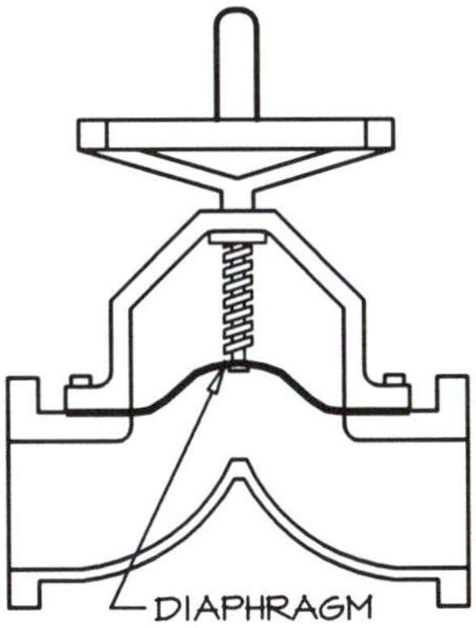 DIAPHRAGM	
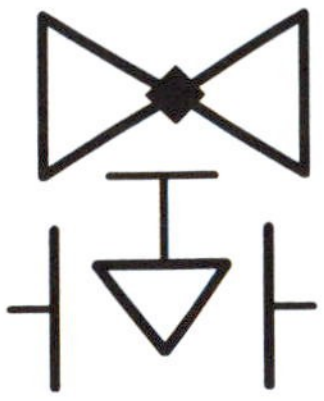	PLUG VALVE	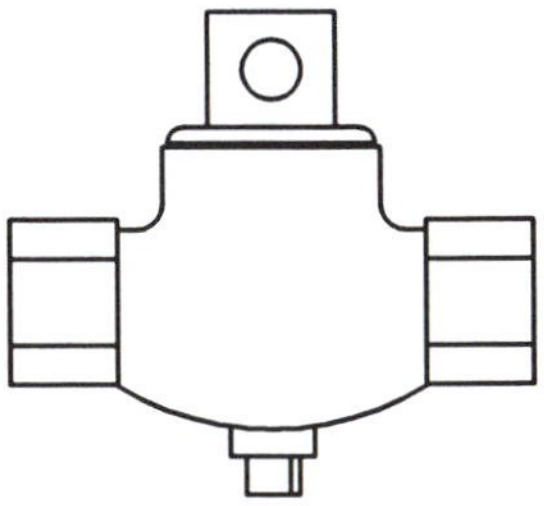	COMMONLY USED ON GAS LINES IN RESIDENTIAL APPLICATIONS.

SYMBOL	NAME	ELEVATION/SECTION	REMARKS
	ROTARY PLUG VALVE	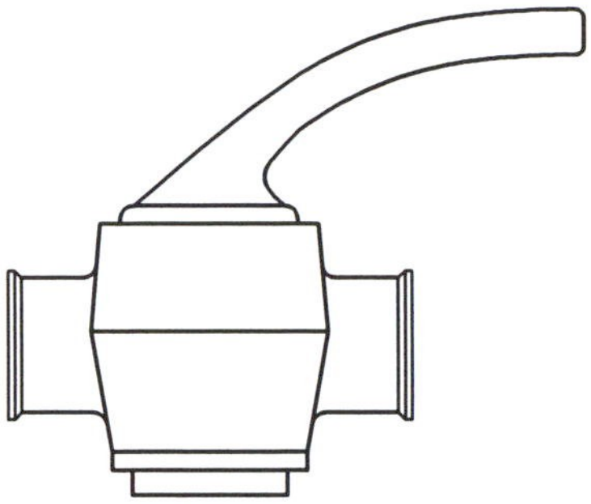	PLUG VALVE WHICH ROTATES THRU 360°.
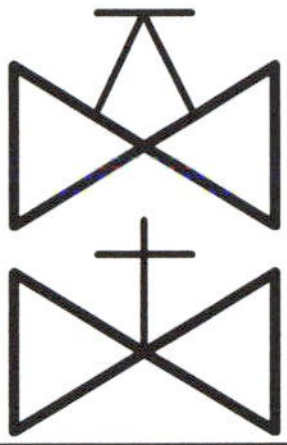	OS&Y VALVE (OUTSIDE STEM & YOKE VALVE)		STEM RISES AND FALLS WHEN OPENED & CLOSED, THUS INDICATING BY SIGHT THE STATE OF THE VALVE.
 NV	NEEDLE VALVE	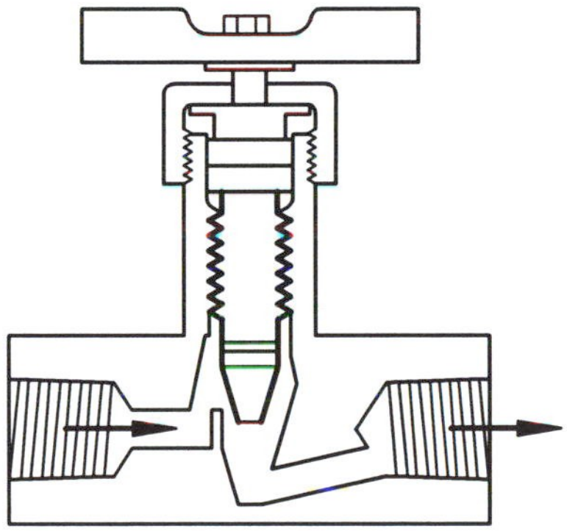	RESTRICTED ORIFICES AND NARROW SEAT PROVIDE FINE FLOW REGULATION.
[A] 	LOCK SHIELD VALVE	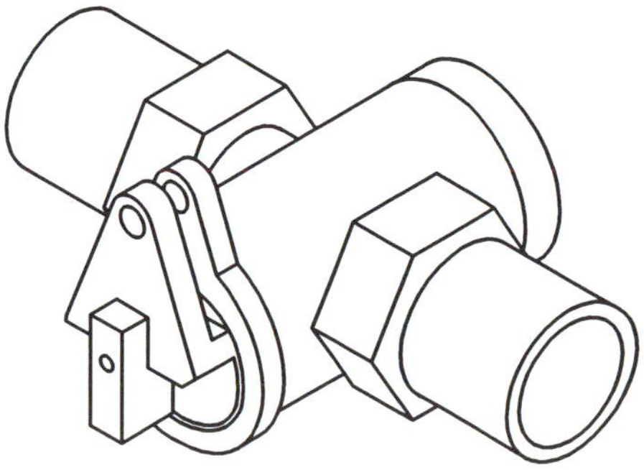	LOCKABLE VALVE.

SYMBOL	NAME	ELEVATION/SECTION	REMARKS
[A] 	QUICK OPENING VALVE		COUNTERWEIGHTS MAKE FOR QUICK VALVE MOVEMENT. COMMON IN PETRO-CHEMICAL INDUSTRY.
	QUICK CLOSING VALVE (FUSIBLE LINK)		
[A] 	SAFETY VALVE		
[A]   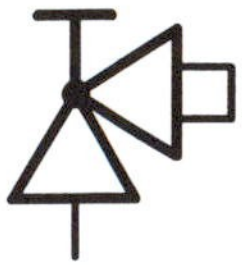	ANGLE, GATE & GLOBE HOSE VALVES		COMMONLY USED ON HOT WATER HEATERS; ATTACH HOSE TO DRAIN.

SYMBOL	NAME	ELEVATION	SECTION	REMARKS
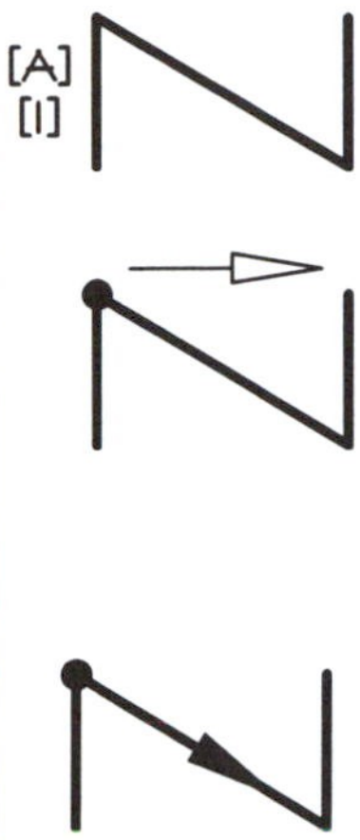 [A] [I]	CHECK VALVE (SWING GATE OR SWING CHECK VALVE)	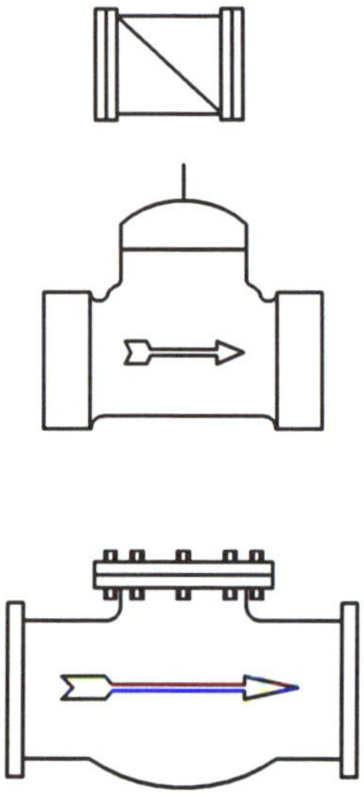	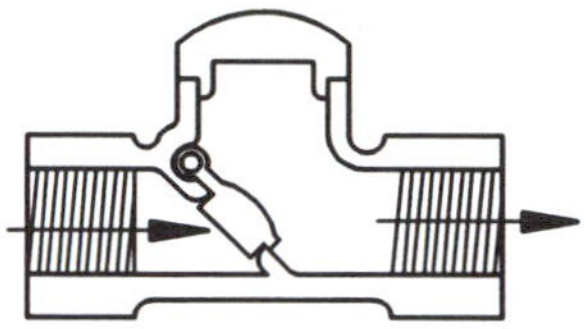	KEEPS FLOW GOING IN ONE DIRECTION. ALSO CALLED BACK -FLOW PREVENTER (BFP).
 S	SPRING CHECK VALVE		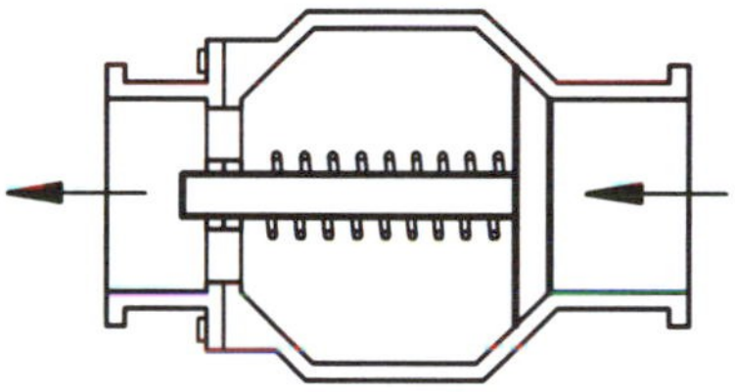	
	BALL TYPE OR AIR BLOW CHECK VALVE		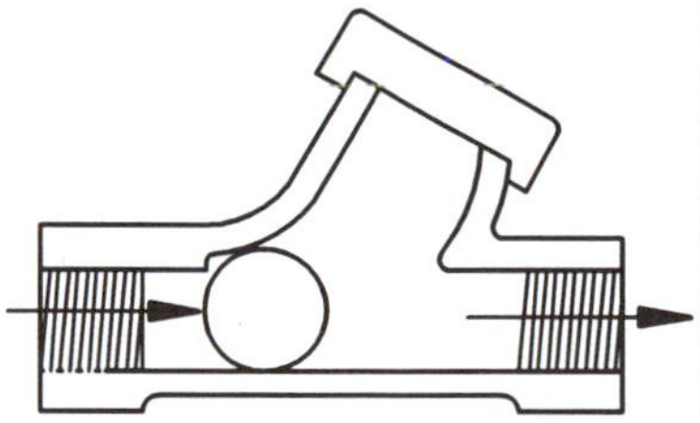	
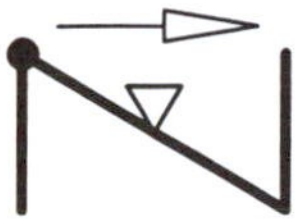	CHECK VALVE, HORIZONTAL GATE		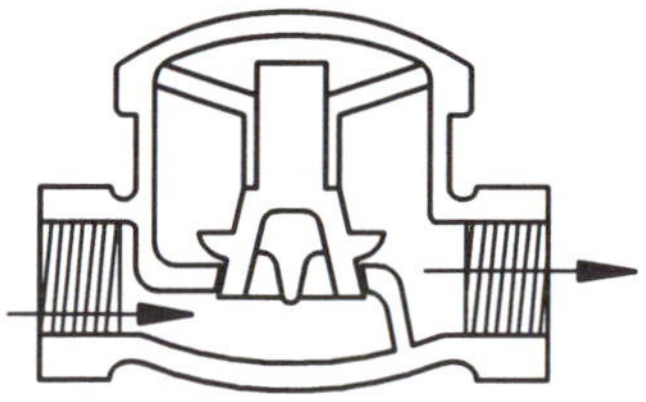	ALSO CALLED LIFT CHECK VALVE.

SYMBOL	NAME	ELEVATION/SECTION	REMARKS
[A] 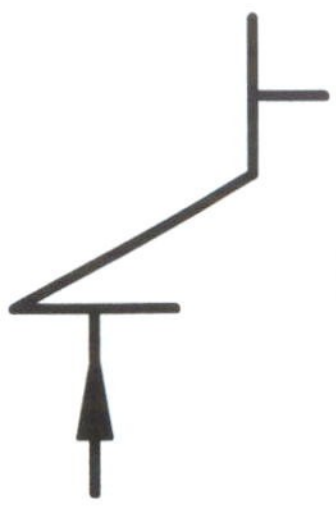	ANGLE CHECK VALVE	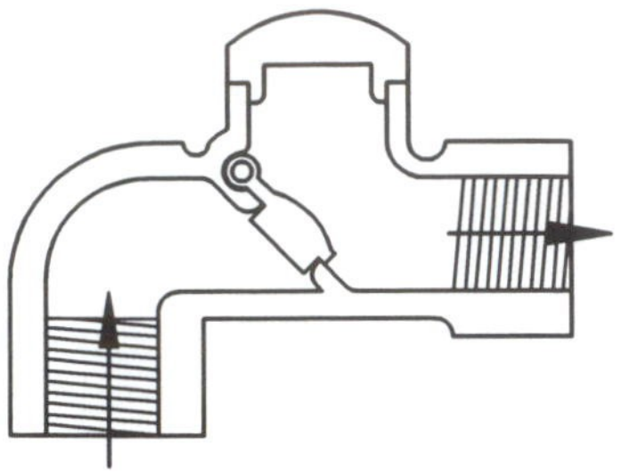	ARCHAIC. REPLACED BY IN-LINE SPRING CHECK VALVE.
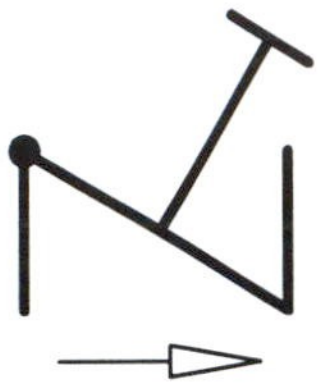	STOP CHECK VALVE	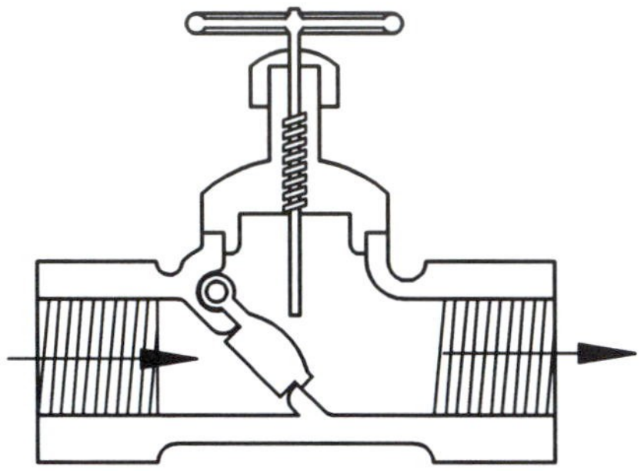	CLAPPER CONTROLLED WITH HANDLE.
	AIR LINE VALVE	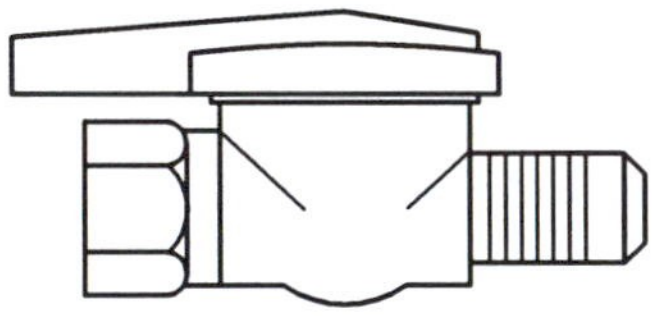	BALL VALVE DESIGNED FOR NATURAL, MFG. GASES AND AIR.
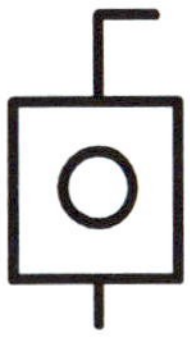	MANUAL AIR VENT		
	UMBRELLA COCK		

SYMBOL	NAME	ELEVATION/SECTION	REMARKS
[A]  R	PRESSURE RELIEF VALVE	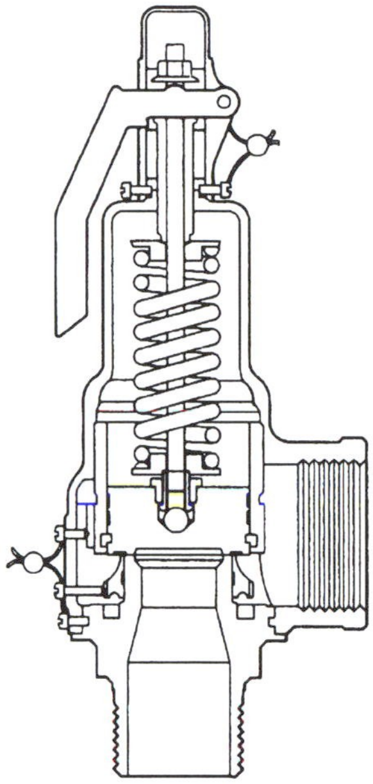	CONTROLLED BY SPRING. MANUAL RELIEF LEVER AT TOP. FLUID EXITS TO DRAIN.  NORMALLY CLOSED VALVE THAT IS ACTUALLY CLOSED.  NORMALLY CLOSED VALVE THAT IS ACTUALLY OPEN.
 T&P (SYMBOLS AS IN PREVIOUS, WITH NOTATION.)	PRESSURE AND TEMPERATURE RELIEF VALVE	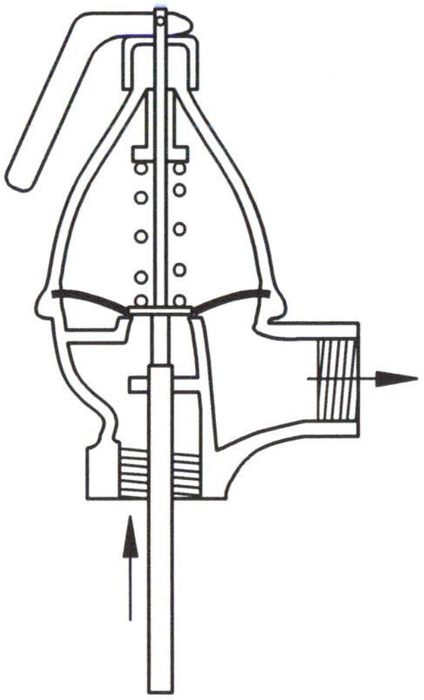	SECTION SHOWS TEMPERATURE SENSOR IN INLET PORT, AND MANUAL RELIEF LEVER ON TOP.
	KEY-OPERATED VALVE		REQUIRES SPECIAL KEY TO OPEN AND CLOSE.

SYMBOL	NAME	ELEVATION/SECTION	REMARKS
	PRESSURE REDUCING VALVE	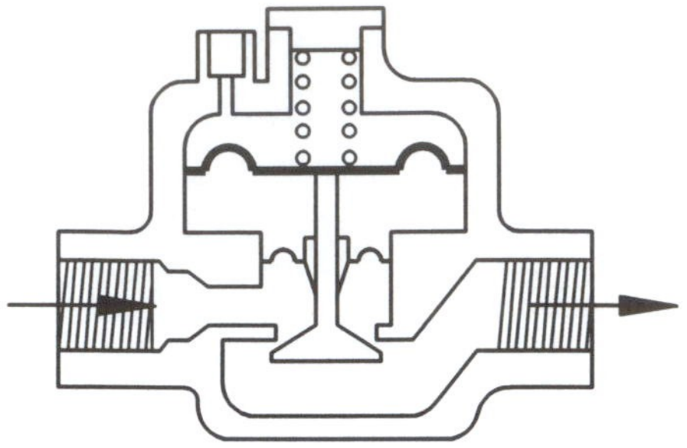	SPRING CONTROL IS ADJUSTABLE.
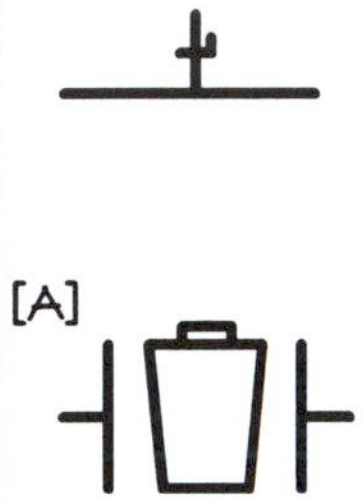	PETCOCK OR COCK OR LABCOCK	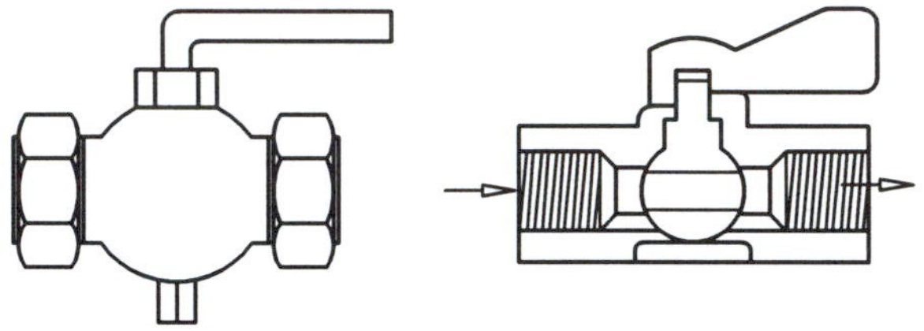	FREQUENTLY USED TO ISOLATE A GAUGE.
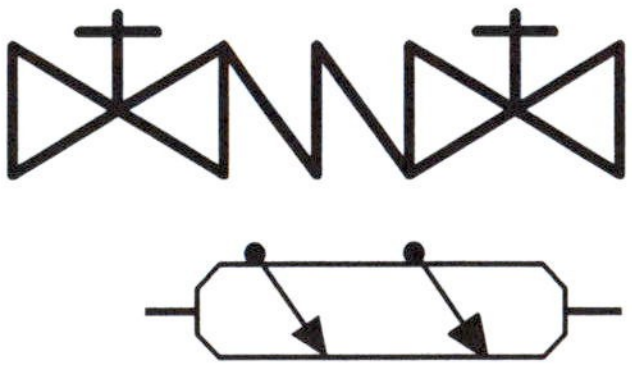	BACKFLOW PREVENTER - DOUBLE CHECK TYPE		
BACKFLOW PREVENTER - REDUCED PRESSURE ZONE TYPE 		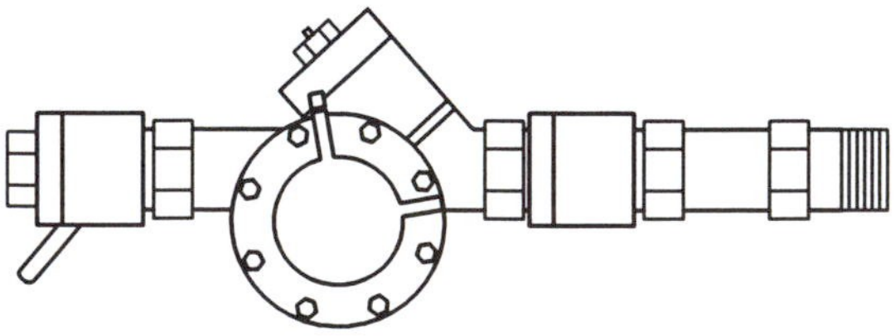	USED WHERE BACK-FLOW PREVENTION IS CRITICAL, SUCH AS POSSIBLE CONTAMIN-ATION.

SYMBOL	NAME	ELEVATION/SECTION	REMARKS
 	VALVE IN PIT OR VALVE BOX	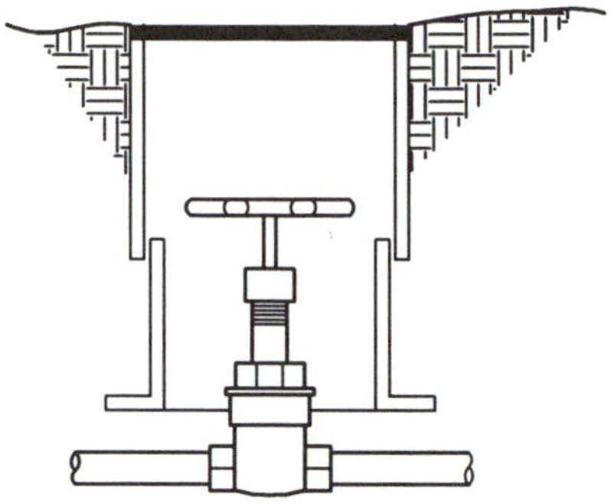	TELESCOPING STYLE VALVE BOX SHOWN WITH BRONZE GATE VALVE.
	WALL BOX VALVE	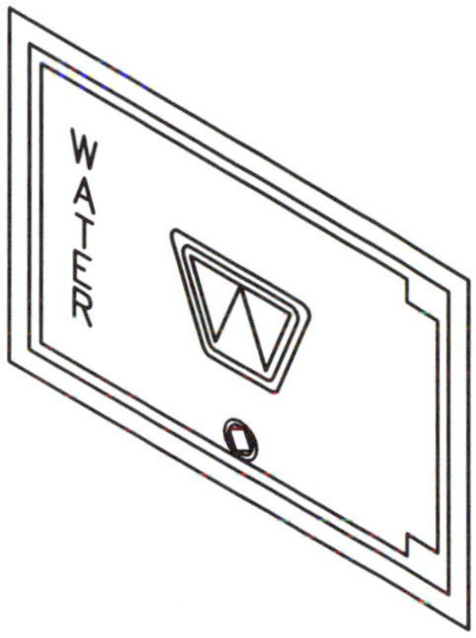 WATER	ILLUSTRATION SHOWS OUTSIDE WATER VALVE BOX.
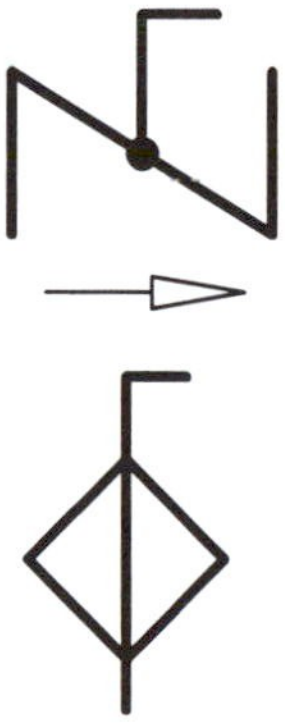	TIPLE DUTY VALVE		PERFORMS THREE FUNCTIONS: ON/OFF BALANCE FLOW PREVENT BACKFLOW
	ANGLED BLOWDOWN VALVE		

SYMBOL	NAME	ELEVATION/SECTION	REMARKS
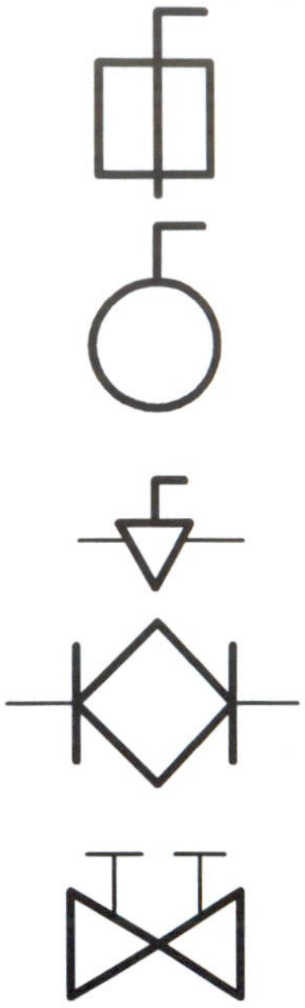	BALANCE VALVE	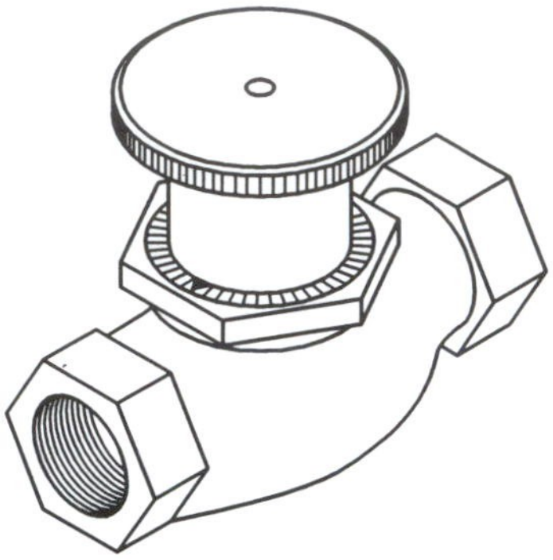	USED IN HOT WATER SYSTEMS TO BLEND WATER AT DIFFERENT TEMPERATURES. SOMETIMES CALLED A 'CIRCUIT SETTER.'
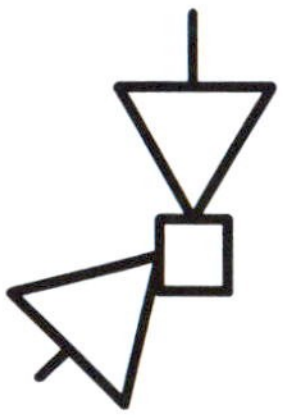	RAM DRAIN VALVE		ARCHAIC. USES AIR PRESSURE TO DRAIN VALVE.
[A] 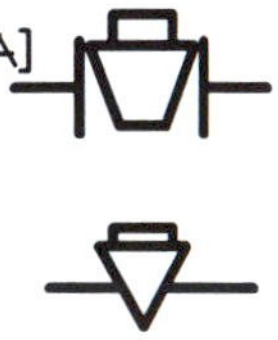	SQUARE HEAD COCK	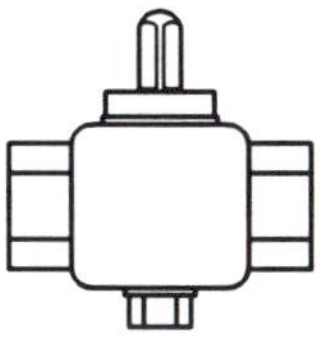	
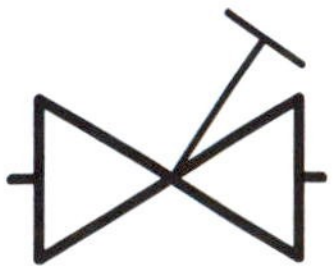	"Y" VALVE	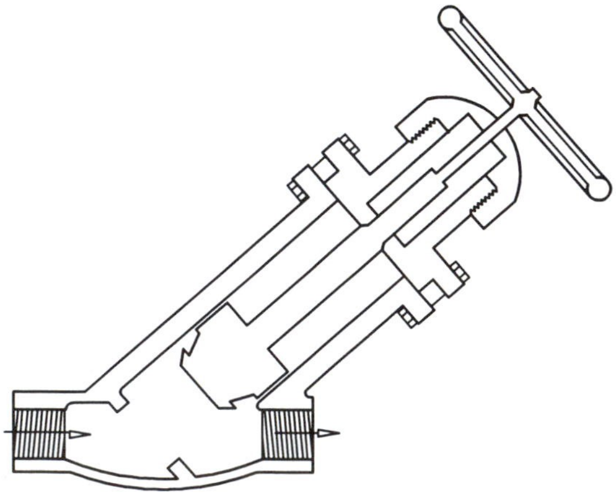	

SYMBOL	NAME	ELEVATION/SECTION	REMARKS
 IhV	INSULATED VALVE BODY		
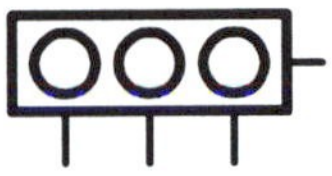	MANIFOLD VALVE	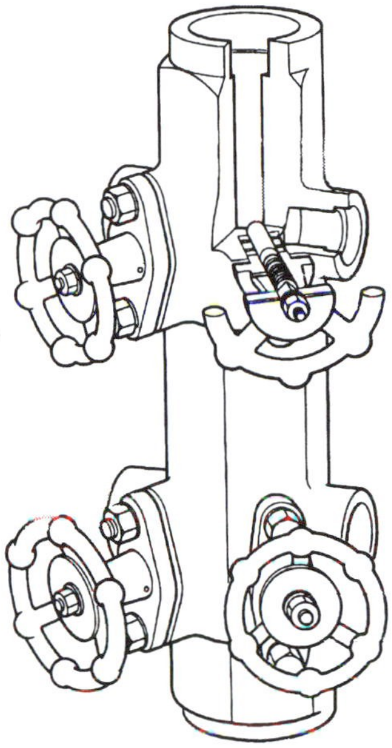	
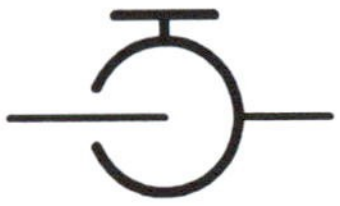	VALVE IN RISER	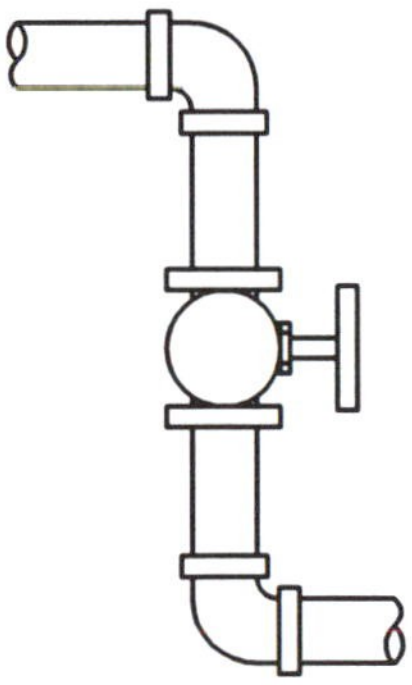	
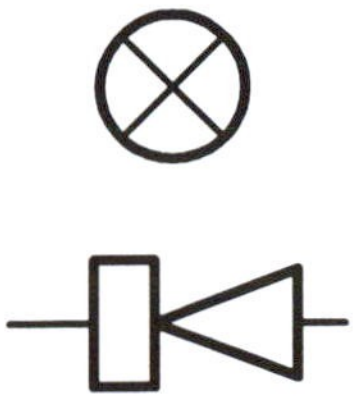	AUTOMATIC EXPANSION VALVE		

SYMBOL	NAME	ELEVATION/SECTION	REMARKS
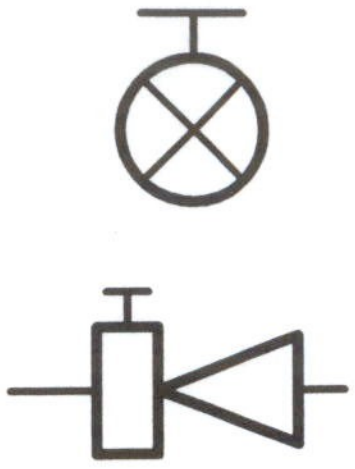	MANUAL EXPANSION VALVE		APPLICATIONS IN REFRIGERATION
	EXPANSION VALVE - THERMOSTATIC		
	HOSE BIBB	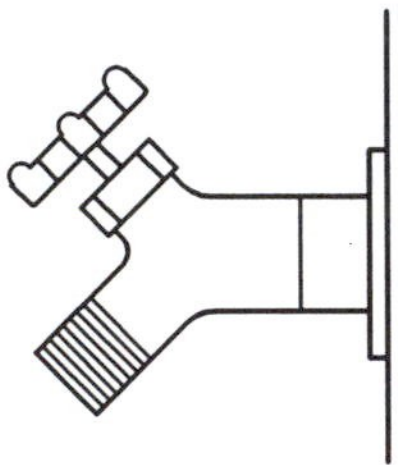	TYPICALLY SOLDERED TO PIPE IN WALL. THREADED FOR HOSE.
	FAUCET		
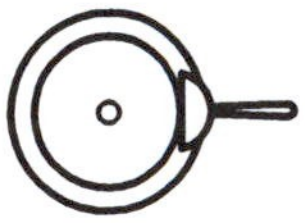	FLUSH VALVE	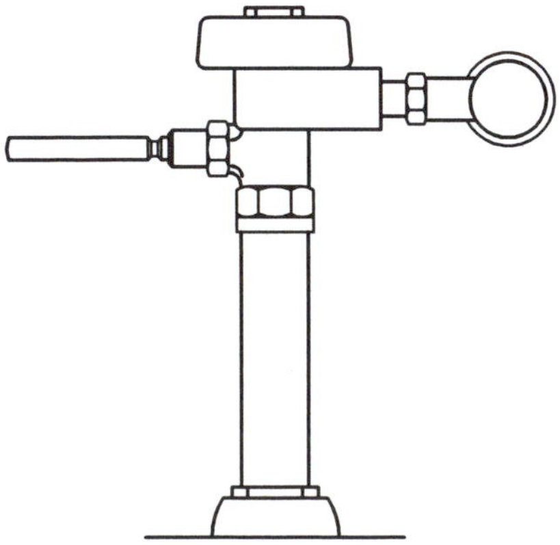	MAY BE FITTED WITH LEVER AS SHOWN, WITH KICK PLATE OR ELECTRONIC SENSOR.

SYMBOL	NAME	ELEVATION/SECTION	REMARKS
	STOP COCK PLUG OR CYLINDER VALVE: 2-WAY		
	STOP COCK PLUG OR CYLINDER VALVE: 3-WAY 2-PORT		FREQUENTLY USED IN TOILET INSTALLATIONS.
	STOP COCK PLUG OR CYLINDER VALVE: 3-WAY 3-PORT		
	STOP COCK PLUG OR CYLINDER VALVE: 4-WAY 4-PORT		

NAME	SYMBOL	OBJECT	REMARKS
HAND CONTROL VALVE		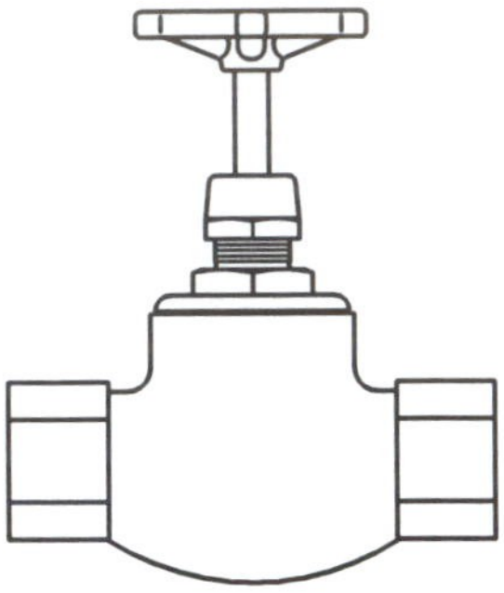	
DIAPHRAGM OPERATED CONTROL VALVE	[A] 	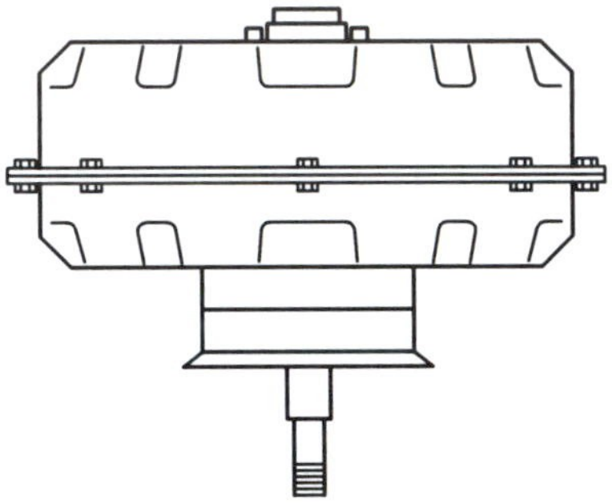	ELEVATION SHOWS CONTROL MECHANISM ONLY
VALVE ACTION (USE ABBREV.)	 FO		
CYLINDER ACTUATED VALVE (SINGLE ACTING)	PNEUMATIC 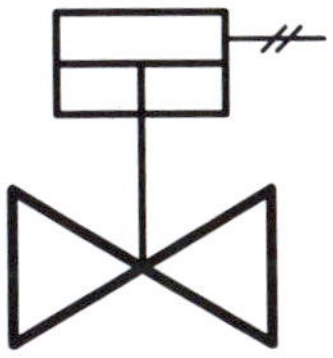 HYDRAULIC 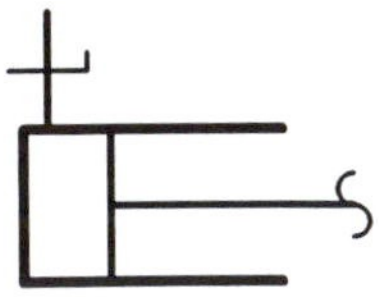		

NAME	SYMBOL	OBJECT	REMARKS
CYLINDER ACTUATED VALVE (DOUBLE ACTING) VALVE POSITIONER	PNEUMATIC  HYDRAULIC 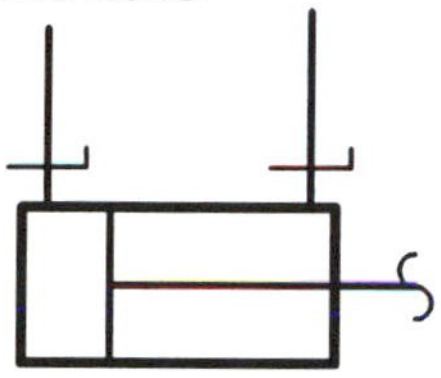 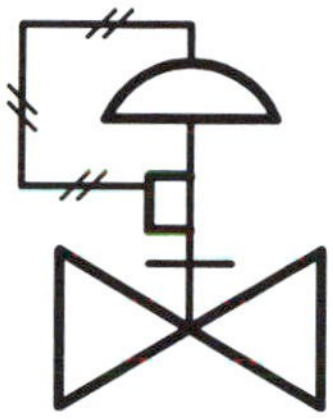	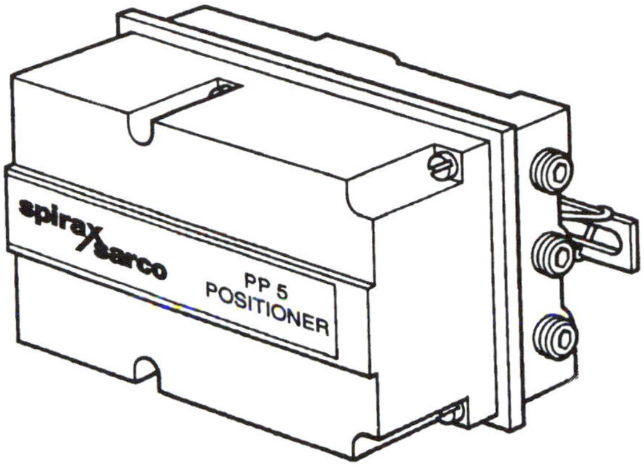 spirax sarco PP 5 POSITIONER	
DOUBLE ACTING DOUBLE END ROD CYLINDER	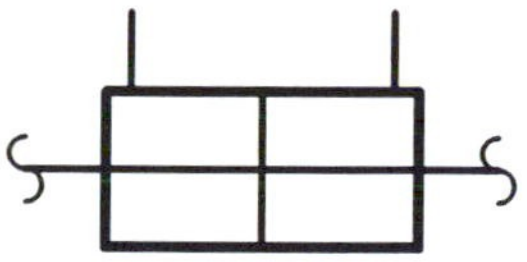		USEFUL FOR SIMULTANEOUSLY OPENING & CLOSING TWO VALVES OR GATES.
REGULATOR SELF-CONTAINED			
SINGLE ACTING-PRESSURE OPENS VALVE	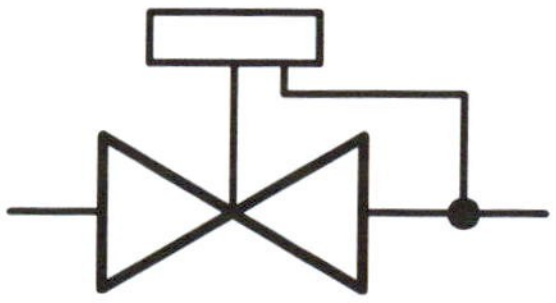		

NAME	SYMBOL	OBJECT	REMARKS
SINGLE ACTING-PRESSURE CLOSES VALVE	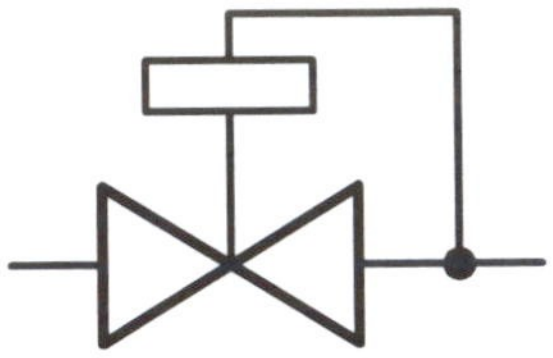	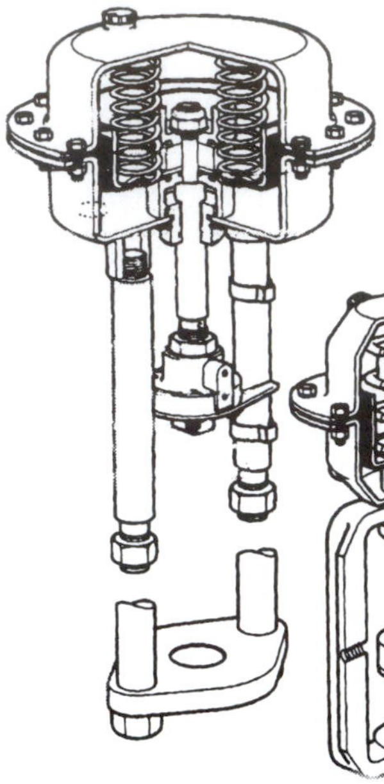	ACTUATORS SHOWN WITHOUT VALVES.
GLOBE VALVE - AIR OPERATED, SPRING CLOSING	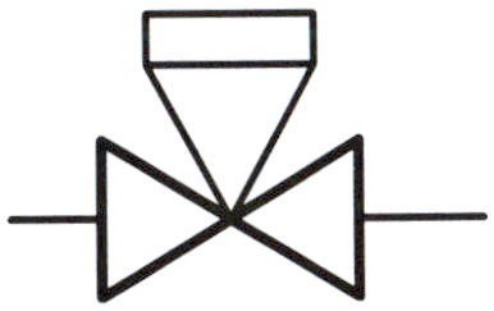		
GLOBE VALVE - DECK OPERATED			
FLOW REGULATOR, SELF-CONTAINED			
FLOW REGULATOR, SELF-CONTAINED	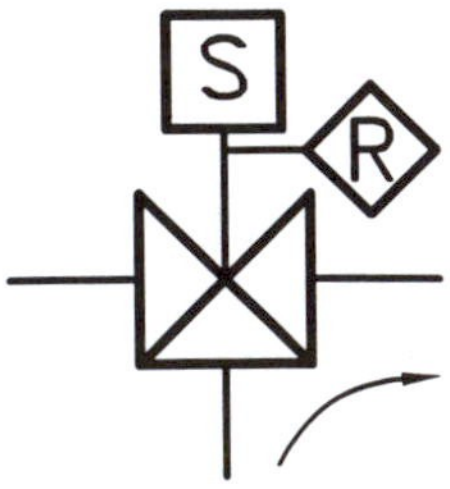 S R		ARROW INDICATES FLOW PATH IN THE DE-ENERGIZED STATE, WITH THE TAIL AT THE COMMON PORT.

NAME	SYMBOL	OBJECT	REMARKS
FLOAT VALVE	[A] 	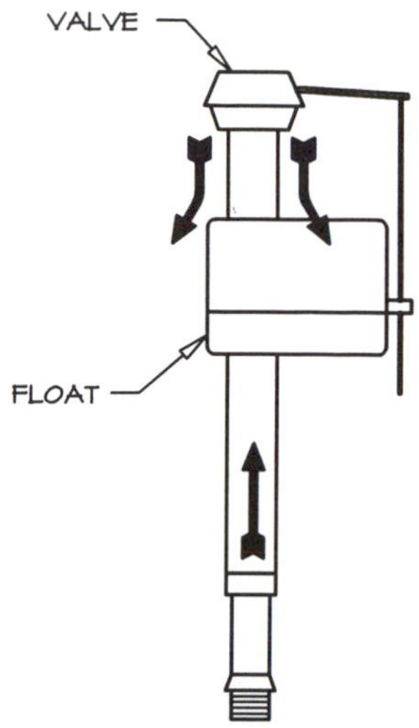 VALVE FLOAT	ILLUSTRATION SHOWS TOILET FLUSH VALVE.
ROTARY ACTUATED VALVE	 M		
MOTOR OPERATED CONTROL VALVE	[A]  M		
SOLENOID OPERATED CONTROL VALVE	 S		
PRESSURE-BALANCED DIAPHRAGM ACTUATED	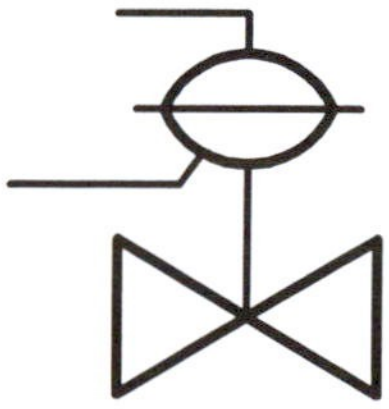		

NAME	SYMBOL	OBJECT	REMARKS
DIFFERENTIAL PRESSURE REGULATOR VALVE			
VALVE POSITIONER	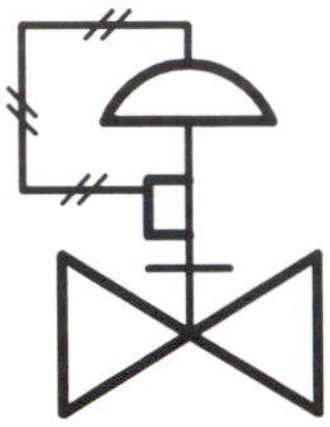		
DIAPHRAGM OPERATED CONTROL VALVE WITH POSITIONER & TRAVEL STOP			
AUTOMATIC 3-WAY VALVE			
PNEUMATIC 3-WAY VALVE	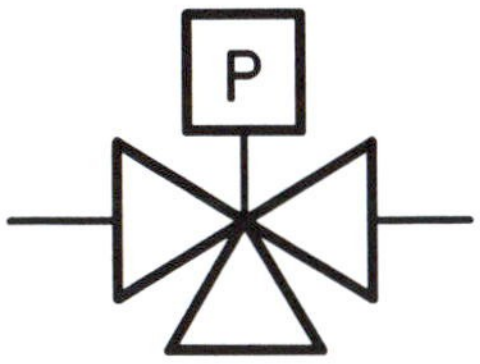 P	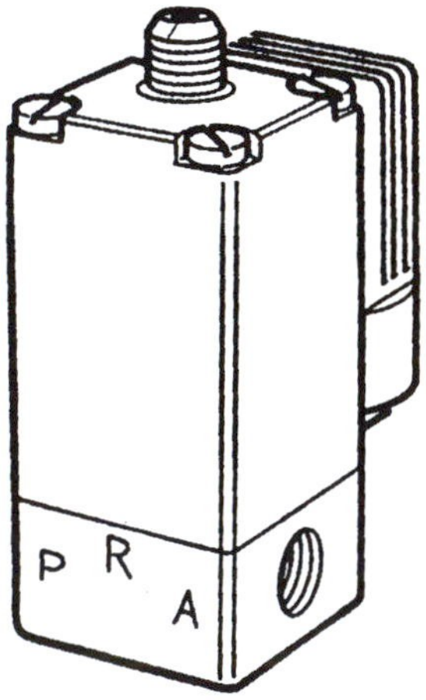 P R A	

NAME	SYMBOL	OBJECT	REMARKS
AUTOMATIC REDUCING VALVE	[A] 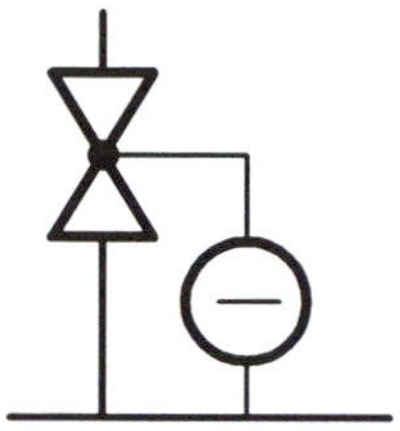		
AUTOMATIC BY-PASS VALVE	[A] 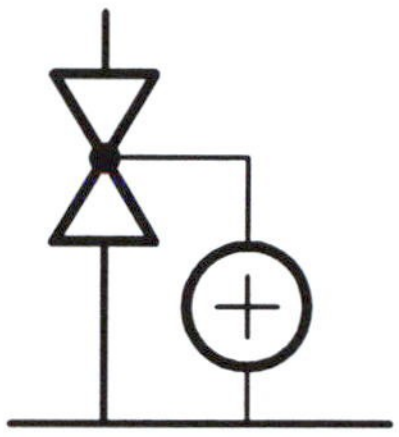		
4-WAY SOLENOID VALVE	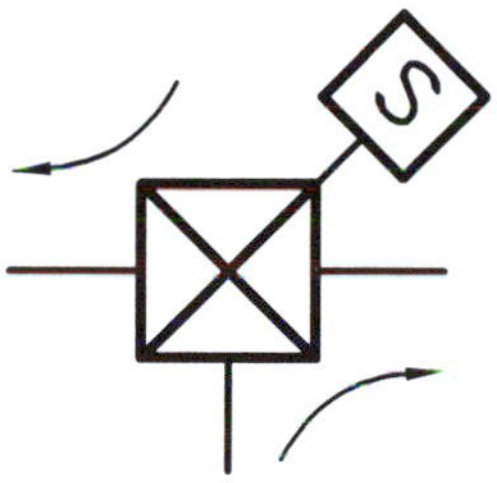 S		ARROW INDICATES FLOW PATH IN THE DE-ENERGIZED STATE, WITH THE TAIL AT THE COMMON PORT.
CONTROL VALVE WITH HAND ACTUATOR			
GOVERNOR-OPERATED VALVE	[A] 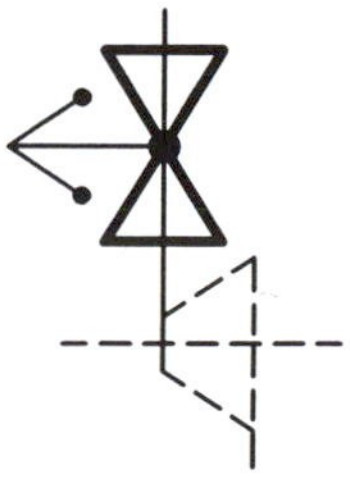		

NAME	SYMBOL	OBJECT	REMARKS
ELECTRO-HYDRAULIC ACTUATOR			AN ELECTRC SIGNAL ACTUATES HYDRAULIC LINES WHICH OPERATE VALVE.
ELECTRO-PNEUMATIC ACTUATOR			AN ELECTRIC SIGNAL ACTUATES PNEUMATIC LINES WHICH OPERATE VALVE.

LINETYPES

DESCRIPTION	RISER	LINE SYMBOL
OXYGEN	O	O
LIQUID OXYGEN	LOX	LOX
NITROGEN	N	N
LIQUID NITROGEN	LN	LN
HYDROGEN	H.	H
HELIUM	H.E.	HG
ARGON	AR	AR
LIQUID PETROLEUM GAS	LPG	LPG
PNEUMATIC TUBE	PN	
OXYGEN	O	O
LIQUID OXYGEN	LOX	LOX
NITROGEN	N	N
LIQUID NITROGEN	LN	LN
HYDROGEN	H.	H
HELIUM	H.E.	HG
ARGON	AR	AR
LIQUID PETROLEUM GAS	LPG	LPG

DESCRIPTION	LINE SYMBOL
PRINCIPAL SYSTEM	————————————
SECONDARY SYSTEM	——— – ——— – ———
TERTIARY SYSTEM	—— – – —— – – ——
QUARTERNARIAL SYSTEM	— – – – —— – – – —
FIFTH SYSTEM IN IMPORTANCE	— — — — — — — —
INVISIBLE LINES (LABEL)	– – – – – – – – – – – – – – – – – –

DESCRIPTION	RISER	LINE SYMBOL
AIR–RELIEF LINE	AR	[A] 1/16" 1/4" 1/2"
BOILER BLOW–OFF	BBO	[A] 3/8" 1/16"
CONTINUOUS BLOW–DOWN	CBD	CBD
BOILER WATER SAMPLE	BWS	BWS
LOW PRESSURE GAS	G	[A] G
MEDIUM PRESSURE GAS	MG	MG
HIGH PRESSURE GAS	HG	HG
NAT. GAS IGNITOR FUEL	G(I)	G(I)
LIQ. PET. GAS IGNITOR FUEL	LPG(I)	LPG(I)
COMPRESSED AIR	A	A
HIGH PRESSURE STEAM	HPS	[A]
HIGH PRESSURE RETURN	HPR	[A]
MEDIUM PRESSURE STEAM	MPS	[A]
MEDIUM PRESSURE RETURN	MPR	[A]
LOW PRESSURE STEAM	LPS	[A]
LOW PRESSURE RETURN	LPS	[A]
MAKE–UP WATER	MUW	[A] 1/2" 1/16" 1/32"
CONDENSATE LINE	C	C [A] ø1/16" 3/8" 1/16" 3/8"

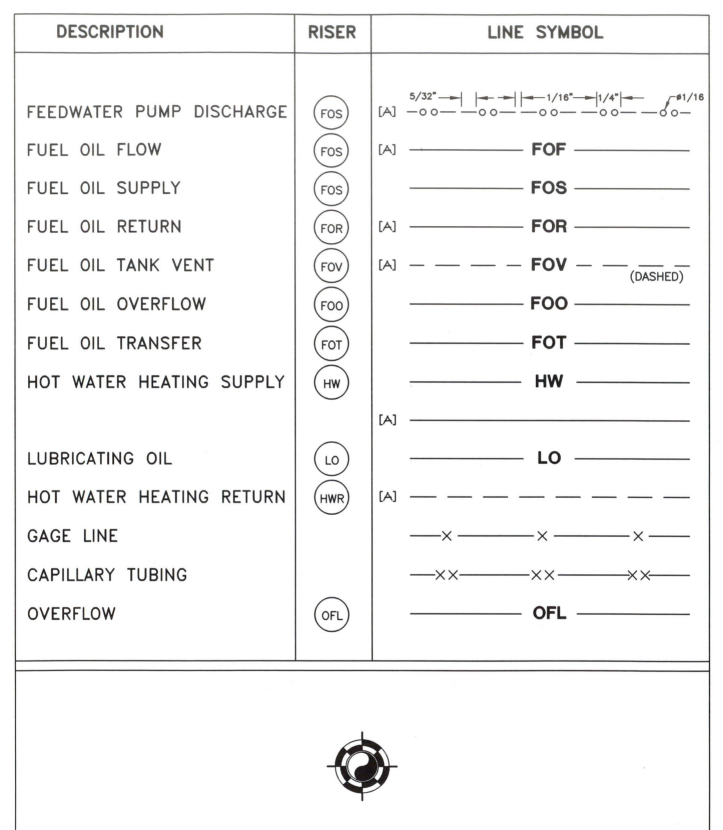

DESCRIPTION	RISER	LINE SYMBOL
FEEDWATER PUMP DISCHARGE	FOS	[A] —o o— —o o— —o o— —o o— —o o— (5/32", 1/16", 1/4", ø1/16)
FUEL OIL FLOW	FOS	[A] ——— FOF ———
FUEL OIL SUPPLY	FOS	——— FOS ———
FUEL OIL RETURN	FOR	[A] ——— FOR ———
FUEL OIL TANK VENT	FOV	[A] — — — — FOV — — — — (DASHED)
FUEL OIL OVERFLOW	FOO	——— FOO ———
FUEL OIL TRANSFER	FOT	——— FOT ———
HOT WATER HEATING SUPPLY	HW	——— HW ———
		[A] ———————
LUBRICATING OIL	LO	——— LO ———
HOT WATER HEATING RETURN	HWR	[A] — — — — — — — —
GAGE LINE		——×——×——×——
CAPILLARY TUBING		——××——××——××——
OVERFLOW	OFL	——— OFL ———

GENERAL RULES:

TWO LINE WIDTHS, HAVING A THICKNESS RATIO OF 2 TO 1, ARE RECOMMENDED FOR USE ON MECHANICAL DRAWINGS. THE THIN LINE SHOULD BE AT LEAST .3 MM (.012 IN) AND THE HEAVY LINE SHOULD BE AT LEAST 0.6 MM (.024 IN). SPACING BETWEEN ADJACENT LINES MAY BE EXAGGERATED TO NOT MORE THAN 3 MM (.12 IN) FOR PURPOSES OF CLARITY WHEN REPRODUCING A DRAWING.

LINETYPE	SAMPLE	REMARKS
VISIBLE LINE	————————	REPRESENTS VISIBLE EDGES AND CONTOURS OF OBJECTS. USE THICK LINE.
HIDDEN LINE	— — — — — — — —	REPRESENTS HIDDEN EDGES & CONTOURS. SHOULD ALWAYS BEGIN AND END WITH A SOLID LINE (CAD LIMITATIONS EXCEPTED). USE THIN LINE. HIDDEN LINES MAY BE OMITTED WHEN NOT ESSENTIAL TO THE DRAWING. FEATURES BEHIND TRANSPARENT MATERIALS SHOULD BE TREATED AS HIDDEN.
CENTER LINE	——— — ———— — ———	CENTER LINES ARE USED TO REPRESENT AXES OR CENTER PLANES OF SYMMETRICAL PARTS & FEATURES, BOLT CIRCLES, AND PATHS OF MOTION. USE THIN LINE. SHOULD START AND END WITH LONG LINE. SHOULD EXTEND EQUALLY AND APPROPRIATELY A SHORT DISTANCE BEYOND THE OBJECT OR FEATURE. SHOULD NOT EXTEND FROM ONE VIEW TO ANOTHER.

MACHINED PARTS LINETYPES

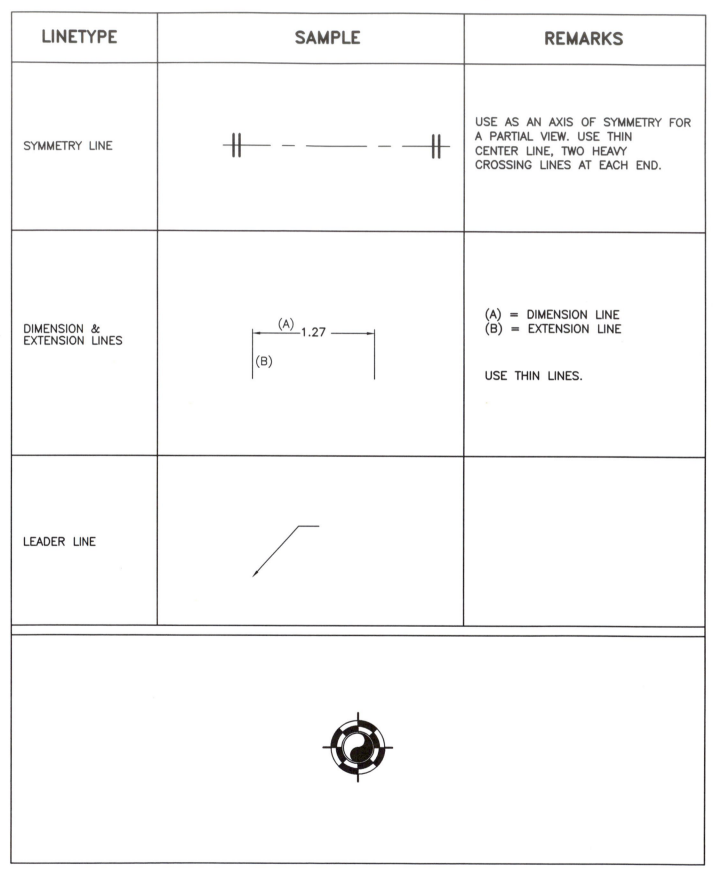

LINETYPE	SAMPLE	REMARKS
SYMMETRY LINE		USE AS AN AXIS OF SYMMETRY FOR A PARTIAL VIEW. USE THIN CENTER LINE, TWO HEAVY CROSSING LINES AT EACH END.
DIMENSION & EXTENSION LINES	(A) 1.27 (B)	(A) = DIMENSION LINE (B) = EXTENSION LINE USE THIN LINES.
LEADER LINE		

LINETYPE	ABBREV.	SAMPLE	REMARKS
WATER LINE	WT LN		
GAS LINE	G LN		
SANITARY SEWER	SAN SW		
SEWER TILE	SW TL		
SEPTIC FIELD LEACH LINE	SP FLD LCH LN		
PROPERTY CORNER	PROP CR		CIRCLE DIAMETER 1/8" TO 3/16" APPROX.

LINETYPE	ABBREV.	SAMPLE	REMARKS
STATE BOUNDARY	BND ST		
COUNTY BOUNDARY	BND CNTY		
TOWN BOUNDARY	BND TWN		
CITY BOUNDARY, INCORPORATED	BND CTY		
NATIONAL BOUNDARY OR STATE RESERVATION	BND NAT OR ST RES		

LINETYPE	ABBREV.	SAMPLE	REMARKS
SMALL AREAS; PARKS, AIRPORTS, ETC.	BND	– – – – – – – – – – – – –	
LAND GRANT BOUNDARY	BND LD GR	—— - - —— - - —— - - ——	
U.S. LAND SURVEY TOWNSHIP BOUNDARY	BND US LD SUR TWN	————————————	
TOWNSHIP BOUNDARY, APPROXIMATED	BND TWN	– – – – – – – – – – – – –	
SECTION LINE BOUNDARY, U.S.	BND SEC LN US LD SUR	————————————	
SECTION BOUNDARY LINE, APPROXIMATE	BND SEC LN	– – – – – – – – – – – – –	

LINETYPE	ABBREV.	SAMPLE	REMARKS
TOWNSHIP BOUNDARY, (NOT U.S. LAND SURVEY)	BND TWN	. . . .	
FENCE		——X——X——X——	

DESCRIPTION	RISER	LINE SYMBOL
WET STANDPIPE	WSP	WSP
DRY STANDPIPE	DSP	DSP
FIRE LINE	F	[A] F
MAIN SUPPLIES SPRINKLER	S	[A] S
BRANCH & HEAD SPRINKLER	BHS	[A] ø1/16
LOW PRESSURE GAS	G	[A] G
MEDIUM PRESSURE GAS	MG	MG
HIGH PRESSURE GAS	HG	HG
COMPRESSED AIR	A	A
VACUUM	V	V
VACUUM CLEANING	VC	VC [A] V
VENT	O	[A] - - - - - - - -
SEWER – CAST IRON	CSI	CSI
SEWER – CLAY TILE (BELL & SPIGOT)	SCT	SCT
COLD WATER – SUPPLY	CW	[A] —— – —— – ——
COLD WATER – RETURN	CWR	[A] —— – – —— – – ——
HOT WATER	HW	[A] —— – – —— – – ——
HOT WATER RETURN	HWR	[A] — – – – — – – – —
SOFTENED WATER	SW	SW

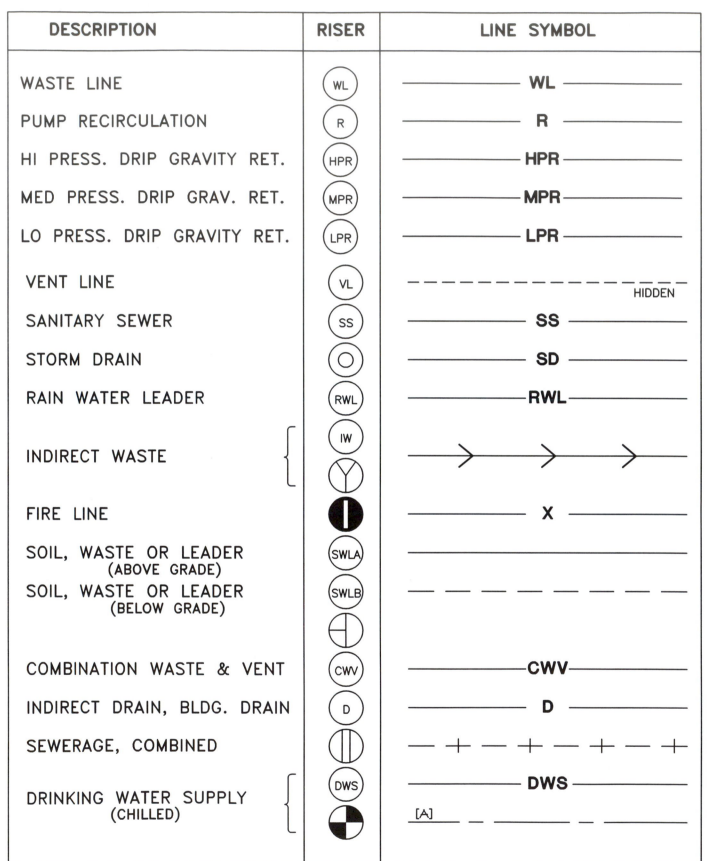

DESCRIPTION	RISER	LINE SYMBOL
WASTE LINE	WL	WL
PUMP RECIRCULATION	R	R
HI PRESS. DRIP GRAVITY RET.	HPR	HPR
MED PRESS. DRIP GRAV. RET.	MPR	MPR
LO PRESS. DRIP GRAVITY RET.	LPR	LPR
VENT LINE	VL	HIDDEN
SANITARY SEWER	SS	SS
STORM DRAIN		SD
RAIN WATER LEADER	RWL	RWL
INDIRECT WASTE	IW	
FIRE LINE		X
SOIL, WASTE OR LEADER (ABOVE GRADE)	SWLA	
SOIL, WASTE OR LEADER (BELOW GRADE)	SWLB	
COMBINATION WASTE & VENT	CWV	CWV
INDIRECT DRAIN, BLDG. DRAIN	D	D
SEWERAGE, COMBINED		
DRINKING WATER SUPPLY (CHILLED)	DWS	DWS [A]

DESCRIPTION	RISER	LINE SYMBOL
SANITIZING HOT WATER SUPPLY – 180°F	SHW	——— – – ——— – – ———
TEMPERED WATER SUPPLY	TS	———— TS ————
COLD WATER		——— – ——— – ———
HOT WATER		——— – – ——— – – ———
HOT WATER RETURN		— – – – ——— – – – —

PROCESS LINETYPES

LINETYPE	SAMPLE	REMARKS
MAJOR FLOW LINE		
OBJECT OR MINOR FLOW LINE		
MATCH LINE, SKID OUTLINE OR BOUNDARY OUTLINE		
MECHANICAL LINK		
ELECTRICAL SIGNAL		
PNEUMATIC SIGNAL		

LINETYPE	SAMPLE	REMARKS
HYDRAULIC SIGNAL		
CAPILLARY LINE		
ELECTRICAL HEAT	ET	
HEAT MEDIUM TRACE	HT	
TUBING LINE		
ACID WASTE	AW; AW; [A] ACID	
ACID VENT	AV; AV	
SALT WATER	SW	

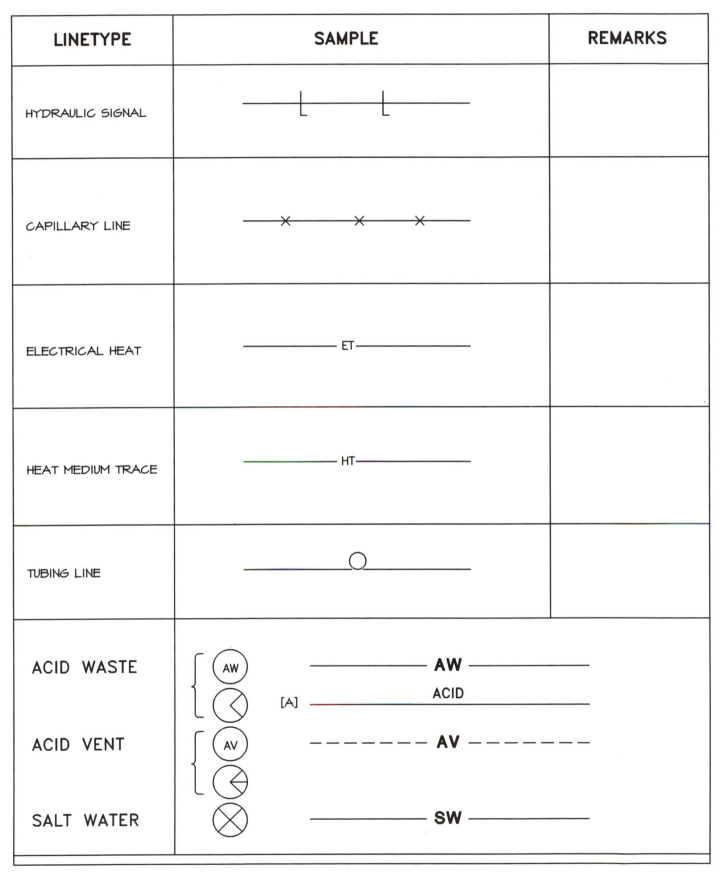

REFRIGERATION LINETYPES

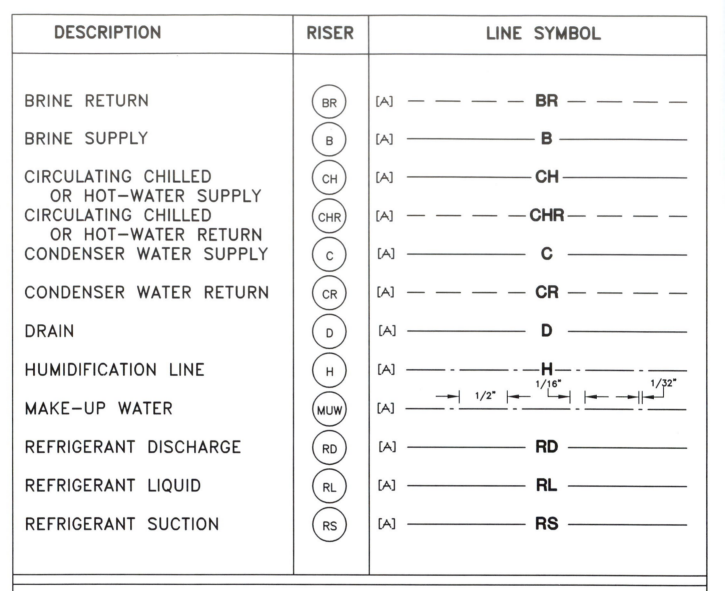

DESCRIPTION	RISER	LINE SYMBOL
BRINE RETURN	BR	[A] — — — — BR — — — —
BRINE SUPPLY	B	[A] ———— B ————
CIRCULATING CHILLED OR HOT–WATER SUPPLY	CH	[A] ———— CH ————
CIRCULATING CHILLED OR HOT–WATER RETURN	CHR	[A] — — — — CHR — — — —
CONDENSER WATER SUPPLY	C	[A] ———— C ————
CONDENSER WATER RETURN	CR	[A] — — — — CR — — — —
DRAIN	D	[A] ———— D ————
HUMIDIFICATION LINE	H	[A] —— - —— - —H— - —— - ——
MAKE–UP WATER	MUW	[A] —— - —— - —— - —— - —— (1/2" 1/16" 1/32")
REFRIGERANT DISCHARGE	RD	[A] ———— RD ————
REFRIGERANT LIQUID	RL	[A] ———— RL ————
REFRIGERANT SUCTION	RS	[A] ———— RS ————

LINETYPE	SAMPLE	REMARKS
POWER TRANSMISSION LINE		SOMETIMES ABBREVIATED: PWR TR LN
CULVERT		
DITCH	(L)	
EDGE OF PAVEMENT	(L)	
CORRUGATED METAL PIPE	(L)	
UNDERGROUND DIRECT BURIAL CABLE		
UNDERGROUND DUCT LINE		INDICATE DUCT AND CONDUCTORS.

ELECTRICAL LINETYPES

LINETYPE	SAMPLE	REMARKS
CONDUIT, CONCEALED IN CEILING OR WALL		
CONDUIT, CONCEALED IN FLOOR		
CONDUIT, EXPOSED		
WIRING TURNED UP		
WIRING TURNED DOWN		
FLEXIBLE METALLIC ARMORED CABLE		

LINETYPE	SAMPLE	REMARKS
HOME RUN TO PANEL	2 1	NUMBER OF ARROWHEADS INDICATES NUMBER OF CIRCUITS. MAY USE NUMBERS TO IDENTIFY CIRCUIT NUMBER. ONE ARROWHEAD MEANS TWO-WIRE CIRCUIT. NUMBER OF HASH MARKS INDICATES NUMBER OF WIRES.
TELEPHONE CONDUIT	T	
TELEVISION ANTENNA CONDUIT	TV	
SOUND SYSTEM CONDUIT	S	NUMBER OF HASH MARKS INDICATES NUMBER OF PAIRS OF CONDUCTORS.

ABBREVIATIONS

ABBREVIATION	MEANING
A	Area
ADA	Americans With Disabilities Act
AFF	Above Finished Floor
AFG	Above Finished Grade
AGA	American Gas Association
AHAM	Association of Home Appliance Manufacturers
AIA	American Institute of Architects
ANSI	American National Standards Institute
ASA	American Standards Association
ASCE	American Society of Civil Engineering
ASHRAE	American Society of Heating, Refrigeration & Air Conditioning Engineers
ASME	American Society of Mechanical Engineers
ASSE	American Society of Sanitary Engineering
ASTM	American Society for Testing & Materials
AUTO	Automatic
BG	Below Grade
BLDG	Building
C TO C	Center to Center
CAB	Cabinet(s)
CAT	Catalog
CCW	Counter Clockwise
CL	Center Line
CLG	Ceiling
CMU	Concrete Masonry Unit
COMM	Communication
CONC	Concrete
CORR	Corridor
CW	Clockwise
DET	Detail
DIA	Diameter
DN	Down
DT	Dust Tight
DWG	Drawing(s)
(E)	Existing
E to C	End to Center
ELEV	Elevation
EM	Emergency
ENCL	Enclosure
EP	Explosion Proof
EQ, EQUIP	Equipment
FF	Finished Floor
FIN	Finished
FLEX	Flexible

ABBREVIATION	MEANING
FLR	Floor
FT	Feet
FUT	Future
GALV	Galvanized
GD OR GND	Ground
GRN	green
GRY	gray
H	Height or High
hp or HP	Horsepower
HT OR HTR	Heater
HVAC	Heating, Ventilating & Air Conditioning
in	Inch(es)
L	Length
lb	Pound(s)
MACH	Machine
MAN	Manual
MAX	Maximum
MIN	Minimum
MT	Empty
MTG	Mounting
(N)	New
NAPHCC	Natl. Assoc. of Plumbing, Heating and Cooling Contractors
NBFU	National Board of Fire Underwriting
NBS	National Bureau of Standards
NC	Normally Closed
NELA	National Electric Light Association
NFPA	National Fire Protection Association
NIC	Not in Contract
NO	Normally Open, or Number
NOM	Nominal
OC	On Center
OD	Outside Diameter
ORN	Orange
OZ	Ounce(s)
psi	Pounds Per Square Inch
PVC	Polyvinyl Chloride (plastic)
(R)	Rough-in Only
RAD	Radius
REV	Reversible
RM	Room
RS	Residential
SCH	Schematic
SCHED	Schedule

ABBREVIATION	MEANING
SPEC	Specification
SQ	Square
STD	Standard
TAW	Throw Away (disposable)
TEMP	Temperature
TYP	Typical or Typically
U	Underground
UL	Underwriter's Laboratories, Inc.
UTY	Utility
V	Volume
W/	With
WD	Wide
WH or HWH	Hot Water Heater
WHT	White
WP	Weatherproof or Waterproof

ABBREVIATION	MEANING
AAF	ABOVE ACCESS FLOOR
AFF	ABOVE FINISH FLOOR
ALUM	ALUMINUM
ARCH	ARCHITECT
AS REQ'D	AS REQUIRED
ASF	ABOVE STRUCTURAL FLOOR
BO	BOTTOM/BASE OF
BOF	BOTTOM/BASE OF FOOTING
BOW	BOTTOM/BASE OF WALL
BRG	BEARING
BTWN	BETWEEN
CJ	CONTROL JOINT
CLG	CEILING
CMU	CONCRETE MASONRY UNIT
CONC	CONCRETE
CT	CERAMIC TILE
DIA	DIAMETER
(E)	EXISTING
EA	EACH
EJ	EXPANSION JOINT
EQ	EQUAL
EQUIP	EQUIPMENT
EST	ESTIMATE/ESTIMATED
EXT	EXTERIOR
FD	FLOOR DRAIN
FE	FIRE EXTINGUISHER
FIN	FINISH
FLR	FLOOR
FNDN	FOUNDATION
FV	FIELD VERIFY
GA	GAUGE
GALV	GALVANIZED
GWB	GYPSUM WALL BOARD
HC	HANDICAPPED
HT	HEIGHT
LAV	LAVATORY
LLH	LONG LEG HORIZONTAL
LLV	LONG LEG VERTICAL
MAS	MASONRY
MATL	MATERIAL
MFG	MANUFACTURER
MO	MASONRY OPENING
MTL	METAL
(N)	NEW
NO	NUMBER
NOM	NOMINAL

ABBREVIATION	MEANING
NTS	NOT TO SCALE
OC	ON CENTER
PL	PROPERTY LINE
PLY	PLYWOOD
PNT	PAINT
PROJ	PROJECT/PROJECTION
PT	PRESSURE TREATED
QT	QUARRY TILE
RAD	RADIUS
RB	RUBBER BASE
RD	ROOF DRAIN
RE	REFER TO/REGARDING
REFL	REFLECTED
REINF	REINFORCE/REINFORCED
REQ	REQUIRED/REQUIREMENT
RO	ROUGH OPENING (FRAMED)
RR	RESTROOM(S)
SAC	SUSPENDED ACOUSTIC CEILING
SF	SQUARE FEET
SIM	SIMILAR
SOG	SLAB ON GRADE
SQ. FT.	SQUARE FEET
STL	STEEL
STN	STAIN
STRUCT	STRUCTURE/STRUCTURAL
T&B	TOP AND BOTTOM
TO	TOP OF
TOF	TOP OF FOOTING
TOW	TOP OF WALL
TYP	TYPICAL
UG	UNDERGROUND
UNO	UNLESS NOTED OTHERWISE
UR	URINAL
VB	VINYL BASE
VERT	VERTICAL
VIF	VERIFY IN FIELD
W/	WITH
WD	WOOD
WG	WIRED GLASS
WWF	WELDED WIRE FABRIC
WWM	WELDED WIRE MESH
@	AT
+/-	PLUS OR MINUS
#	NUMBER/NUMBER OF
℄	CENTERLINE

	ABBREVIATION	MEANING
A	A OR AMP	Amperes
	AC	Alternating Current
	AF	Ampere Frame
	AH	Ampere Hour
	AIC	Ampere Interrupting Capacity
	AL	Aluminum
	ALM	Alarm
	ANT	Antenna
	AT	Ampere Trip
	ATS	Automatic Transfer Switch
	AUTO	Automatic
	AWG	American Wire Gauge
	A/C	Air Conditioner
B	B&S	Brown & Sharpe, Wire Gauge (AWG)
	BLK	Black
	BLU	Blue
	BRN	Brown
	BR—_	Bridge rectifier — (Number)
	BSG	British Standard Gauge, Wire Gauge
	BTY	Battery
	BX	Flexible Armored Cable
	B+	Best Best, Iron Telephone Wire
C	C	Conduit
	CAB	Cabinet
	CAT	Catalog

	ABBREVIATION	MEANING
	CB	Circuit Breaker
	CMC	Contact Making or Breaking Ammeter
	CMA	Contact Making or Breaking Clock
	CCT or CKT	Circuit
	CMS	Combination Motor Starter
	CMV	Contact Making or Breaking Voltmeter
	COND	Conductor
	CPT	Control Power Transformer
	CRO	Oscilloscope
	CSP	Central Switch Panel
	CT	Current Transformer
	CU	Copper
	C–_	Capacitor – (Number)
D	dBm	dBm Meter
	dB	Decibel, Decibel Meter
	DC	Direct Current
	DCP	Dimmer Control Panel
	DISC	Disconnect
	DIST	Distribution
	DP	Double Pole
	DPDT	Double Pole Double Throw
	DPST	Double Pole Single Throw
	DT	Double Throw
	DTR	Demand–totalizing Relay

	ABBREVIATION	MEANING
	D–_	Diode – (Number)
E	E	Voltage
	EC	Electrical Contractor
	EG	Earth Ground
	ELECT	Electrical
	EM	Emergency
	EMT	Electrical Metallic Tubing
	EP	Explosion Proof
	EQ OR EQUIP	Equipment
	ESP	Emergeny Switch Panel
	E.H.P.	Electrical Horsepower
F	F	Frequency Meter
	F OR FLUOR	Fluorescent
	f	Frequency
	FE	Iron
	FSS	Fused Safety Switch
	F–_	Fuse – (Number)
G	G	Galvanometer
	G, GD, GND	Ground
	GA	Gauge
	GALV	Galvanized
	GD	Grounded
	GEN	Generator

	ABBREVIATION	MEANING
H	HP or hp	Horsepower
	HID	High Intensity Discharge
	HPS	High Pressure Sodium
	HTR or HT	Heater
	HZ or cps	Hertz (cycles)
	H_	Hot In (Number)
I	I	Indicating
	I or INC	Incandescent
	I	Current
	IG	Isolated Ground
	INT	Integrating
	ISC	Short Circuit Current
J	J–_	Jack (Number)
	JB	Junction Box
K	KVA	Kilo Volt Ampere
	KW	Kilowatt
	K–_ or RY–_	Relay– (Number)
L	LED	Light Emitting Diode
	LP–_	Lamp – (Number)
	LTG	Lighting
	LTS	Lights
	L–_	Inductor – (Number)
	L_	Line in (Number)

	ABBREVIATION	MEANING
	LV	Low Voltage
M	MCB	Main Circuit Breaker
	MDF	Main Distribution Frame, Telephone
	MDP	Main Distribution Panel
	MERC	Mercury Vapor
	MH	Metal Halide
	MLO	Main Lug Only
	M–G	Motor Generator
	M–_	Motor – (Number)
	MV	Medium Voltage
N	N or NEU	Neutral
	NELA	National Electric Light Assoc.
	NFSS	Non Fused Safety Switch
	NL	Night Light
	(NL)	New Location
	NM	Noise Meter
	N/C	Normally Closed
	N/O	Normally Open
O	O	Overload Contactor
	OHM	Ohmeter
	OL	Overload
	ORN	Orange
	OSCG	Oscillograph String

	ABBREVIATION	MEANING
P	P	Power or Pole
	_P	(Number of) Poles
	PB	Push Button
	PBX	Private Branch Telephone Exchange
	PC	Pull Chain
	PCB	Printed Circuit Board
	PF	Phase or Phasemeter
	PH	
	PNL	Panel
	PRI	Primary
	PROT	Protector
	P–_	Plug – (Number)
Q	Q–_	Transistor – (Number)
R	R	Resistance or Recessed
	RD	Recording Demand Meter
	REC	Recording
	RECEPTS	Receptacles
	RED	Red
	RFI	Radio Frequency Interference
	RMC	Rigid Metallic Conduit
	RY	Relay
	R–_	Resistor – (Number)

	ABBREVIATION	MEANING
S	S OR SW	Switch
	SCR	Silicon Controlled Rectifier
	SE	Service Entry
	SI	Silver
	SPK OR SPKR	Speaker
	SWBD	Switchboard
	SY	Synchroscope
T	T	Temperature Meter or Thermometer
	TC	Thermocouple
	TELE	Telephone
	THC	Thermal Converter
	TP	Test Point
	TRFMR	Transformer
	TS	Terminal Strip
	TSTAT	Thermostat
	TV	Television
	T–_	Transformer – (Number)
	TT	Total Time
U	UF	Underground Feeder
	uL	Micro Logic (IC)
	USE	Underground Service Entrance

	ABBREVIATION	MEANING
V	V	Volts, Voltmeter
	VA	Volt—ampere, Volt—ammeter
	VAR	Varmeter
	VARH	Varhour Meter
	VCR	Video Cassette Recorder
	VI	Volume Indicator or Meter; Audio Level
	VIO	Violet
	VU	Standard Volume Indicator or Meter
W	W	Watts
	WH	Watthour Meter
	WP	Weatherproof
	WHT	White
X	XFMR	Transformer
	XTAL	Crystal
Y	YEL	Yellow

ABBREVIATION	MEANING
AD	Area Drain
AE	Air Extractor
AFD	Area Floor Drain
AHU	Air Handling Unit
AP	Access Panel
BS	Bar Sink
Btu	British Thermal Unit(s)
Btuh	Btu per hour
BU	Blower Unit
BWV	Back Water Valve
c	Capita (person)
CB	Catch Basin
CD	Ceiling Diffuser
CFM	Cubic Feet Per Minute
CG	Ceiling Grille
CI	Cast Iron
CL	Condensate Line
CO	Cleanout
CODP	Cleanout Deck Plate
CR	Ceiling Register
CSR	Ceiling Supply Register
cu.ft.	Cubic Feet
cu.in	Cubic Inches
CV	Check Valve
CW	Cold Water
CWM	Clothes Washing Machine
CWR	Cold Water Return
DL	Ductliner in Duct
EA	Exhaust Air
F/L	Full Load
FCU	Fan Coil Unit
FCO	Floor Cleanout
FD	Fire Damper
FG	Floor Grille
FR	Floor Register
GPM	Gallons Per Minute
HSWG	High Sidewall Grille
HSWR	High Sidewall Register
HWC	Hot Water Convector
LSWR	Low Sidewall Register
MBH	Thousand Btu per Hour
MD	Manual Damper
OA	Outside Air
PF	Propeller Fan

ABBREVIATION	MEANING
PROT	Protection
PRV	Power Roof Ventilator
RA	Return Air
SA	Supply Air
SD	Splitter Damper
SLV	Sleeve
TEAO	Totally Enclosed, Air Over
TEAC	Totally Enclosed, Air Cooled
TENV	Totally Enclosed, Non-ventilated
TV	Turning Vanes
UN	Unit Heater
UV	Unit Ventilator
UVS	Utility Vent Set
WCO	Wall Cleanout
WF	Wall Fin Radiator
WH	Wall Heater

INDEX

D

E

F

G

H

I

J

K

L

M

N

O

P

R

S

T

U

V

W

About the Author

Doug Wolff owns and operates his own AutoCAD drafting studio near Boulder, Colorado. With six years of intensive AutoCAD drafting experience, he also has 15 additional years' background in many phases of the construction industry, including design and building of custom and production wooden cabinetry and furniture.